NUTRITIONAL AND SAFETY ASPECTS OF FOOD PROCESSING

FOOD SCIENCE

A Series of Monographs

Series Editor
OWEN R. FENNEMA
Department of Food Science
University of Wisconsin
College of Agriculture
Madison, Wisconsin

Volume 1. FLAVOR RESEARCH: Principles and Techniques
by A. Teranishi, I. Hornstein, P. Issenberg, and E. L. Wick (Out of Print)

Volume 2. PRINCIPLES OF ENZYMOLOGY FOR THE FOOD SCIENCES
by John R. Whitaker

Volume 3. LOW-TEMPERATURE PRESERVATION OF FOODS AND LIVING MATTER
by Owen R. Fennema, William D. Powrie, and Elmer H. Marth

Volume 4. PRINCIPLES OF FOOD SCIENCE
Edited by Owen R. Fennema

Part I: FOOD CHEMISTRY *Edited by Owen R. Fennema*

Part II: PHYSICAL METHODS OF FOOD PRESERVATION *by Marcus Karel, Owen R. Fennema, and Daryl B. Lund*

Part III: FOOD FERMENTATIONS *by E. H. Marth, N. F. Olson, and C. H. Amundson*

Volume 5. FOOD EMULSIONS
Edited by Stig Friberg

Volume 6. NUTRITIONAL AND SAFETY ASPECTS OF FOOD PROCESSING
Edited by Steven R. Tannenbaum

Other Volumes in Preparation

NUTRITIONAL AND SAFETY ASPECTS OF FOOD PROCESSING

Edited by STEVEN R. TANNENBAUM

Department of Nutrition and Food Science
Massachusetts Institute of Technology
Cambridge, Massachusetts

MARCEL DEKKER, INC. New York and Basel

Library of Congress Cataloging in Publication Data

Main entry under title:

Nutritional and safety aspects of food processing

(Food science ; v. 6)
Includes indexes.
1. Food industry and trade. 2. Nutrition
I. Tannenbaum, Steven R., [DATE]
TP372.5.N87 664 78-31276
ISBN 0-8247-6723-3

MARCEL DEKKER, INC.
270 Madison Avenue, New York, New York 10016

Current printing (last digit):
10 9 8 7 6 5 4 3 2 1

PRINTED IN THE UNITED STATES OF AMERICA

To John T. R. Nickerson (Nick),
who taught me food technology in the laboratory and classroom,
and professional ethics by example

PREFACE

The history of this book is entwined with the history of my professional career. Both walk the boundary of nutrition, food safety, and food science.

I became a food scientist because of the influence of Sam Goldblith. Particularly, I have always remembered a story he told me, that as a POW in a Japanese camp in World War II he and his men sustained themselves by consuming the juice of common grasses to obtain vitamin and mineral supplements.

While I was taking my first course in nutrition with Bob Harris, he had just received the first galleys of Nutritional Evaluation of Food Processing, copies of which were passed out to the class. Thus, the subject of my first course in nutrition included the subject of this book. My inclinations to understand and teach the chemical aspects of foods led to the concept of developing a book that treated the principles of the effect of processing of foods on the nutrients therein.

I owe all of my colleagues and teachers a debt of gratitude: To Sandy Miller for teaching me to do biological research, Mark Karel for teaching me to approach a problem with rigor and organization, to Gerry Wogan for leading me into the pastures of food safety, to Nick (to whom the book is dedicated), and to my other friends who have contributed directly and indirectly to this book.

I also owe special thanks to Marie Ludwig for incisive editorial assistance, and to Carol, Lisa, and Mark for reasons only they will know.

Steven R. Tannenbaum

CONTRIBUTORS

MICHAEL C. ARCHER, Department of Nutrition and Food Science, Massachusetts Institute of Technology, Cambridge, Massachusetts

BENJAMIN BORENSTEIN,* Roche Chemical Division, Hoffmann-La Roche Inc., Nutley, New Jersey

NICHOLAS CATSIMPOOLAS, Department of Nutrition and Food Science, Massachusetts Institute of Technology, Cambridge, Massachusetts

J. CLAUDE CHEFTEL, Laboratoire de Biochimie et Technologie Alimentaires, Université des Sciences et Techniques, Montpellier, France

SAMUEL A. GOLDBLITH, Department of Nutrition and Food Science, Massachusetts Institute of Technology, Cambridge, Massachusetts

MARCUS KAREL, Department of Nutrition and Food Science, Massachusetts Institute of Technology, Cambridge, Massachusetts

FRED A. KUMMEROW, Department of Food Sciences, Burnsides Research Laboratory, University of Illinois, Urbana, Illinois

EDWARD G. PERKINS, Department of Food Sciences, Burnsides Research Laboratory, University of Illinois, Urbana, Illinois

NEVIN S. SCRIMSHAW, Department of Nutrition and Food Science, Massachusetts Institute of Technology, Cambridge, Massachusetts

ANTHONY J. SINSKEY, Department of Nutrition and Food Science, Massachusetts Institute of Technology, Cambridge, Massachusetts

DAVID J. SISSONS, Unilever Research, Colworth Laboratory, Colworth House, Sharnbrook, Bedfordshire, U.K.

*Current affiliation: CPC International Inc., Englewood Cliffs, New Jersey.

STEVEN R. TANNENBAUM, Department of Nutrition and Food Science, Massachusetts Institute of Technology, Cambridge, Massachusetts

GEOFFREY M. TELLING, Unilever Research, Colworth Laboratory, Colworth House, Sharnbrook, Bedfordshire, U.K.

LLOYD A. WITTING, Supelco, Inc., Supelco Park, Bellefonte, Pennsylvania

GERALD N. WOGAN, Department of Nutrition and Food Science, Massachusetts Institute of Technology, Cambridge, Massachusetts

VERNON R. YOUNG, Department of Nutrition and Food Science, Massachusetts Institute of Technology, Cambridge, Massachusetts

CONTENTS

NUTRITIONAL AND SAFETY ASPECTS OF FOOD PROCESSING

CHAPTER 1

THE CONSUMER, THE PRODUCT, AND THE PROMISE

Samuel A. Goldblith

Department of Nutrition and Food Science
Massachusetts Institute of Technology
Cambridge, Massachusetts

I. INTRODUCTION

Industry, since ancient times, has felt a responsibility to the consumer and has either regulated itself or been regulated through governmental intervention.

Perhaps the oldest story of regulation deals with a building code which was enacted about 1800 B.C. in Babylonia by Hammurabi, who decreed that, if a dwelling collapsed, the penalty was death to the builder.

Although we may not agree with the harshness of the sentence, it was a model code in that it stressed performance rather than details of construction.

Today in the food industry we are still faced with performance of products; that is, the product must match its promise to the consumer in terms of content, storage life, etc. In addition, there are various legalistic criteria to be met, such as proper labeling, ingredient listing, and meeting the standards of identity.

Moreover, there are numerous additional constraints upon the ability of the food industry to produce a particular product in response to consumer demand or government regulation. These include availability and costs of energy and nonrenewable resources, spiraling costs of commodities and labor, as well as limits of productivity.

Thus, quality control is a complicated procedure involving consumers, manufacturers, retailers, and distributors, as well as legal implications and governmental regulations. Its aim is to see that the consumer gets what is advertised and paid for, in terms of quality, quantity, safety, and nutrition of foods.

II. CONSUMERISM

Consumerism is a social movement seeking to protect the rights and powers of buyers in relation to sellers. Its concerns in relation to the food industry include safety and nutritional value of foods, food additives, food costs, and preservation of the environment.

Unfortunately, consumerism seems all too frequently to be a matter of right ends, but wrong means, since it is the consumer who must ultimately pay the price of governmental regulations which are often full of duplications and inefficiently administered.

Thomas Jefferson promised "a wise and frugal government," but, as a country grows, regulations become necessary. A system without regulations can lead to anarchy, whereas overregulation can be so costly that, as Alfred W. Eames, Chairman of Del Monte, wrote: "The well-informed consumer might decide that he or she cannot afford all the governmental regulation that he or she is getting."

For example, while it costs $4 billion for the annual budget of the regulatory agencies, it costs industry $42 billion to meet these regulations, and the consumer must pay the cost of regulatory mismanagement, overregulation, and governmental inefficiencies.

However, although we have methodologies for dealing with the generation of numbers, or the exact sciences, we do not have adequate methodologies for dealing with consumer behavior. We are living in turbulent times wherein food companies must take into account the consumer's focus on ecological goals, as well as the economic goals of the company.

If we look into the history of quality control and the rights of the consumer we find some interesting historical developments.

III. FOOD QUALITY AND GUARANTEES

Nicholas Appert, who invented canned foods in 1831, recognized that the consumer needed to be educated about this new process. The consumer had no way of evaluating the quality of his wares before purchase, since the contents of his tins could not be examined.

Faced with the problem of giving the consumer a satisfactory guarantee, and yet protecting himself from false fraudulent claims, Appert warned against bulging tins, presumed to be defective, which were not to be opened, but returned for replacement. He also gave instructions on proper storage of tins during sea voyages.

Thus, Appert educated the consumer concerning his product, in addition to guaranteeing its quality.

IV. TRUTH IN LABELING

Louis Pasteur, as part of his public service, was an adviser to the French Board of Hygiene and Sanitation, which had the responsibility for food laws.

In 1877, Pasteur was asked by the Board to determine whether canned peas were being treated with copper salts to render them green. He went to the marketplace, obtained 14 tins of peas, and found 10 to contain copper.

With much insight, as he was ignorant of the food chemistry involved, Pasteur stated, "Canned peas not treated with copper salts have a yellowish color. There does not exist in the food preservation industry a method which permits the manufacture of preserved peas with a more or less green color without the addition of copper salts." He further stated that no tolerance should be set for copper without a label indicating its usage and the carrier's reason for its addition.

Similarly, in 1879, Pasteur was called upon to investigate a process for the preservation of meats, which consisted of dipping or pumping them with benzoic acid. Although the process worked, he questioned it in relation to public health, pointing out that benzoic acid was known to transform to hippuric acid in the body, and was even used for the treatment of certain bladder diseases. He could not answer as to its safety "for all constitutions" and urged the Board to be "conservative." He felt that the authorities could permit the use of benzoic acid solution if the food to be preserved had "as a label of declaration that the preservation was due to a weak and measured dose of benzoic acid."

In his report to the Board on November 15, 1879, Pasteur summarized his views on copper salts in canned peas, as well as on the preservation of other foods with chemical agents such as boric, salicyclic, and benzoic acids. He cautioned against eating foods containing the acids over extended periods of time, until more was known about them. He also stated: "...there is only one way for the administration and for us to teach industry to be honorable in its responsibilities, that is, to require a true declaration of the nature of foreign substances added to food products..." He also emphasized that the label declaration of additives should be printed in "legible characters." As he said, "The public— the entire public—has the right to know."

V. QUALITY CONTROL AND THE MANUFACTURER

Manufacturers of consumer goods are notoriously poor at handling customer complaints. Perhaps only one out of 100 defects is reported to the food stores. The consumer just buys a different brand the next time or shops elsewhere. The manufacturer only hears about complaints when they become numerous, or a public health hazard arises, such that a recall is instituted or forced upon him.

The net result is that, whereas 65% of the consumers had confidence in business in 1965, in 1974 it had dropped to 15%.

One way of regaining this confidence in the food industry is to realize that quality control cannot be assured by examining an occasional finished product in a governmental or industrial laboratory. Instead quality assurance is achieved by manufacturing quality into the product by developing and adhering to a code of good manufacturing practices, with the hazard analysis at critical control points in force. Adherence to these principles should also reduce the overregulation pressures of the consumer advocates.

VI. TOWARD THE FUTURE

In 1962 President Kennedy presented to Congress a consumer message with four basic consumer rights:

1. The right to safety
2. The right to be informed
3. The right to choose (from a diverse number of products)
4. The right to be heard (by producers of goods and services)

President Nixon's "Buyer's Bill of Rights" was similar to President Kennedy's message. Thus, the legacy of Appert and Pasteur continues to the present time.

Using the modern techniques of "manufacturing quality into the product," food products may be made to match their promise to the consumer, in terms of nutrition, storage life, and gustatory delight. This quality assurance should be achievable at minimal regulatory expense and minimal control costs to the manufacturer, and, therefore, at minimum cost to the consumer.

The succeeding chapters of this book will indicate how modern processing has preserved the nutrient content of the food and protected it from contamination by harmful substances.

REFERENCES

1. S. A. Goldblith, Pro Bona Publico—In the Best Scientific Tradition, Red Devil 3(2):7 (1970).
2. S. A. Goldblith, Quality Guarantees—An Historical Note, Food Technol. 26(10):40 (1972).
3. J. B. Wilkinson, The Promise and the Product, J. Soc. Cosmet. Chem. 26:497 (1975).

CHAPTER 2

HUMAN NUTRIENT REQUIREMENTS AND DIETARY ALLOWANCES

Vernon R. Young
Nevin S. Scrimshaw

Department of Nutrition and Food Science
Massachusetts Institute of Technology
Cambridge, Massachusetts

I. INTRODUCTION

The earliest forms of life were presumably simple bacteria-like organisms which were capable of synthesizing all of the molecules needed for growth and reproduction from salts, simple carbon compounds, nitrogen and water [1].

About one billion years ago the first single-celled animals developed, which differed from plant cells in their need to obtain various carbon compounds from the environment. All animals need to ingest organic compounds, whether they be single-celled protozoa or complex multicellular organisms. Whereas plants have retained a metabolic capacity to synthesize all of the 20 amino acids present in cell protein, animals are dependent on their diets to supply about half of these amino acids. The precise number and relative amounts depend upon the species and developmental stage of the organism [2-4]. In terms of their nutritional requirements, it is as if all animals suffer from an inborn error of metabolism, or a hereditary disease, to which plants are immune.

More than 30 years ago, Beadle and Tatum [5] demonstrated with neurospora that gene mutation can bring about alterations in the needs of cells and organisms for exogenous sources of compounds. Neurospora like other plants do not normally require a preformed source of vitamins and amino acids. However, exposure to X rays produced a series of mutations that caused a loss in ability to synthesize vitamins such as thiamine, pyridoxine and p-aminobenzoic acid, as well as the amino acids histidine, lysine, and tryptophan. It has been suggested [6] that there may be a selective advantage associated with these mutations, due to elimination of "unnecessary" enzymatic processes.

An evolutionary occurrence of significance for human nutrition is the variation among animal species in ability to synthesize ascorbic acid (vitamin C), as depicted in Fig. 1. Chatterjee et al. [7,8] have suggested that the capacity for synthesizing this vitamin began in the amphibians approximately 350 million years ago, and that about 25 million years ago, in a progenitor of man and other primates, a gene mutation led to the loss of the enzyme L-gulonolactome oxidase. As shown in Fig. 1, this enzyme catalyzes the terminal step in the conversion of glucose to ascorbic acid. The mutation was not lethal because an adequate supply of dietary vitamin C was present in the environment.

These genetic changes, producing a dependence on food sources, have profoundly influenced the course of evolution and development in animals. Thus, for example, an organization of muscular tissues and a complex neuroendocrine system were required in the higher animals to enable them to obtain an adequate diet. However, in addition, with reliance on exogenous sources of essential nutrients, the development of specific nutritional deficiencies became possible. Such deficiencies have affected the health of earlier generations and continue to affect the lives of populations throughout the world today [9].

In view of a rapidly expanding world population, and the importance of maintaining a balance between it and food supply, as well as the trends in technically developed regions toward use of more highly processed foods and foods of unconventional origin, it is important to study the pattern and amounts of nutrients required by humans to maintain adequate health and function. In this chapter we will survey these nutrient requirements from both qualitative and quantitative standpoints. However, the emphasis will be on the broad issues of estimating requirements and dietary allowances, rather than on a detailed account of the roles of individual nutrients.

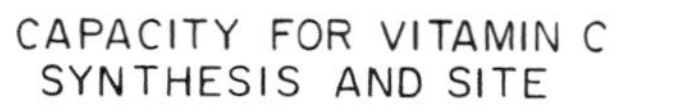

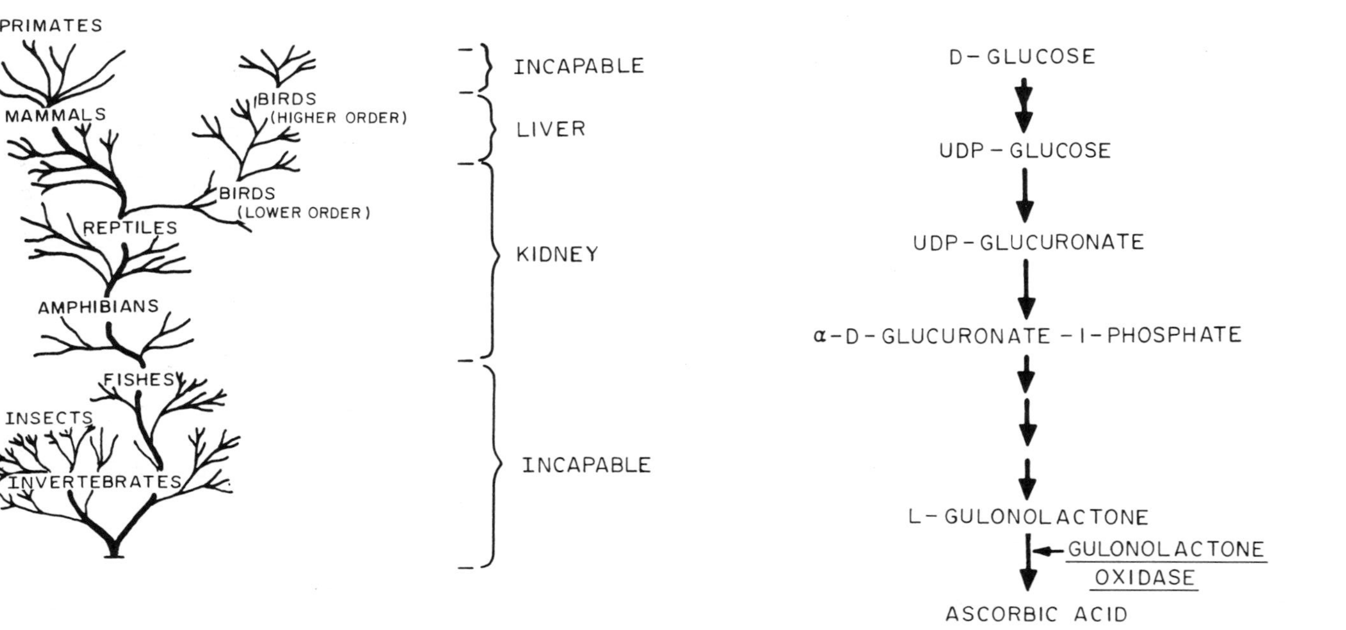

Fig. 1. Evolutionary changes in the synthesis of vitamin C, related to the presence (or absence) of gulonolactone oxidase (from Chatterjee [7]).

II. THE ESSENTIAL NUTRIENTS

Nutrients are used for growth, maintenance, tissue repair, and reproduction, and foods are the vehicles for them. An individual food may contain only a few nutrients, or it may supply many, but no single food provides all nutrients in amounts and proportions necessary for adequate health [10]. However, if the total diet supplies all essential nutrients, the cells and body organs can synthesize many thousands of additional metabolically important substances.

Foods in their native state cannot be used by the cell. They must first undergo digestion within the intestinal tract, resulting in release of nutrients, which are then transported across the mucosal wall of the intestine. Ultimately the nutrients enter the blood stream and are transported to tissues, where they are utilized for various physiological and metabolic functions. To prevent accumulation to toxic levels in organs and body fluids of nutrients and/or their metabolites, the phases of absorption, utilization, and catabolism are integrated at the whole body level. Intakes in excess of cellular needs are excreted or enter catabolic pathways to be removed as breakdown products via routes such as urine, feces, bile, and sweat (see Fig. 2).

However, very high nutrient intakes may exceed the metabolic capacity of the organism, resulting in pathological effects and a deterioration of health. This state occurs relatively frequently in certain diseases, such as maple syrup urine disease and phenylketonuria, where infants cannot adequately dispose of the branched-chain amino acids (leucine, isoleucine, and valine) and phenylalanine, respectively. The free amino acids rise to high levels in blood and tissues, including the brain, and if these elevated concentrations are maintained mental retardation may occur. The treatment of these conditions includes diets supplying low levels of the "overabundant" nutrients [11]. Another example of the accumulation to toxic levels of an essential nutrient is hypervitaminosis A, due to excessive dietary intake of vitamin A. This condition is characterized by a drying and desquamation of the skin, headaches, loss of hair, and bone and joint pain.

On the other hand, if the intake of a nutrient is insufficient to meet the normal needs of cells, metabolic responses occur within the cells and organs to conserve their limited supply. These changes include a more effective absorption of nutrients from the intestine, as in iron deficiency [12], and/or the activation of biochemical mechanisms to improve retention and utilization of the nutrient by body tissues. Thus, the oxidation of essential amino acids is low when dietary intake is inadequate and high when it is more than sufficient to meet amino acid requirements [13]. In this way the organism can adjust, at least over short periods, to a range of nutrient intakes. However, when utilization and metabolic control of nutrients are altered by genetic disease, infection, or administration of

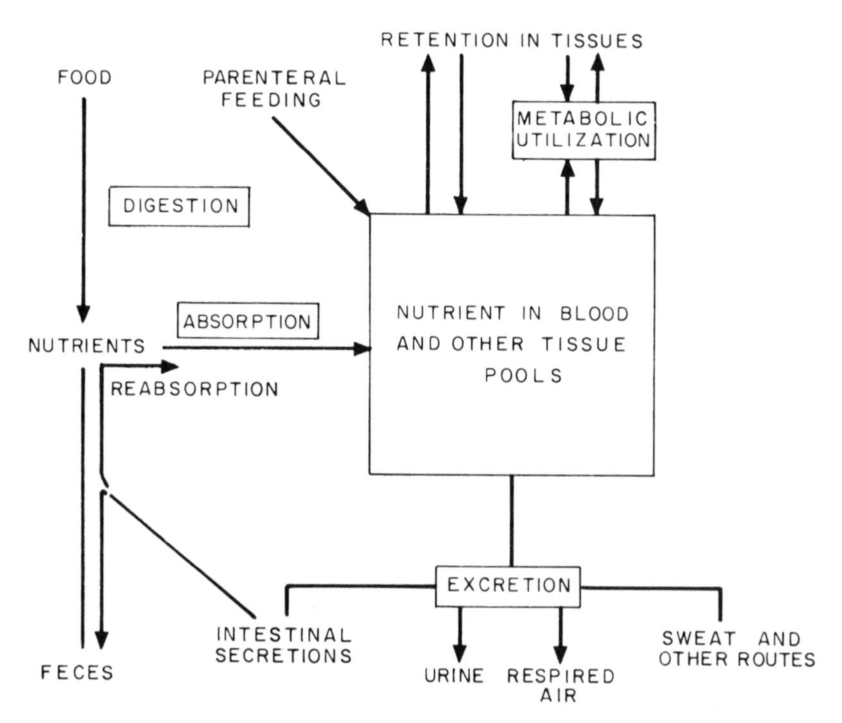

Fig. 2. Diagrammatic representation of the major stages in the utilization of nutrients.

drugs, the requirements for nutrients will also be affected. Unless the dietary supply is appropriate to compensate for these effects, health will deteriorate.

A detailed review of the 45 to 50 essential dietary constituents is beyond the scope of this discussion, but a brief general summary of their function and utilization should serve as a background for the topics that follow. Table 1 gives a compilation of major features for each essential nutrient.

Carbohydrates, fats, proteins, vitamins, and minerals comprise the general classes of nutrients. The first three classes serve as major sources of fuel. In addition, proteins supply essential amino acids and serve as a source of utilizable nitrogen for cell and tissue protein synthesis. Amino acids may also serve as precursors for metabolically active compounds. Examples are: tryptophan and tyrosine for neurotransmitter; tryptophan for niacin; tyrosine for iodothyronines; lysine for carnitine formation. These uses of the amino acids may be of little quantitative importance, in terms of total daily intake, but they have considerable physiological significance. Thus it has been estimated that about 0.1% of

TABLE 1 A Summary of the Dietary Sources, Functions, and Effects of Deficient or Excess Intakes of the Individual Essential Nutrients

Essential amino acids	Dietary sources	Major body functions	Deficiency	Excess
Aromatic Phenylalanine Tyrosine Basic Lysine Histidine Branched chain Isoleucine Leucine Valine Sulfur containing Methionine Cystine Other Tryptophan Threonine	From proteins Good sources Legumes Dairy products Meat Fish Adequate sources Rice Corn Wheat Poor sources Cassava Sweet potato	Precursors of structural protein, enzymes, antibodies, hormones, metabolically active compounds Certain amino acids have specific functions: (a) Tyrosine is a precursor of epinephrine and thyroxine (b) Arginine is a precursor of polyamines (c) Methionine is required for methyl group metabolism (d) Tryptophan is a precursor of serotonin	Deficient protein intake leads to development of kwashiorkor and, coupled with low energy, intake to marasmus.	Excess protein intake possibly aggravates or potentiates chronic disease states.

Essential fatty acids Arachidonic Linoleic Linolenic	Vegetable fats (corn, cottonseed, soy oils) Wheat germ Vegetable shortenings	Involved in cell membrane structure and function. Precursors of prostaglandins (regulation of gastric function, release of hormones, smooth muscle activity)	Poor growth Skin lesions	Not known

Vitamin	Dietary sources	Major body functions	Deficiency	Excess
Water-soluble				
Vitamin B-1 (thiamine)	Pork, organ meats, whole grains, legumes	Coenzyme (thiamine pyrophosphate) in reactions involving the removal of carbon dioxide	Beriberi (peripheral nerve changes, edema, heart failure)	None reported
Vitamin B-2 (riboflavin)	Widely distributed in foods	Constituent of two flavin nucleotide coenzymes involved in energy metabolism (FAD and FMN)	Reddened lips, cracks at corner of mouth (cheilosis), lesions of eye	None reported

TABLE 1 (continued)

Vitamin	Dietary sources	Major body functions	Deficiency	Excess
Niacin	Liver, lean meats, grains, legumes (can be formed from tryptophan)	Constituent of two coenzymes involved in oxidation-reduction reactions (NAD and NADP)	Pellagra (skin and gastrointestinal lesions, nervous, mental disorders)	Flushing, burning and tingling around neck, face and hands
Vitamin B-6 (pyridoxine)	Meats, vegetables, whole-grain cereals	Coenzyme (pyridoxal phosphate) involved in amino acid metabolism	Irritability, convulsions, muscular twitching, dermatitis near eyes, kidney stones	None reported
Pantothenic acid	Widely distributed in foods	Constituent of coenzyme A, which plays a central role in energy metabolism	Fatigue, sleep disturbances, impaired coordination, nausea (rare in man)	None reported
Folacin	Legumes, green vegetables, whole-wheat products	Coenzyme (reduced form) involved in transfer of single-carbon units in nucleic acid and amino acid metabolism	Anemia, gastrointestinal disturbances, diarrhea, red tongue	None reported

Vitamin B-12	Muscle meats, eggs, dairy products, (not present in plant foods)	Coenzyme involved in transfer of single-carbon units in nucleic acid metabolism	Pernicious anemia, neurological disorders	None reported
Biotin	Legumes, vegetables, meats	Coenzyme required for fat synthesis, amino acid metabolism and glycogen (animal-starch) formation	Fatigue, depression, nausea, dermatitis, muscular pains	Not reported
Choline	All foods containing phospholipids (egg yolk, liver, grains, legumes)	Constituent of phospholipids. Precursor of putative neurotransmitter acetylcholine	Not reported in man	None reported
Vitamin C (ascorbic acid)	Citrus fruits, tomatoes, green peppers, salad greens	Maintains intercellular matrix of cartilage, bone and dentine. Important in collagen synthesis	Scurvy (degeneration of skin, teeth, blood vessels, epithelial hemorrhages)	Relatively nontoxic. Possibility of kidney stones
Fat-soluble				
Vitamin A (retinol)	Provitamin A (beta-carotene) widely distributed in green vegetables. Retinol present in milk, butter, cheese, fortified margarine	Constituent of rhodopsin (visual pigment). Maintenance of epithelial tissues. Role in mucopolysaccharide synthesis.	Xerophthalmia (keratinization of ocular tissue), night blindness, permanent blindness	Headache, vomiting, peeling of skin, anorexia, swelling of long bones

TABLE 1 (continued)

Vitamin	Dietary sources	Major body functions	Deficiency	Excess
Vitamin D	Cod-liver oil, eggs, dairy products, fortified milk and margarine	Promotes growth and mineralization of bones. Increases absorption of calcium	Rickets (bone deformities) in children. Osteomalacia in adults	Vomiting, diarrhea, loss of weight, kidney damage
Vitamin E (tocopherol)	Seeds, green leafy vegetables, margarines, shortenings	Functions as an antioxidant to prevent cell-membrane damage	Possibly anemia	Relatively nontoxic
Vitamin K (phylloquinone)	Green leafy vegetables. Small amount in cereals, fruits and meats	Important in blood clotting (involved in formation of active prothrombin)	Conditioned deficiencies associated with severe bleeding. Internal hemorrhages	Relatively nontoxic. Synthetic forms at high high doses may cause jaundice

Mineral	Dietary sources	Major body functions	Deficiency	Excess
Calcium	Milk, cheese, dark-green vegetables, dried legumes	Bone and tooth formation Blood clotting Nerve transmission	Stunted growth Rickets, osteoporosis Convulsions	Not reported in man
Phosphorus	Milk, cheese, meat, poultry, grains	Bone and tooth formation. Acid base balance	Weakness, demineralization of bone Loss of calcium	Erosion of jaw (fossy jaw)

Sulfur	Sulfur amino acids (methionine and cystine) in dietary proteins	Constituent of active tissue compounds, cartilage and tendon	Related to intake and deficiency of sulfur amino acids	Excess sulfur amino acid intake leads to poor growth
Potassium	Meats, milk, many fruits	Acid-base balance Body water balance Nerve function	Muscular weakness Paralysis	Muscular weakness Death
Chlorine	Common salt	Formation of gastric juice Acid-base balance	Muscle cramps Mental apathy Reduced appetite	Vomiting
Sodium	Common salt	Acid-base balance Body water balance Nerve function	Muscle cramps Mental apathy Reduced appetite	High blood pressure
Magnesium	Whole grains, green leafy vegetables	Activates enzymes. Involved in protein synthesis	Growth failure Behavioral disturbances Weakness, spasms	Diarrhea
Iron	Eggs, lean meats, legumes, whole grains, green leafy vegetables	Constituent of hemoglobin and enzymes involved in energy metabolism	Iron-deficiency anemia (weakness, reduced resistance to infection)	Siderosis Cirrhosis of liver
Fluorine	Drinking water, tea, seafood	May be important in maintenance of bone structure	Higher frequency of tooth decay	Mottling of teeth Increased bone density Neurological disturbances

TABLE 1 (continued)

Mineral	Dietary sources	Major body functions	Deficiency	Excess
Zinc	Widely distributed in foods	Constituent of enzymes involved in digestion	Growth failure, small sex glands, skin lesions	Fever, nausea, vomiting, diarrhea
Copper	Meats, drinking water	Constituent of enzymes associated with iron metabolism	Anemia, bone changes (rare in man)	Rare metabolic condition (Wilson's disease)
Silicon Vanadium Tin Nickel	Widely distributed in foods	Function unknown (essential for animals)	Not reported in man	Industrial exposures: Silicon - silicosis Vanadium - lung irritation Tin - vomiting Nickel - acute pneumonitis
Selenium	Seafood, meal, grains	Functions in close association with vitamin E	Anemia (rare)	Gastrointestinal disorders, lung irritation
Manganese	Widely distributed in foods	Constituent of enzymes involved in fat synthesis	In animals: poor growth, disturbances of nervous system, reproductive abnormalities	Poisoning in manganese mines: generalized disease of nervous system

Iodine	Marine fish and shellfish, dairy products, many vegetables	Constituent of thyroid hormones	Goiter (enlarged thyroid)	Very high intakes depress thyroid activity
Molybdenum	Legumes, cereals, organ meats	Constituent of some enzymes	Not reported in man	Inhibition of enzymes
Chromium	Fats, vegetable oils, meats	Involved in glucose and energy metabolism	Impaired ability to metabolize glucose	Occupational exposures: skin and kidney damage
Cobalt	Organ and muscle meats, milk	Constituent of vitamin B-12	Not reported in man	Industrial exposure: dermatitis and diseases of red blood cells
Water	Solid foods, liquids, drinking water	Transport of nutrients Temperature regulation Participates in metabolic reactions	Thirst, dehydration	Headaches, nausea Edema High blood pressure

Source: Slightly modified from Scrimshaw and Young [14].

dietary lysine is converted to carnitine in the rat [15], whereas diets low in lysine lead to reduced levels of carnitine in heart and skeletal muscle [16].

Fats are concentrated sources of dietary energy and also serve as carriers of the fat-soluble vitamins. Lipids play essential roles in many enzyme reactions and in the maintenance of cell membrane structure and function. Dietary fats are also the source of essential fatty acids, such as linoleic (18:2n-6), linolenic (18:3n-3), and arachidonic (20:4n-6) acids which are involved in the maintenance of normal membrane structure and function [17]. In addition, they are precursors of a class of "local" hormones, the prostaglandins. These fatty acid derivatives exhibit diverse actions, involving secretion, blood platelets, smooth muscle metabolism and modulation of nervous system activity. Thus it has recently been shown that an increase in dietary linoleic acid can influence prostaglandin biosynthesis, decrease the thrombotic tendency of blood platelets and improve heart function [18].

Although dietary fats are important in the maintenance of normal metabolism and health, excessive intakes can promote disease. It is judged from epidemiological studies that excessive intake of saturated fats is one of the risk factors associated with the development of coronary heart disease [19] and possibly with various types of cancer [20]. However, there is considerable debate over the extent to which these diseases are due to specific dietary factors as opposed to other environmental variables.

Dietary carbohydrates are also important sources of energy intake. The nutritionally important carbohydrates are: (a) the monosaccharides (glucose or fructose), (b) disaccharides (sucrose, lactose), (c) polysaccharides (starch, glycogen). In human nutrition the celluloses and other plant polysaccharides are not used for energy because they are not hydrolyzed by secretions of the human digestive tract. However, they contribute to dietary fiber intake (unavailable carbohydrates and lignin, a phenyl-propane derivative), which now appears to play an important role in the maintenance of normal gastrointestinal function, metabolism, and health. However, the significance of dietary fiber in relation to human health and disease deserves further study [21], and this is currently an active area in nutrition research.

Available carbohydrates such as sucrose, lactose, and starch are not specifically required for human health, as far as is known. However, they contribute about 40% of energy intake, with sucrose consumption comprising from 15-20% of the total energy intake in the American diet [22].

Vitamins and minerals comprise the remaining two classes of essential nutrients. Both vitamins and minerals appear in foods in a variety of forms, the chemical nature of which may have considerable nutritional importance, since it determines whether the nutrient is made available to the body during the process of digestion and absorption. Thus, a measure of the total amount of various nutrients in a food may not be useful information for the nutritionist; it is important to know the fraction of total nutrient content available to the body. The subject of mineral availability is discussed in more detail in Chapter 5.

There are 13 vitamins necessary for normal growth, reproduction, and maintenance of health. Their metabolic functions are diverse; for example, thiamine, pyridoxine, riboflavin, and niacin act as coenzymes, and a low dietary intake of these vitamins causes reduced activities of enzymes associated with energy transfer and utilization in amino acid metabolism.

Vitamins A, D, E, and K have attracted considerable research interest in recent years, and significant advances have been made in understanding their metabolism and mechanisms of action. For example, the recent discovery of a retinol-phosphate-mannose-glycolipid intermediate suggests that vitamin A may play a primary role in glycoprotein synthesis [23]. Vitamin K now appears to be required for the conversion of a precursor to prothrombin in a step involving the γ-carboxylation of glutamic acid residues [24]. The identification of $1\alpha,25(OH)_2$-D_3(1,25-dihydroxyvitamin-D_3) as the active form of vitamin D [25,26], and its exclusive formation in the kidney, provides a final example of this new knowledge. Indeed, recent work leads to the conclusion that, in biochemical and physiological terms, vitamin D, and possibly vitamin A, are better defined as hormones [27]. Furthermore, other developments have led to the successful application of vitamin D metabolites and analogs in the treatment of various disorders of mineral metabolism [28].

The mineral elements constitute the final class of nutrients to be discussed. The so-called macroinorganic elements (sodium, potassium, chlorine, calcium, magnesium, phosphorus) are present in the body in significant quantities, their combined mass being about 3 kg in adult men. The microelements, or trace minerals, are required only in small amounts, usually less than 30 mg per day. Their body content is about 30 g.

The major mineral elements, such as sodium, potassium, calcium, magnesium and chloride, fulfill electrochemical functions. Calcium, potassium, magnesium, copper, and zinc participate as catalysts in enzyme systems, and calcium, phosphorus, and fluorine play a role in maintaining the structure of hard tissues, bones and teeth. Other essential mineral elements include iodine for formation of thyroid hormones (tri- and tetraiodothyronine), and iron as a constituent of heme as well as enzymes concerned with oxidation-reduction reactions [29]. Dietary iron inadequacy leads to low blood levels of hemoglobin, which is characterized clinically by weakness and a reduced capacity to carry out hard physical labor [30].

Improved methods for the quantitative analysis of trace elements in body fluids and tissues, and development of systems for rearing experimental animals under highly controlled conditions, have led to expansion of the list of essential mineral nutrients. Elements previously considered to be hazardous, such as nickel, tin, vanadium, and silicon, now appear to be essential for animal health [31,32]. These minerals are widely distributed in nature and in the human food supply, so that a primary dietary deficiency appears to be unlikely in humans. However, changes in the balance of intake of trace elements may have important consequences for long-term health and be among the factors responsible for increases in the incidence of heart disease, arteriosclerosis, and various forms of cancer in the affluent populations of North America, Europe, and Japan [33].

III. FACTORS AFFECTING NUTRIENT REQUIREMENTS

Various factors affecting the nutrient requirements of individuals and populations are listed below.

1. Dietary Factors
   - Chemical form of nutrient
   - Presence or level of other dietary constituents
   - Food processing
2. Host Factors
   - Age
   - Sex
   - Genetic
   - Physiological States
     - Growth
     - Pregnancy
     - Lactation
     - Aging
   - Pathological States
     - Metabolic disease
     - Trauma
     - Neoplasia
     - Other stress and drugs
3. Environmental Factors
   - Physical
     - Temperature
     - Altitude
   - Biological
     - Infectious agents
   - Social
     - Dietary habits
     - Environmental sanitation
     - Personal hygiene
     - Physical activity

A fundamental component of variation in requirements is that introduced by genetic differences (host category). The inborn errors of metabolism [11, 34] are extreme examples of the nutritional implications of genetic variation in humans. Although many of these errors have important nutritional implications, and in certain diseases [11] successful management depends upon appropriate dietary modification to restrict or greatly increase intake of a particular nutrient, these conditions are of medical importance rather than of broad public health significance.

The issue of genetic differences in minimum nutrient requirements among normal human populations has not yet been examined extensively, and much more comparative information is needed on nutrient requirements of groups of differing ethnic and cultural backgrounds. Sex differences are also important in relation to minimum nutrient needs, as males and females differ in energy expenditure [35] and in response to undernutrition [36]. However, apart from the recognition of increased iron loss in premenopausal women, there has been insufficient investigation of sex differences in relation to minimum physiological needs for nutrients.

Those host factors associated with growth, development, pregnancy, and lactation are generally taken into account in the estimation of human nutrient requirements and recommended dietary allowances. Nutrient needs are relatively high during the growth and developmental period of life, and fall with attainment of adulthood. Furthermore, pregnancy and lactation increase nutrient needs. Age considerations in discussions of human requirements are usually limited to the period of growth and development, but comprehensive revisions of national and international dietary allowances should include specific recommendations for the elderly, especially because their relative numbers are increasing in the populations of the technically developed regions.

Another host factor to be considered is the effect of stress on nutrient requirements, particularly that arising from infection or trauma. The metabolic consequences of acute systemic infections have been extensively documented [37]. First, there is an increased rate of synthesis of immunoglobulins and other proteins characteristic of the acute phase reaction. This is followed by catabolic responses that result in increased losses of body nitrogen, vitamins A and C, iron, zinc, and other nutrients, as well as decreases in blood levels of these nutrients. Acute or chronic infections may also interfere with absorption of nutrients whenever the gastrointestinal tract is significantly involved, as in cases of diarrhea and heavy concentrations of intestinal parasites.

The net result of these processes is depletion of tissue nutrients during the acute phase of the infection, followed by a physiological increase in the need for nutrients during the recovery phase. However, the quantitative data are inadequate to determine how much nutrient intakes should be increased to meet the additional demands created by infections and other stressful stimuli, including fever, anxiety, pain, and physical trauma.

Dietary factors that affect nutrient requirements are numerous and cannot be discussed extensively here. For example, the form of dietary iron, type of diet, and composition of individual meals affect the utilization of iron and partially determine the iron intake necessary to meet individual requirements. Figure 3 shows the effect of the presence of bread rolls and meat on the absorption of a test dose of tagged ferrous sulfate. It is now well established that meat improves overall iron absorption [38,39]. Furthermore, Cook et al. [38] found that bread reduced the availability of iron, since ferrous sulfate consumed alone was absorbed about four times more effectively than when it was incorporated into bread. Thus, a better understanding of the factors affecting iron absorption and utilization is needed to determine minimum intakes and to understand variation in requirements among apparently similar individuals.

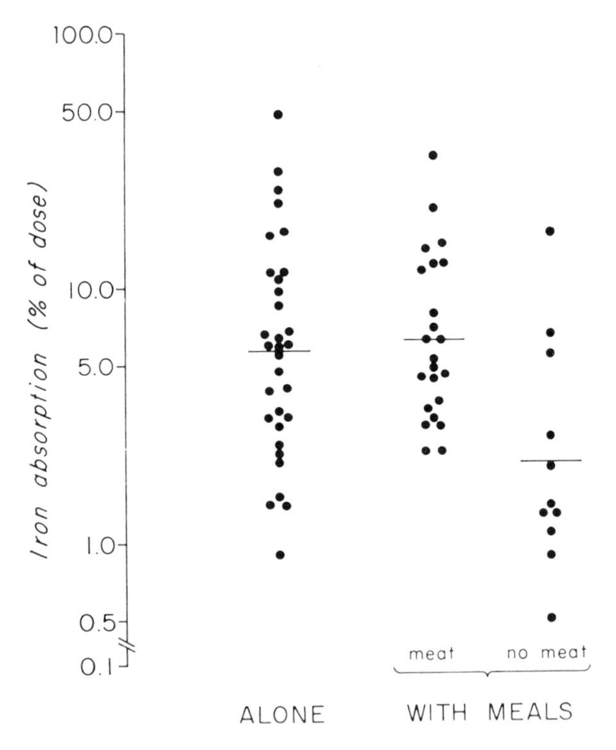

Fig. 3. Absorption of 3 mg supplement of tagged ferrous sulfate in 60 g dinner roll. Rolls were consumed alone or with meals, and in latter case with or without meat. Horizontal bar is geometric mean (from Cook et al. [38]).

A second example of the effect of dietary factors on estimated nutrient requirements is found in studies of protein needs in adult subjects. It has long been recognized that the level of energy intake influences the efficiency of utilization of dietary protein [40,41]. Inoue et al. [42], Calloway [43], and ourselves [44,45] have observed that relatively small changes in the level of energy intake in humans have marked effects on utilization of dietary nitrogen, with higher energy intakes increasing nitrogen retention. The net effect is a reduction of minimum protein intake estimated as necessary for body nitrogen maintenance.

Previous dietary history, as well as current diet, may have an important effect on the nutrient requirements of an individual or population, as suggested by studies on calcium requirements. Table 2 summarizes results of two studies, one by Hegsted et al. [46] with Peruvian adult male prisoners, and the other by Steggerda and Mitchell [47] with U.S. adult subjects. In these two studies, markedly different estimates of the mean daily calcium requirement were obtained, those determined by Hegsted et al. [46] with Peruvians being much lower than those obtained with healthy U.S. subjects, whose intakes of calcium prior to the experiments were probably generous.

Because low calcium intakes result in enhanced absorption of calcium and reduce endogenous loss [48], the difference in the estimates obtained in the two studies may be due to adaptive changes in calcium metabolism. However, it is not known for certain that the estimates reflect true differences in calcium needs as opposed to methodological differences. Furthermore, Walker and Linkswiler [49] have demonstrated marked effects of the level of dietary protein intake on calcium retention and balance, and the extent to which differences in protein intake affected the outcome of the calcium balance studies on Peruvian and U.S. subjects is also unknown.

TABLE 2 Comparison of Estimated Mean Calcium Requirement in Two Groups of Adult Men

Group	Mean daily calcium requirement (mg/kg)	Author and reference
10 Peruvian men (30-56 yr.)	0.5 (0-7.1)[a]	Hegsted et al. [46]
13 U.S. men (20-44 yr.)	7.4	Steggerda and Mitchell [47]

[a]Range of requirements indicated in parentheses.

Environmental factors, including sanitation and personal hygiene, also affect nutrient needs of a population, since they have an influence on the prevalence of infection which in turn affects nutrient utilization. In addition, climate will partially determine the dietary intake levels of nutrients which are adequate for a given population, since differences exist in the capacity of foods to meet nutrient needs, and climate determines types of foods which are available.

All of the factors affecting nutrient requirements (see above) make it difficult to be precise about the nutrient needs of humans. Furthermore, the quantitative significance of many of these factors is unknown, even though this knowledge is critical for the development of recommended dietary allowances.

IV. METHODS OF ESTIMATING NUTRIENT REQUIREMENTS IN MAN

It is difficult to estimate quantitative requirements for essential nutrients in humans, and the status of current knowledge in this area is generally unsatisfactory. The choice of the appropriate criteria for assessing these requirements is particularly challenging [50].

For animals, the requirements for food or individual nutrients can be judged in relation to certain economic and production criteria that are relatively easy to measure. However, the requirements for human health and well-being cannot be so readily assessed. Maximal resistance to disease, optimal physical fitness or healthy long life would be desirable criteria, but they cannot be measured in experimental situations. Therefore, other criteria must be used.

Most of the quantitative requirements for essential nutrients in humans are based on the concept of preventing clinically evident nutritional disease. Figure 4 depicts schematically the stages of development of nutritional disease with eventual appearance of clinical symptoms that are frequently nonspecific. The sequence begins with an inadequate supply of a nutrient to cells and body tissues. The inadequacy may arise from a primary dietary deficiency or other factors that reduce absorption or retention of a nutrient, or increase the relative requirement.

One approach to quantification of nutrient requirements is the epidemiological method, which involves assessing the nutrient intakes of populations free of nutritional disease and comparing them with intakes in populations where specific nutritional deficiency occurs. This is not an exact procedure, however, as it is difficult to determine precisely the nutrient intakes of free subjects, as well as to assess nutritional status.

Another type of study is based on the metabolic or balance technique, which involves assessment of minimum intake necessary to prevent, on the one hand, clinically evident signs and symptoms of nutrient deficiency [51,

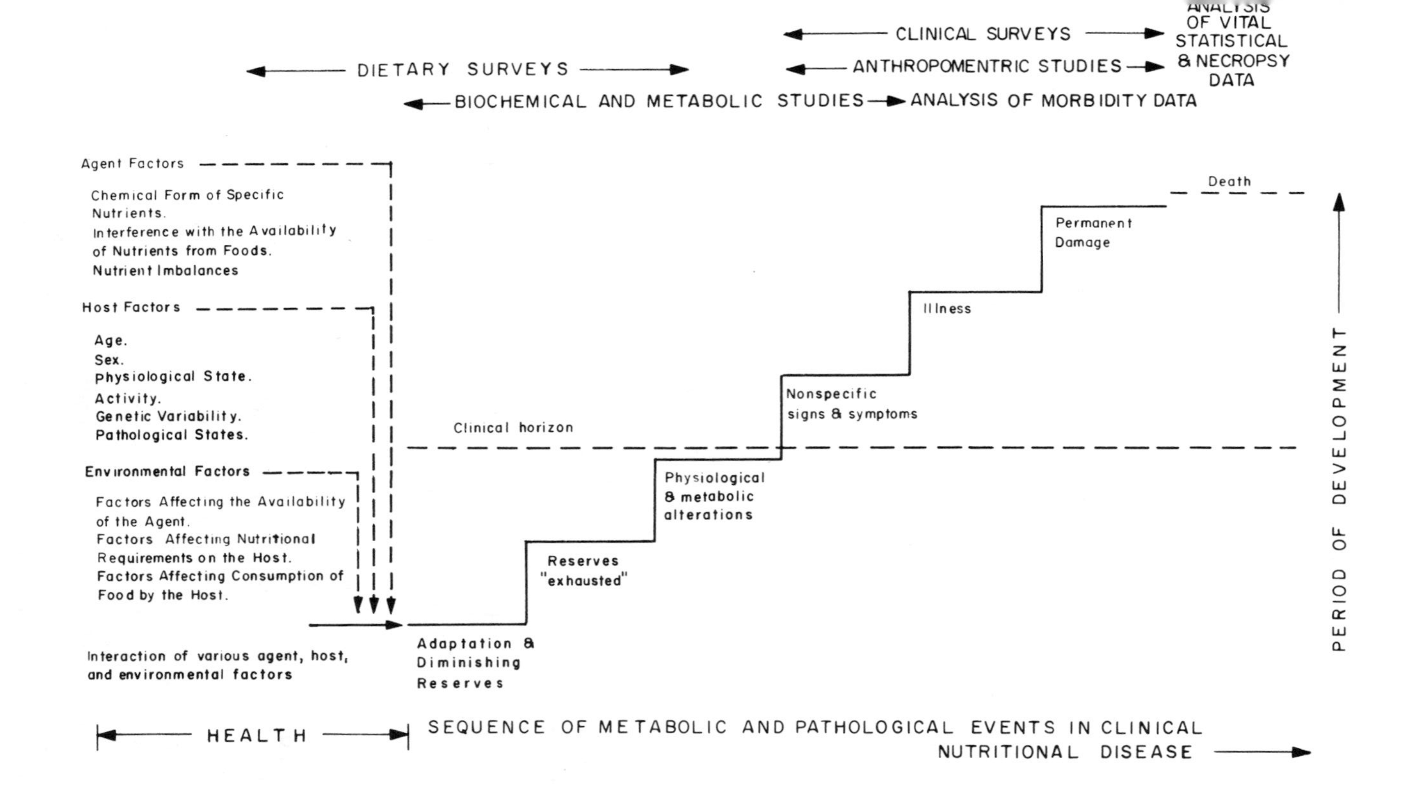

Fig. 4. Sequence of events in the development of clinically evident nutritional disease. The agent, host, and environmental factors which determine requirements and/or nutritional state are shown. Furthermore, this figure gives an indication of how information obtained from surveys and metabolic studies relates to the various nutritional states. (Figure kindly provided by G. Arroyave [INCAP]).

52], such as failure to support growth in the young or maintain body composition in adults, or on the other hand, changes at the subclinical stage of nutritional disease.

At the subclinical level, when intake or availability of a nutrient is inadequate, the content of the nutrient or its products will decrease in body fluids and tissues. In time, biochemical and pathological lesions will develop, which are expressed as altered enzyme activity, reduced rates of formation of tissue protein and other metabolically active compounds, and changes in efficiency of energy utilization [53]. To study these changes, volunteers are fed highly controlled diets, which may be very different in form and composition from free-choice diets. Under the study conditions, the level of the test nutrient intake is altered to determine the minimum level of intake necessary to meet a given criterion, such as the maintenance of a specific biochemical function. In the study of thiamine requirements the index might be the activity of erythrocyte transketolase [54], whereas in the case of protein requirements of children, the level of plasma albumin might serve as the biochemical index of intake adequacy [55].

Although the metabolic approach is more precise than the dietary-epidemiological method, such studies with humans are laborious, time-consuming, and expensive. In addition, their design and conduct is often restricted by ethical considerations. For these reasons, metabolic studies usually involve the participation of a limited number of volunteers. Nevertheless, the metabolic balance technique, or modifications of it, has been used extensively to determine nutrient requirements of humans. For example, this approach has been used for measuring the requirements for protein [56] and essential amino acids [57], and for calcium [58], zinc [59], magnesium [60], and iron [61].

However, the limitations of this technique should be recognized, and have been discussed previously by us in reference to protein and amino acid requirements [62], as well as by others [63-65]. Thus, this technique presents a problem in not providing information about where in the body the nutrient is retained or how it is used. For example, body nitrogen balance can be achieved at various intakes of protein which are below usual dietary intakes [14]. Yet, there is no direct evidence that tissues and body organs are functioning adequately, and that health will be maintained at the minimum nitrogen intake necessary to support body nitrogen balance, a criterion used in many studies of protein and amino acid needs of adults. Thus, sole reliance on nutrient balance determinations is an inadequate approach to the definition of human requirements, and our [44,45,66] recent long-term metabolic nitrogen balance studies conducted in healthy young men further emphasize this point.

The duration of dietary periods in metabolic balance studies is usually short, often 10 days or less in infants and children, and not over two to three weeks in adults. The long-term nutritional and health significance of these relatively brief experimental diet periods has not been critically examined. Thus, the present estimates of nutrient requirements based on results of short-term metabolic and balance experiments must be accepted with caution.

All of these methodological problems must be considered in evaluating published data on human nutrient requirements. For some of the known essential nutrients, no quantitative studies have been made, and only epidemiological data are available. In some cases, it has been necessary to extrapolate data obtained in animal experiments in order to estimate adequate dietary intakes for humans.

Finally, it is important to point out that the accuracy and precision of the various methods for estimating human requirements are usually not known. Rarely are replicate studies conducted within the same individual to determine the reproducibility of the estimated requirement for a given nutrient.

V. RECOMMENDED DIETARY ALLOWANCES

A. Purpose

Determination of the quantitative needs for essential nutrients based on the above approaches serves as a data base for the development of recommended dietary allowances (RDAs). The latter are needed as guidelines in the planning of diets and food supplies and in nutritional labeling of foods, as well as for evaluating nutritional adequacy from food consumption data.

As discussed earlier, the minimum requirement for a nutrient varies among apparently similar individuals, and RDAs are designed to meet the nutritional needs of practically all healthy persons within a population [67-69]. Current dietary standards represent judgments based on available studies, and since the data on human nutrient requirements are still limited, it is not surprising that the recommendations of different committees do not always agree. A comparison of some of the dietary allowances proposed by various committees is shown in Table 3 to illustrate this point.

B. Variation in Relation to the RDA

The issue of variation in nutrient needs among individuals is of particular importance in relation to development of dietary standards, and it is worthwhile to discuss the problem briefly in relation to RDAs.

In any healthy population, variation in nutrient requirements among individuals is usually assumed to be normally distributed. However, the data available on this aspect of human nutrition are very limited. As shown in Fig. 5, we have observed that the obligatory urinary nitrogen losses of a large group of M.I.T. students followed a Gaussian distribution. The assumption was made here that the minimum protein requirement of similar individuals was normally distributed to a quantitatively similar extent [73], but this assumption is not yet proven. More significant perhaps is the fact

TABLE 3 Comparisons of Recommended Intakes for Selected Nutrients in Young Men, as Proposed by Different Authorities

Nutrient, Daily Requirement	United States, 1974 [68]	Canada, 1974 [71]	West Germany, 1975 [72]	FAO/WHO, 1974 [69]
Vitamin C (mg)	45	30	75	30
Iron (mg)	10		12	5-9[a]
Folic acid (μg)	400[b]	200[c]	400	200[c]
Calcium (mg)	800		800	400-500
Protein (g/kg)	0.8[d]		0.9	0.81[e]

Source: Adapted from Truswell [70].
[a]For diets in which 10-25% energy is derived from animal sources or soybean.
[b]Total food folacin as determined by Lactobacillus casei assay.
[c]Expressed as free folate.
[d]Assuming NPU (net protein utilization) of diet 0.75.
[e]Assuming NPU of diet 0.70, compared with egg or milk.

that data comparable to those shown in Fig. 5 are not available for any other essential nutrients.

Nevertheless, the mean requirement and dietary allowance for a population group can be estimated (see Fig. 6), assuming normal distribution. However, it is impractical to develop recommendations appropriate to all members of a population, since a few normal individuals in the group will need much more of a nutrient than can be recommended for the others. For nutrients other than energy, the mean requirement plus two standard deviations is regarded as a reasonable recommendation, since this intake level would be sufficient to satisfy the needs of 97.5% of the population.

Therefore, it is important to know the variance in requirements for a given nutrient within the target population. For example, in the report of the FAO/WHO Committee on Energy and Protein Requirements [74], it was assumed that the coefficient of variation in protein requirements was about 15%. Thus, the estimated mean requirement plus 30% of this value should be sufficient to cover 97.5% of the population.

We have recently applied four statistical approaches to estimating dietary protein allowances, using nitrogen balance data obtained in young men [75]. These approaches are schematically depicted in Fig. 7.

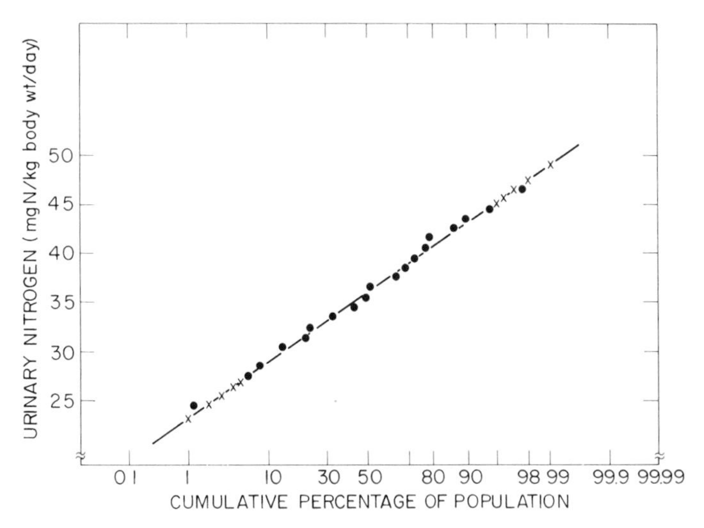

Fig. 5. Probability plot of obligatory urinary nitrogen excretion in 83 young male M.I.T. students. Solid dots are actual values and crosses indicate estimates of highest and lowest percentiles (from Scrimshaw et al. [73]).

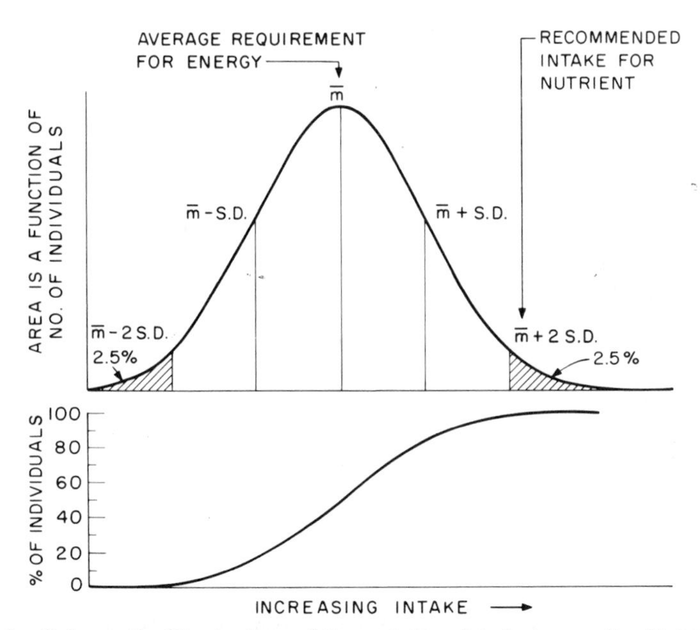

Fig. 6. Schematic illustration of the relationship between the distribution of nutrient requirements among individuals within a population and intake of a nutrient necessary to meet the needs of nearly all members of the population group. (Note: unlike RDAs for nutrients, the energy allowance is based on the mean requirement for that population.)

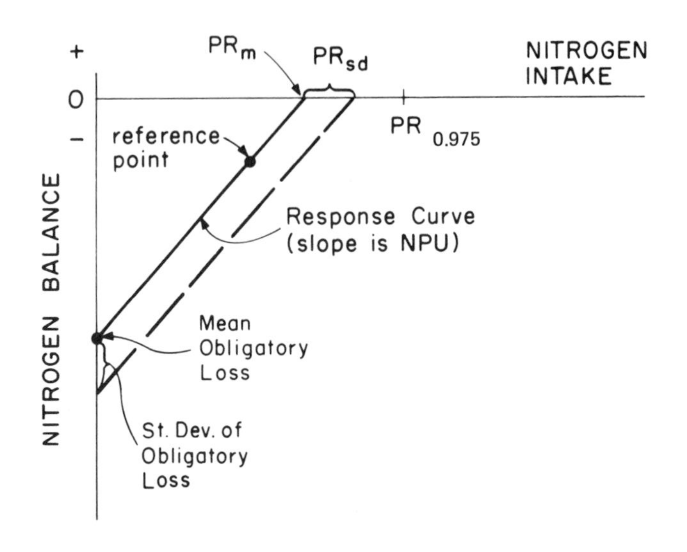

(A) Single-level method

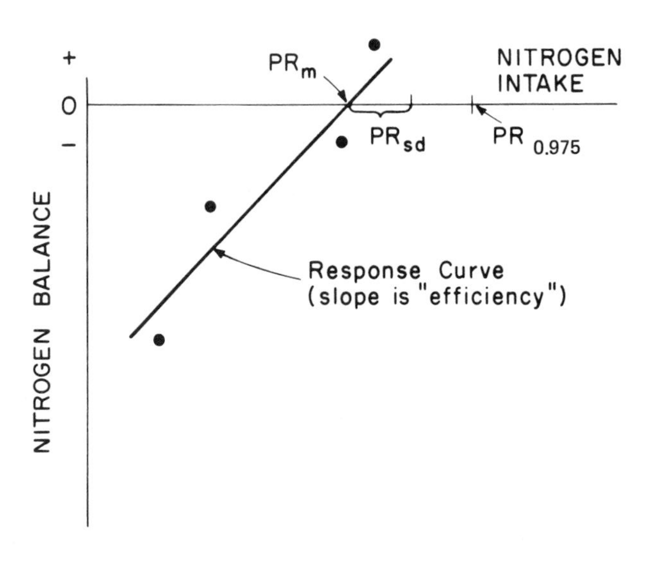

(B) Multiple level - constant variance method

Fig. 7. Four methods for estimating a protein allowance from nitrogen balance data. (PR_m = mean protein requirement, PR_{sd} = standard deviation of protein requirement, and $PR_{0.975}$ = 97.5% protein allowance) (from Rand et al. [75]).

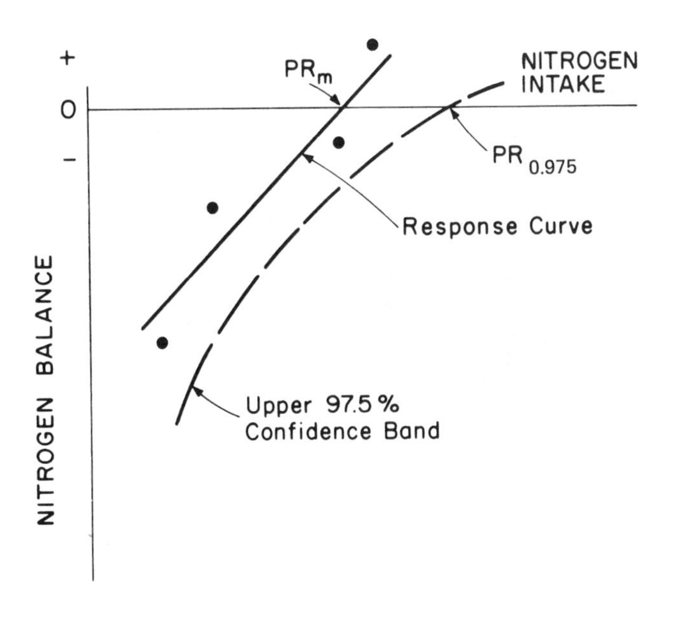

(C) Multiple level - confidence band method (pooled data)

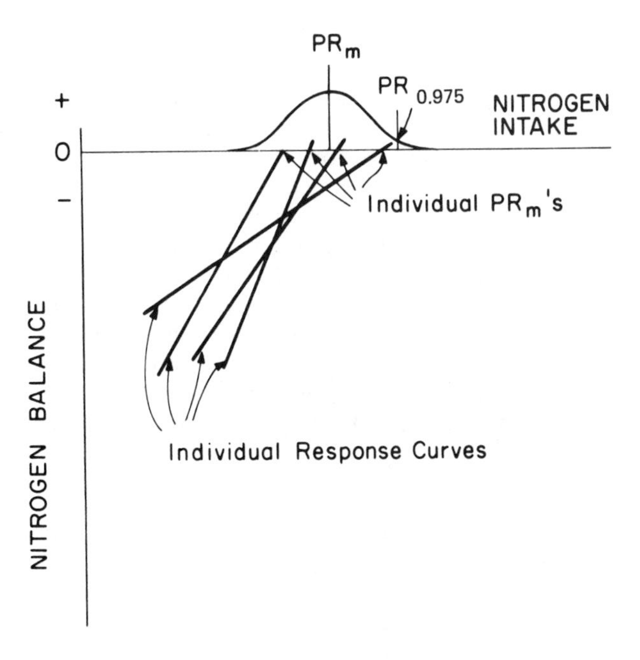

(D) Multiple level - confidence band method (individual responses)

Two separate statistical problems are involved. First, the mean or average requirement for that population (PR_m) must be estimated. With respect to protein requirements, the usual approach is to determine the relationship between nitrogen balance and nitrogen intake, and to define PR_m as the minimum intake necessary to achieve zero nitrogen balance in adults, or a given level of nitrogen retention in growing infants and children. Secondly, it is necessary to estimate the variability in requirements within a population; the symbol PR_{sd} is used here for the standard deviation of the population protein requirement. In Fig. 7, four approaches to estimating this quantity are indicated. The two quantities, PR_m and PR_{sd}, can then be used to estimate the protein allowance for a population.

In the first two methods for estimating variability in requirements, the single- and multiple-level constant variance methods (Figs. 7a and 7b), the standard deviation of the population requirement is assumed to be equivalent to the coefficient of variability of total obligatory nitrogen losses, i.e., nitrogen losses in subjects after brief adaptation to a protein-free diet. The difference between these two approaches lies only in how the mean requirement is estimated. If the nitrogen balance response curve is linear throughout the submaintenance to maintenance range of nitrogen intake, the population allowances derived from the two methods are identical. However, for both methods the coefficient of variation in nitrogen utilization is assumed to be constant over the entire submaintenance to maintenance range of nitrogen intake.

In the third method (multiple level, confidence band method) the above assumption is eliminated and the variation in requirements is calculated directly from the regression line of nitrogen balance at graded intakes of protein. The advantage of this method (Fig. 7c) is that the variability of nitrogen needs is calculated directly from the nitrogen balance data obtained in the submaintenance to near-maintenance level of nitrogen intake.

Finally, the fourth method (multiple level, individual response) is comparable to the third method, consisting of analysis of nitrogen balance data for each individual and calculations of tolerance intervals (Fig. 7d).

To illustrate these four techniques, we have applied them to nitrogen balance data obtained from seven young adult men consuming graded levels of beef protein [76]. Table 4 shows the predicted mean requirements and the allowances sufficient to satisfy 97.5% of the population, using the four different techniques. Notice that these methods are similar in terms of the estimated mean requirement (PR_m). However, for the 97.5% protein allowances ($PR_{0.975}$), the estimates of requirements increase from methods 1. through 4., reflecting the additional sources of variation taken into account in successive methods. In addition, the variability in nitrogen losses at protein-free intakes does not adequately predict variability in utilization of proteins where intake is close to requirement levels. Based on this analysis, the size of protein allowance will depend in part upon the method and approaches used to determine the variability in the nitrogen balance results.

TABLE 4 Estimates of the Protein Allowance for Young Men Given Beef Protein Using Four Methods of Analysis of Nitrogen Balance Data[a]

Method of estimation	Mean daily requirement	Daily population allowance
	(mg N/kg body wt)	
Single test level	71	91
Multiple level, constant variance	82	104
Multiple level, confidence band (pooled data)	82	110
Multiple level, confidence band (individual responses)	83	121

Source: From Rand et al. [75].
[a]A description of the four methods listed is given in Fig. 7. The nitrogen balance data used in these calculations were those reported by Young et al. [76].

These data also help to illustrate the problems that arise in formulating population allowances for dietary nutrients on the basis of a limited number of subjects. For the purposes of determining allowances applicable to entire populations of countries, it is of great importance to study large numbers of subjects. Many different national populations should be represented in these investigations, if these estimates are to be safely applied to the world at large. Unfortunately, for most allowances so far proposed, the data are dependent upon a limited body of information.

C. Nutrient Availability in Relation to the RDA

An additional factor that needs to be considered in setting RDAs is the source and availability of nutrients in foods. For some nutrients the dietary requirement may be met by a precursor, as for example in the case of vitamin A, where an alternative dietary source is provided by the carotenoids. The most common source of "vitamin A" activity is β-carotene, which in the diets of populations of developing countries may provide 60 to 100% of vitamin A needs.

The Joint FAO/WHO Expert Group on Requirements of Vitamin A, Thiamine, Riboflavine and Niacin [77] concluded, for guideline purposes, that on a weight basis β-carotene has only one-sixth of the activity of retinol when ingested by man. However, the relative activity varies considerably with different foods and methods of preparation. The RDA for vitamin A is expressed, therefore, as "retinol equivalents," where

1 retinol equivalent = 1 μg of retinol
= 6 μg of β-carotene
= 12 μg of other vitamin A active compounds

The availability of nutrients from different foodstuffs depends on the efficiency of absorption. For iron this efficiency is low, only 5-25%, depending upon the composition of the diet, where availability is greatest in meat and lowest in cereals (see Table 5). Therefore, the level of the RDA will depend upon the nature of the diet. Recommended intakes of iron, as proposed by the FAO/WHO, are related to three types of diet (see Table 6).

Similarly, in the RDAs for protein, provision is made for differences in the quality of the protein. Thus, the 1973 FAO/WHO [74] recommendations are given in terms of egg or milk protein, and a method is described for adjusting the recommendation for proteins of lower quality. Similarly, in the recommendations for protein allowances by the U.S. Food and Nutrition Board [68], it is assumed that the mixed proteins in the U. S. diet are utilized with only 75% efficiency compared to higher quality proteins used in experimental studies.

TABLE 5 Absorption of Iron from Different Types of Diet

Type of diet	Assumed upper limit for iron absorption by normal individuals, %
Less than 10% of energy from foods of animal origin or from soybeans[a]	10
10-25% of energy from foods of animal origin or from soybeans	15
More than 25% of energy from foods of animal origin or from soybeans	20

Source: Beaton and Patwardhan [67].
[a]Soybean is grouped with foods of animal origin owing to the high bioavailability of its iron content.

D. Current Recommendations for Energy and Nutrient Intakes

Table 6 summarizes the recommendations of the various FAO and FAO/WHO expert committees on nutrient allowances, and Table 7 summarizes the most recent recommendations of the U. S. Food and Nutrition Board [68]. The recommendations made by other national committees have been summarized in a recent report by the International Union of Nutritional Sciences [78], which may be consulted for comparative purposes. The reader is also referred to the original publications for detailed information concerning the derivation of allowances and a discussion of practical applications.

It must be emphasized that the recommended allowances, or "safe levels of intake" [74], are amounts considered sufficient for the maintenance of health in nearly all people. These recommendations are concerned with health maintenance and they are not intended to be sufficient for therapeutic purposes. They are not designed to cover the additional requirements that may arise during and following recovery from infection, or under conditions of malabsorption, trauma, and metabolic disease. Nor are they adequate for rehabilitation of undernourished individuals. Thus, the possible benefits that might occur with considerably higher intakes of vitamin C in a variety of clinical situations, such as the improvement of wound healing in surgical patients or prevention or amelioration of certain types of infection, are not relevant in considerations of RDAs [67]. The doses of vitamin C used in these situations are so large in comparison with normal intake that their effects may be regarded as pharmacological rather than as "vitamin" effects [67], and recommending such high intakes as a usual dietary practice for healthy populations cannot be justified [79-80].

As may be noted from Tables 6 and 7, not all of the essential nutrients appear in the main tables of the reports on recommended allowances. Table 8 lists the nutrients that appear in the main tables of these reports and indicates other nutrients that are mentioned in the texts.

E. Use of the RDAs in the Evaluation of Dietary Intake Data

Finally, the use of RDAs for evaluation of dietary intake data and nutritional status of populations will be briefly considered.

From the earlier discussion (Sec. V.B), it should be clear that most healthy individuals actually have requirements below the recommended levels, except in the case of energy allowances (see Fig. 6). In the latter instance there is good reason to believe that almost all individuals would have intakes that approximated their requirements, if the average intake of a population group were equal to the allowance. For all other nutrients where the RDA is set toward the upper level of requirements, the likelihood of an inadequate nutrient intake is low, when all members of a popu-

TABLE 6 FAO and FAO/WHO Recommendations for the Daily Intake of Energy, Protein, Iron, Calcium and Several Vitamins.

Age and sex	Weight (kg)	Expression of requirement	Energy (kcal)	Energy (MJ)	Protein (g)[a] A	Protein (g)[a] B	Protein (g)[a] C	Protein (g)[a] D
Infants								
6-8 months	8.2	per day	900	3.77	14	17	19	23
		per kg per day	110	0.46	1.62	2.02	2.30	2.71
9-11 months	9.4	per day	990	4.14	14	17	19	23
		per kg per day	105	0.44	1.44	1.80	2.04	2.40
Children, both sexes								
1-3 years	13.4	per day	1360	5.69	16	20	23	27
		per kg per day	101	0.42	1.19	1.48	1.68	1.99
4-6 years	20.2	per day	1830	7.66	20	26	29	34
		per kg per day	91	0.38	1.02	1.26	1.43	1.67
7-9 years	28.1	per day	2190	9.16	25	31	35	41
		per kg per day	78	0.33	0.88	1.10	1.25	1.47
Adolescents, male								
10-12 years	36.9	per day	2600	10.88	30	37	43	50
		per kg per day	71	0.30	0.80	1.01	1.15	1.36
13-15 years	51.3	per day	2900	12.13	37	46	53	62
		per kg per day	57	0.24	0.72	0.90	1.02	1.20
16-17 years	62.9	per day	3070	12.84	38	47	54	63
		per kg per day	49	0.21	0.62	0.75	0.85	1.00
Adolescents, female								
10-12 years	38.0	per day	2350	9.83	29	36	41	48
		per kg per day	62	0.26	0.77	0.95	1.08	1.27
13-15 years	49.9	per day	2490	10.42	31	39	45	52
		per kg per day	50	0.21	0.63	0.79	0.89	1.05
16-17 years	54.4	per day	2310	9.67	30	37	43	50
		per kg per day	43	0.18	0.58	0.69	0.78	0.92
Adults								
Reference man	65.0	per day	3000	12.55	37	46	53	62
		per kg per day	46	0.19	0.57	0.71	0.81	0.95
Reference woman	55.0	per day	2200	9.20	29	36	41	48
		per kg per day	40	0.17	0.52	0.65	0.74	0.87
During last half of pregnancy		per day	add 350	1.46	add 9	add 11	add 13	add 15
During first 6 months of lactation		per day	add 550	2.30	add 11	add 21	add 24	add 28

Source: Beaton and Patwardhan [67].

[a]"Safe levels of intake" of protein are described in terms of four different sources:

A Milk or egg protein (relative NPU = 100)
B Mixed diet rich in animal source foods (relative NPU = 80)
C Mixed cereal-legume diet with small amounts of animal source foods (relative NPU = 70)
D Staple cereal diets with few other sources of protein (relative NPU = 60)

[b]Iron requirements are described in terms of three types of diets classified by the proportion of the energy derived from animal sources or soybean:

A 25% or more
B 10-25%
C less than 10%

Iron (mg)[b] A	B	C	Calcium[c] (mg)	Vitamin A[d] (μg)	Vitamin D[e] (μg)	Thiamine (mg)	Riboflavin (mg)	Niacin (niacin equiva-lents)[f]	Folic acid[g] (μg)	Vitamin B_{12} (μg)	Ascorbic acid (mg)
5	7	10	500-600	300	10	0.4	0.5	5.9	60	0.3	20
5	7	10	500-600	300	10	0.4	0.6	6.5	60	0.3	20
5	7	10	400-500	250	10	0.5	0.8	9.0	100	0.9	20
5	7	10	400-500	300	10	0.7	1.1	12.1	100	1.5	20
5	7	10	400-500	400	2.5	0.9	1.3	14.5	100	1.5	20
5	7	10	600-700	575	2.5	1.0	1.6	17.2	100	2.0	20
9	12	18	600-700	725	2.5	1.2	1.7	19.1	200	2.0	30
5	6	9	500-600	750	2.5	1.2	1.8	20.3	200	2.0	30
5	7	10	600-700	575	2.5	0.9	1.4	15.5	100	2.0	20
12	18	24	600-700	725	2.5	1.0	1.5	16.4	200	2.0	30
14	19	28	500-600	750	2.5	0.9	1.4	15.2	200	2.0	30
5	6	9	400-500	750	2.5	1.2	1.8	19.8	200	2.0	30
14	19	28	400-500	750	2.5	0.9	1.3	14.5	200	2.0	30
			1000-1200	750	10	add 0.15	add 0.20	add 2.3	400	3.0	50
			1000-1200	1200	10	add 0.20	add 0.35	add 3.6	300	2.5	50

These sources range from typical North American diets to the cereals diets of Asia.
[c]"Safe practical allowances" of calcium.
[d]Expressed as retinol.
[e]Expressed as cholecalciferol; 2.5 μg = 100 I.U.
[f]Niacin equivalent = 1 mg niacin or 60 mg tryptophan.
[g]Expressed as free folic acid. Requirements expressed as "total folate" have been estimated to be twice as high as those shown.

TABLE 7 1974 Recommended Daily Dietary Allowances of the U.S. Food and Nutrition Board, National Academy of Sciences-National Research Council

								Fat-soluble vitamins			
		Weight		Height				Vitamin A Activity		Vitamin D	Vitamin E Activity[c]
Subject	Age (years)	(kg)	(lb)	(cm)	(in.)	Energy (kcal)	Protein (g)	(RE)[a]	(IU)	(IU)	(IU)
Infants	0.0-0.5	6	14	60	24	kg x 117	kg x 2.2	420[c]	1400	400	4
	0.5-1.0	9	20	71	28	kg x 108	kg x 2.0	400	2000	400	5
Children	1-3	13	28	86	34	1300	23	400	2000	400	7
	4-6	20	44	110	44	1800	30	500	2500	400	9
	7-10	30	66	135	54	2400	36	700	3300	400	10
Males	11-14	44	97	158	63	2800	44	1000	5000	400	12
	15-18	61	134	172	69	3000	54	1000	5000	400	15
	19-22	67	147	172	69	3000	54	1000	5000	400	15
	23-50	70	154	172	69	2700	56	1000	5000		15
	51+	70	154	172	69	2400	56	1000	5000		15
Females	11-14	44	97	155	62	2400	44	800	4000	400	12
	15-18	54	119	162	65	2100	48	800	4000	400	12
	19-22	58	128	162	65	2100	46	800	4000	400	12
	23-50	58	128	162	65	2000	46[b]	800	4000		12
	51+	58	128	162	65	1800	46	800	4000		12
Pregnant						+300	+30	1000	5000	400	15
Lactating						+500	+20	1200	6000	400	15

[a]Retinol equivalents.

[b]Assumed to be all as retinol in milk during the first six months of life. All subsequent intakes are assumed to be half as retinol and half as β-carotene when calculated from international units. As retinol equivalents, three fourths are as retinol and one fourth as β-carotene.

[c]Total vitamin E activity, estimated to be 80% as α-tocopherol and 20% other tocopherols.

Water-soluble vitamins							Minerals					
Ascorbic Acid (mg)	Folacin[d] (μg)	Niacin[e] (mg)	Riboflavin (mg)	Thiamin (mg)	Vitamin B_6 (mg)	Vitamin B_{12} (μg)	Calcium (mg)	Phosphorus (mg)	Iodine (μg)	Iron (mg)	Magnesium (mg)	Zinc (mg)
35	50	5	0.4	0.3	0.3	0.3	360	240	35	10	60	3
35	50	8	0.6	0.5	0.4	0.3	540	400	45	15	70	5
40	100	9	0.8	0.7	0.6	1.0	800	800	60	15	150	10
40	200	12	1.1	0.9	0.9	1.5	800	800	80	10	200	10
40	300	16	1.2	1.2	1.2	2.0	800	800	110	10	250	10
45	400	18	1.5	1.4	1.6	3.0	1200	1200	130	18	350	15
45	400	20	1.8	1.5	2.0	3.0	1200	1200	150	18	400	15
45	400	20	1.8	1.5	2.0	3.0	800	800	140	10	350	15
45	400	18	1.6	1.4	2.0	3.0	800	800	130	10	350	15
45	400	16	1.5	1.2	2.0	3.0	800	800	110	10	350	15
45	400	16	1.3	1.2	1.6	3.0	1200	1200	115	18	300	15
45	400	14	1.4	1.1	2.0	3.0	1200	1200	115	18	300	15
45	400	14	1.4	1.1	2.0	3.0	800	800	100	18	300	15
45	400	13	1.2	1.0	2.0	3.0	800	800	100	18	300	15
45	400	12	1.1	1.0	2.0	3.0	800	800	80	10	300	15
60	800	+2	+0.3	+0.3	2.5	4.0	1200	1200	125	18+[f]	450	20
80	600	+4	+0.5	+0.3	2.5	4.0	1200	1200	150	18	450	25

[d]The folacin allowances refer to dietary sources as determined by *Lactobacillus casei* assay. Pure forms of folacin may be effective in doses less than one fourth of the recommended dietary allowance.

[e]Although allowances are expressed as niacin, it is recognized that on the average 1 mg of niacin is derived from each 60 mg of dietary tryptophan.

[f]This increased requirement cannot be met by ordinary diets; therefore, the use of supplemental iron is recommended.

TABLE 8 A Listing of Reports Concerned with Recommended Intakes and an Indication of the Nutrients Discussed in these Reports

Reports		Ref.
Technical Commission on Nutrition, League of Nations (1938)	(12 nutrients)[a]	81
Calories, calcium, phosphorus, iron, iodine, vitamin A, vitamin B_1, riboflavin, vitamin C, vitamin D, protein (fats).		
British Medical Association (1950)	(11 nutrients)	82
Calories, protein, Ca, Fe, vitamin A, vitamin D, thiamin, riboflavin, niacin, ascorbic acid, iodine.		
US National Research Council (1964)	(10 nutrients)	83
Calories, protein, Ca, Fe, vitamin A, thiamin, riboflavin, niacin, ascorbic acid, vitamin D.		
US National Research Council (1968)	(17 nutrients)	84
Energy, protein, vitamin A, vitamin D, vitamin E, ascorbic acid, folacin, niacin, riboflavin, thiamin, vitamin B_6, vitamin B_{12}, Ca, P, I, Fe, magnesium.		
Department of Health and Social Security (1969)	(10 nutrients + 16)	85
Energy, protein, thiamin, riboflavin, nicotinic acid, ascorbic acid, vitamin A, vitamin D, Ca, Fe (vitamin B_6, folic acid, vitamin B_{12}, pantothenic acid, vitamin E, essential fatty acids (EFA), sodium, potassium, chlorine, Mg, P, fluorine, Cu, zinc, manganese, iodine).		
FAO/WHO (Passmore, Nicol, Rao, Beaton & DeMaeyer, 1974)	(12 nutrients + 9)	69
Energy, protein, vitamin A, vitamin D, thiamin, riboflavin, niacin, folic acid, vitamin B_{12}, ascorbic acid, Ca, Fe (pyridoxine, I, F, Zn, Mg, Cu, chromiun, selenium, molybdenum).		
US National Research Council (1974)	(18 nutrients + 12)	68
Energy, protein, vitamin A, vitamin D, vitamin E, ascorbic acid, folacin, niacin, riboflavin, thiamin, vitamin B_6, vitamin B_{12}, Ca, P, I, Fe, Mg, Zn (EFA, vitamin K, pantothenate, biotin, Na, K, Cl, Cu, F, Cr, Mn, Mo).		
Deutsche Gesellschaft für Ernährung (1975)	(24 nutrients + 10)	72
Energy, protein, EFA, water, Na, Cl, K, Ca, P, Mg, Fe, I, F, vitamin A, vitamin D, vitamin E, thiamin, riboflavin, niacin, vitamin B_6, folic acid, pantothenic acid, vitamin B_{12}, vitamin C (biotin ,vitamin K, Cu, Cr, cobalt, Mn, Se, Zn, carbohydrate).		

Source: Truswell [70].

[a]Nutrients shown in parentheses are those mentioned in the text of the respective reports, but for which recommendations were not given in the main tables of the reports.

lation have intakes equal to or in excess of the RDA. However, even if the average intake of a population exceeds the allowance, all members of the group cannot be assumed to have adequate intakes, since this will depend upon intake distribution and variation in requirements among individuals. Similarly, if the dietary intake of an individual is substantially below the allowance, it does not necessarily mean that the individual is receiving an inadequate intake, although the risk of inadequacy is clearly increased [86, 87].

In addition, a further problem in evaluating the nutritional and health significance of dietary intake data is the inadequate status of information on the nutrient composition of foods. Many of the data on food composition are old, and although they are currently under revision, it is frequently unclear how the values were derived and whether the data are sufficiently representative for use in assessing dietary survey data. Furthermore, many important nutrients are not listed in the published tables on food composition, and there is uncertainty regarding the availability of many of the nutrients in foods.

VI. SUMMARY

Knowledge of the nutrient needs of man provides an essential basis for the rational design of national food and nutrition programs and for assessment of nutritional adequacy of food sources and diets. In this chapter, we have reviewed the approaches available for estimating nutrient requirements of individuals and population groups. The limitations of these approaches and factors affecting requirements were discussed.

The differences between minimum physiological requirements and recommended dietary allowances (RDAs) were considered, together with use of the RDA in interpretation of dietary data.

Although there is a continuing need for RDAs, the data on which they are based are highly inadequate. In particular, more quantitative information is needed about the effects of physiological state, dietary, and environmental factors on nutrient requirements for long-term health.

REFERENCES

1. H. N. Munro, in Mammalian Protein Metabolism (H. N. Munro, ed.), Vol. III, Academic Press, New York, 1969, pp. 3-19.
2. G. E. Gaull, D. K. Rossin, N. C. R. Raiha, and K. Heinonen, J. Pediat. 90:348 (1977).
3. G. E. Gaull, J. A. Sturman, and N. C. R. Raiha, Pediat. Res. 6:538 (1972).
4. H. N. Munro, in Mammalian Protein Metabolism (H. N. Munro, ed.), Vol. III, Chap. 25, Academic Press, New York, 1969, p. 133.

5. G. W. Beadle and E. L. Tatum, Proc. Natl. Acad. Sci. U.S. 27:499 (1941).
6. L. Pauling, Science 160:265 (1968).
7. I. B. Chatterjee, Science 182:1271 (1973).
8. I. B. Chatterjee, A. K. Majumder, B. K. Nandi, and N. Subramanian, Ann. N.Y. Acad. Sci. 258:24 (1975).
9. G. H. Beaton and J. M. Bengoa (eds.), Nutrition in Preventive Medicine, World Health Organization, Geneva, 1976.
10. D. M. Hegsted and L. M. Ausman, Nutrition Today 8:22 (1973).
11. C. R. Scriver and L. E. Rosenberg, Amino Acid Metabolism and Its Disorders, W. B. Saunders Co., Philadelphia, 1973.
12. A. Turnbull, in Iron in Biochemistry and Medicine (A. Jacobs and M. Worwood, eds.), Chap. 10, Academic Press, New York, 1974, p. 369.
13. J. C. Waterlow, Lancet ii:1091 (1968).
14. N. S. Scrimshaw and V. R. Young, Sci. Am. 235:50 (1976).
15. V. Tanphaichitr and H. P. Broquist, J. Biol. Chem. 248:2176 (1973).
16. V. Tanphaichitr and H. P. Broquist, J. Nutr. 103:80 (1973).
17. E. Aaes-Jorgensen, Physiol. Rev. 41:1 (1961).
18. A. J. Vergroessen, Nutr. Rev. 35:1 (1977).
19. D. M. Hegsted, R. B. McGandy, M. L. Myers, and F. J. Stare, Am. J. Clin. Nutr. 17:281 (1965).
20. E. L. Wynder, Cancer Res. 35:3388 (1975).
21. A. I. Mendeloff, Nutr. Rev. 33:321 (1975).
22. C. H. Ford, R. B. McGandy, and F. J. Stare, Prevent. Med. 1:426 (1972).
23. L. DeLuca, N. Maestri, G. Rosso, G. Wolf, J. Biol. Chem., 248:641 (1973).
24. J. W. Suttie and C. M. Jackson, Physiol. Rev. 57:1 (1977).
25. H. F. DeLuca, Am. J. Clin. Nutr. 29:1258 (1976).
26. M. R. Haussler, in: Present Knowledge in Nutrition (D. M. Hegsted, C. O. Chichester, W. J. Darby, K. W. McNutt, R. Stalvey, and E. H. Stotz, eds.), Chap. 10, Nutrition Foundation Inc., Washington, D.C., 1976, pp. 82-97.
27. H. F. DeLuca, Fed. Proc. 33:2211 (1974).
28. M. R. Haussler, Nutr. Rev. 32:257 (1974).
29. E. Frieden, Nutr. Rev. 31:41 (1973).
30. F. E. Viteri, in Proc. Int. Symp. on Malnutrition and Functions of Blood Cells, U.S.-Japan Cooperative Medical Science Program, 1972, pp. 559-583.
31. E. Carlisle, Fed. Proc. 33:1758 (1974).
32. K. Schwarz, Fed. Proc. 33:1748 (1974).
33. R. Masrioni, Bull. World Health Org. 40:305 (1969).
34. C. R. Scriver, Metabolism 22:1319 (1973).
35. J. V. G. A. Durnin, Proc. Nutr. Soc. 35:145 (1976).
36. E. M. Widdowson, Proc. Nutr. Soc. 35:175 (1976).

37. W. R. Beisel, in Nutrient Requirements in Adolescents (J. I. McKigney and H. N. Munro, eds.), Chap. 14, Academic Press, New York, 1976, p. 257.
38. J. D. Cook, V. Minnich, C. V. Moore, A. Rasmussen, W. B. Bradley, and C. A. Finch, Am. J. Clin. Nutr. 26:861 (1973).
39. M. Layrisse and C. Martinez-Torres, Progr. Hematol. 4:137 (1971).
40. D. H. Calloway and H. Spector, Am. J. Clin. Nutr. 2:405 (1954).
41. H. N. Munro, Physiol. Rev. 31:449 (1951).
42. G. Inoue, Y. Fujida, and Y. Niiyama, J. Nutr. 103:1673 (1973).
43. D. H. Calloway, J. Nutr. 105:914 (1975).
44. C. Garza, N. S. Scrimshaw, and V. R. Young, Am. J. Clin. Nutr. 29:2801 (1976).
45. C. Garza, N. S. Scrimshaw, and V. R. Young, Brit. J. Nutr. 37:403 (1977).
46. D. M. Hegsted, I. Moscoso, and C. L. C. Collazos, J. Nutr. 46:181 (1952).
47. F. R. Steggerda and H. H. Mitchell, J. Nutr. 45:201 (1951).
48. O. J. Malm, in The Transfer of Calcium and Strontium Across Biological Membranes (R. H. Wasserman, ed.), Academic Press, New York, 1963, p. 143.
49. R. M. Walker and H. Linkswiler, J. Nutr. 102:297 (1972).
50. J. Yudkin, Sci. J. 4:48 (1968).
51. W. H. Bartley, A. Krebs, and J. R. P. O'Brien, Med. Res. Council Spec. Rept., Ser. No. 280, H. M. Stationery Office, London, 1953.
52. R. E. Hodges, E. M. Baker, J. Hood, and H. F. Sauberlich, Am. J. Clin. Nutr. 22:535 (1969).
53. M. Kleiber, Nutr. Abst. Rev. 15:207 (1945-46).
54. M. Brin, in Newer Methods of Nutritional Biochemistry (A. A. Albanese, ed.), Vol. III, Chap. 9, Academic Press, New York, 1967, pp. 407-445.
55. R. G. Whitehead and G. A. O. Alleyne, Brit. Med. Bull. 28:72 (1972).
56. M. I. Irwin and D. M. Hegsted, J. Nutr. 101:385 (1971).
57. M. I. Irwin and D. M. Hegsted, J. Nutr. 101:539 (1971).
58. M. I. Irwin, and E. W. Keinholz, J. Nutr. 103:1019 (1973).
59. J. A. Halsted, C. J. Smith, and M. I. Irwin, J. Nutr. 104:345 (1974).
60. M. S. Seelig, Am. J. Clin. Nutr. 14:342 (1964).
61. J. Bowering, A. M. Sanchez, and M. I. Irwin, J. Nutr. 106:985 (1976).
62. V. R. Young and N. S. Scrimshaw, in Evaluation of Proteins for Humans (C. E. Bodwell, ed.), Chap. 2, Avi Publishing Co., Westport, Conn., 1977, pp. 11-54.
63. D. L. Duncan, Nutr. Abstr. Rev. 28:695 (1958).
64. D. M. Hegsted, J. Nutr. 106:307 (1976).
65. W. M. Wallace, Fed. Proc. 18:1125 (1959).
66. C. Garza, N. S. Scrimshaw, and V. R. Young, J. Nutr. 107:335 (1977).

67. G. H. Beaton and V. N. Patwardhan, in Nutrition in Preventive Medicine (G. H. Beaton and J. M. Bengoa, eds.), WHO, Geneva, 1976, p. 445.
68. National Research Council, Recommended Dietary Allowances, 8th rev. ed., National Academy of Sciences, Washington, 1974.
69. World Health Organization, Handbook of Human Nutritional Requirements, (R. Passmore, B. M. Nicol, M. Narayana Rao, G. H. Beaton, and E. M. DeMaeyer, eds.), Geneva, 1974. [Also issued as FAO Nutrition Studies, No. 28.]
70. A. S. Truswell, Proc. Nutr. Soc. 35:1 (1976).
71. Committee for Revision of the Canadian Dietary Standard, Bureau of Nutritional Sciences, 1974. Quoted by Nutr. Rev. 33:156 (1975).
72. Deutsche Gesellschaft für Ernährung, Empfehlungen für die Nahrstoffzufuhr, Umschau Verlag, Frankfurt am Main, 1975.
73. N. S. Scrimshaw, M. A. Hussein, E. Murray, W. M. Rand, and V. R. Young, J. Nutr. 102:1595 (1972).
74. FAO/WHO, Energy and Protein Requirements, Tech. Rep. Ser. No. 522, World Health Organization, Geneva, 1973.
75. W. M. Rand, N. S. Scrimshaw, and V. R. Young, Am. J. Clin. Nutr., 30:1129 (1977).
76. V. R. Young, L. Fajardo, E. Murray, W. M. Rand, and N. S. Scrimshaw, J. Nutr. 105:534 (1975).
77. FAO/WHO, Report of a Joint FAO/WHO Expert Group on Requirements of Vitamin A, Thiamine, Riboflavin and Niacin, WHO Tech. Rep. Ser. No. 362, Geneva, 1967.
78. Committee on International Dietary Allowances of the International Union of Nutritional Sciences (IUNS), Nutr. Abstr. Rev. 45:89 (1975).
79. A. E. Harper, J. Am. Dietet. Assoc. 64:151 (1974).
80. T. H. Jukes, Proc. Natl. Acad. Sci. 71:1949 (1974).
81. Technical Commission on Nutrition, League of Nations, Bull. Health Org. 7:470 (1938).
82. British Medical Association, Report of the Committee on Nutrition, British Medical Association, London, 1950.
83. National Research Council, Recommended Dietary Allowances, 6th ed., National Academy of Sciences, Washington, 1964.
84. National Research Council, Recommended Dietary Allowances, 7th ed., National Academy of Sciences, Washington, 1968.
85. Department of Health and Social Security, Recommended Intakes of Nutrients for the United Kingdom, H. M. Stationery Office, London, 1969.
86. G. H. Beaton and L. D. Swiss, Am. J. Clin. Nutr. 27:485 (1974).
87. L. D. Swiss and G. H. Beaton, Am. J. Clin. Nutr. 27:373 (1974).

CHAPTER 3

VITAMINS

Michael C. Archer
Steven R. Tannenbaum

Department of Nutrition and Food Science
Massachusetts Institute of Technology
Cambridge, Massachusetts

I. INTRODUCTION

The purpose of this chapter is to summarize the information available on the chemistry of vitamin losses in processed and stored foods. This subject has been treated in many review articles, some of which are listed under References. Special mention must be made of Nutritional Evaluation of Food Processing, which contains a large amount of detailed information on various foods in both the raw and finished state [1]. Rather than recapitulate and summarize the many studies carried out on individual foods and processes, the approach in this chapter is to review analytically the chemistry of the individual vitamins and the general factors leading to nutrient losses.

Information on the fate of vitamins in processed foods is reasonably adequate for only a few vitamins. Considerable information is available on ascorbic acid, provitamins A, and thiamine; less on riboflavin and vitamin B_6; and very little on folic acid and vitamin B_{12}. This is the result partially of the relative importance past investigators have attached to these nutrients and also of analytical problems. The amount of space devoted to an individual nutrient in this chapter is influenced both by the complexity of its chemistry and the amount of available information. At some future time, it is hoped that sufficient information will accumulate to fill the more obvious gaps.

II. GENERAL CAUSES FOR LOSS OF VITAMINS

All foods which undergo processing are subject to some degree of loss in vitamin and mineral content, even though availability of a nutrient is occasionally increased or some antinutritional factor inactivated. In general, food processing is carried out in a manner which attempts to minimize nutrient losses and maximize safety of the product. In addition to losses from processing, there are significant preprocess influences on nutrient content. These include genetic variation, degree of maturity, conditions of soil, use and type of fertilizer, climate, availability of water, light (length of day and intensity), and postharvest or postmortem handling.

A. Genetics and Maturity

Numerous examples of genetic variation are given in Harris and Karmas [1]. Data on the effect of maturity are more difficult to find, but an excellent example is that of tomatoes (Table 1). Not only does the ascorbic acid content vary over the period of maturity, but maximum vitamin content occurs when the tomato is in an immature state.

TABLE 1 Influence of Degree of Maturity on Ascorbic Acid Content of New Yorker Variety Tomatoes

Weeks from anthesis	Average wt (g)	Color	Ascorbic acid (mg %)
2	33.4	Green	10.7
3	57.2	Green	7.6
4	102.5	Green-yellow	10.9
5	145.7	Yellow-red	20.7
6	159.9	Red	14.6
7	167.6	Red	10.1

Source: Malewski and Markakis [2].

B. Preplant Handling

Most of the other agricultural variables listed above are known to influence content of some vitamin or mineral, but results are poorly documented or come from poorly designed experiments. The last category, which involves the history of the food from time of harvest or slaughter to time of processing, causes considerable variation in nutritional value. Other chapters of this volume deal with various biochemical processes in plant and animal tissues. Since many of the vitamins are also cofactors for enzymes or may be subject to degradation by endogenous enzymes, particularly those released after death of the plant or animal, it is fairly obvious that postharvest or postmortem practices will cause substantial fluctuations in nutrient content.

C. Trimming

Plant tissues in particular are subject to trimming and subdividing practices which lead to losses in nutrient content of the edible portion compared to the whole plant. The skins and peels of fruits and vegetables are usually removed. It has been reported that the level of ascorbic acid is higher in the apple peel than in the flesh, and the waste core of the pineapple is also higher in vitamin C than the edible portion. Similarly, niacin is reported to be richer in the epidermal layers of the carrot root than in the root which remains after processing [3]. It is likely that similar concentration differences can be found in foods such as potatoes, onions, beets, etc. When peeling is accomplished by such chemically drastic procedures as lye treatment, significant losses of nutrients will also occur in the outer fleshy layer. Trimming of vegetables such as spinach, broccoli, green beans, and asparagus involves discarding bits of stems or tougher portions of the plant and undoubtedly causes losses of some nutrients.

D. Milling

A special category of trimming involves the milling of cereals. All cereals which are milled undergo a significant reduction of nutrients, the extent of the loss being governed by the efficiency with which the endosperm of the seed is separated from the outer seed coat (bran) and germ. The loss of each nutrient follows its own characteristic pattern, as shown for wheat in Fig. 1.

The loss of certain vitamins and minerals from milled cereals was deemed so relevant to health in the United States populace that the concept of adding nutrients back in the final stages of processing was proposed in the 1940s. After a long series of hearings, the Food and Drug Administration

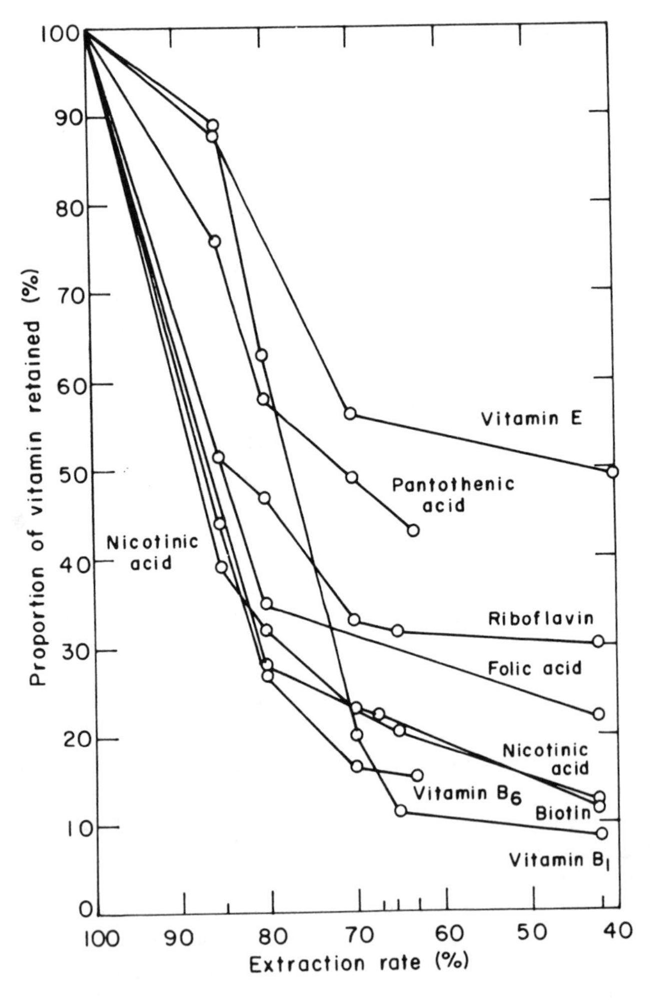

Fig 1. Relation between extraction rate and proportion of total vitamins of the grain retained in flour (from Ref. 4).

issued regulations for a standard of identity for enriched bread. These standards required that four nutrients, namely thiamine, niacin, riboflavin, and iron, be added to flour, with calcium and vitamin D considered as optional. Currently, if bread is to be labeled "enriched," it must meet these standards, but mandatory enrichment is required only by certain state laws.

E. Leaching and Blanching

One of the most significant routes for the loss of water-soluble nutrients is via extraction from cut or sensitive surfaces. Food processing operations which lead to losses of this type include washing, flume conveying, blanching, cooling, and cooking. The nature and extent of the loss will of course depend on pH, temperature, ratio of water to food and of surface to volume, maturity, etc.

Operations of this type may also lead to secondary influences on nutrient content, such as contamination with trace metals and additional exposure to oxygen. In some foods an improvement in mineral content can occur, e.g., increased calcium from exposure to hard water.

Of the operations listed above, blanching leads to the most important nutrient losses. A discussion of blanching, including blanching methods and the influence of blanching time and temperature, was published by Lee [5]. Blanching is normally accomplished with steam or hot water, the choice depending upon the type of food and subsequent process. Steam blanching generally results in smaller losses of nutrients, since leaching is minimized in this process. Leaching during blanching or cooking can be almost entirely eliminated by using microwave cookers [6], since the need for a heating medium is eliminated.

An example of typical vitamin losses from cooked broccoli and the distribution of vitamins between the solid and liquid portions is shown in Table 2. If both liquid and solid portions are considered, there is no measurable loss of B-vitamins and a 10-15% loss of vitamin C. If the

TABLE 2 Vitamin Content and Distribution in Cooked Broccoli

	Vitamin content (%)[a]					
	Solid portion			Liquid portion		
Cooking method	C	B_1	B_2	C	B_1	B_2
Microwave	64	76	71	23	31	31
Pressure (steam)	72	90	94	6	8	8
Boiling	60	75	69	25	33	33

Source: Thomas et al. [7].
[a]Expressed as a percent of broccoli's original vitamin content.

liquid is discarded, as is the usual practice, substantial losses are found for both boiling and microwave cooking. Losses during microwave cooking are untypically high and might have been minimized by using less cooking water.

Under conditions of good manufacturing practice, leaching, blanching, and cooking losses in the food plant should be no greater, and possibly even smaller, than those found for average practice in the home. A voluminous quantity of data on the vitamin content of canned foods provides verification for this statement [8].

F. Processing Chemicals

A number of different chemicals are added to foods as preservatives or processing aids, a subject which is treated in greater depth in Chap. 10. Some of these compounds have a detrimental effect on the content of certain vitamins. For example, oxidizing agents are generally destructive to vitamins A, C, and E. Therefore, the use of bleaching or improving agents for flour may lead to a loss of vitamin activity. However, the older process of aging flour through natural oxidative processes would undoubtedly lead to similar losses.

Sulfite (SO_2) is used to prevent both enzymic and nonenzymic browning of fruits and vegetables. As a reducing agent it protects ascorbic acid, but as a nucleophile it is detrimental to thiamine (see thiamine).

Nitrite is used to preserve meats and may either be added as such or formed by microbial reduction of nitrate. Certain vegetables which contain high concentrations of nitrate, such as spinach and beets, may also contain nitrites because of microbial activity. Nitrite reacts rapidly with ascorbic acid [9] and can also lead to the destruction of carotenoid provitamins, thiamine, and possibly folic acid. Nitrite can act either as an oxidizing agent,

$$NO + H_2O = H^+ + HNO_2 + e^- \qquad E_o = -0.99 \text{ V}$$

or by nucleophilic substitution on N or S, or by addition to double bonds. These reactions are pH sensitive since the reacting species is N_2O_3, formed as follows:

$$H^+ + NO_2^- \rightleftharpoons HNO_2 \qquad pK_a = 3.4$$

$$H^+ + NO_2^- + HNO_2 \rightleftharpoons N_2O_3 + H_2O$$

Thus, the reaction with ascorbic acid is pH sensitive, occurring at a negligible rate above pH 6 and proceeding very rapidly at a pH close to or below 3.4.

Ethylene and propylene oxides are used as sterilizing agents. They are biologically active because of their abilities to alkylate proteins and nucleic acids and can react with vitamins such as thiamine via a similar mechanism. Their primary use in the food industry is for sterilizing spices.

Alkaline conditions are often employed to extract proteins. Conditions of high pH are also encountered where alkaline baking powders are used. Special instances are found where cooked foods, such as eggs, have a pH in the vicinity of 9 because of loss of CO_2. The destruction of some vitamins (see individual vitamins) including thiamine, ascorbic acid, and pantothenic acid is greatly increased under alkaline conditions. Strong acid conditions are only rarely encountered in foods and few vitamins are sensitive to this condition.

G. Deteriorative Reactions

There are a number of general reactions which impair sensory properties of processed and stored foods, and which also cause loss of nutrients. Enzymic reactions have already been mentioned, but enzymic contamination by added ingredients should also be considered. Examples might include ascorbic acid oxidase from plant materials or thiaminase from fishery products, when foods containing these enzymes are added to foods which normally would not contain the enzymes.

Lipid oxidation causes the formation of hydroperoxides, peroxides, and epoxides, which will in turn oxidize or otherwise react with carotenoids, tocopherols, ascorbic acid, etc. The fate of other readily oxidizable vitamins such as folic acid, B_{12}, biotin, and vitamin D has not been adequately investigated, but serious losses would not be unexpected. The decomposition of hydroperoxides to reactive carbonyl compounds could lead to further losses, particularly for thiamine, some forms of B_6, pantothenic acid, etc.

Nonenzymic browning reactions of carbohydrates also lead to highly reactive carbonyl compounds, which may react with certain vitamins in a fashion similar to that proposed for carbonyls derived from lipid oxidation.

III. WATER-SOLUBLE VITAMINS

A. Ascorbic Acid

1. Structure

Ascorbic acid is a highly soluble compound that has both acidic and strong reducing properties. These qualities are attributable to its enediol structure which is conjugated with the carbonyl group in a lactone ring. The natural form of the vitamin is the L-isomer; the D-isomer has about 10% of the activity of the L-isomer and is added to foods for nonvitamin purposes.

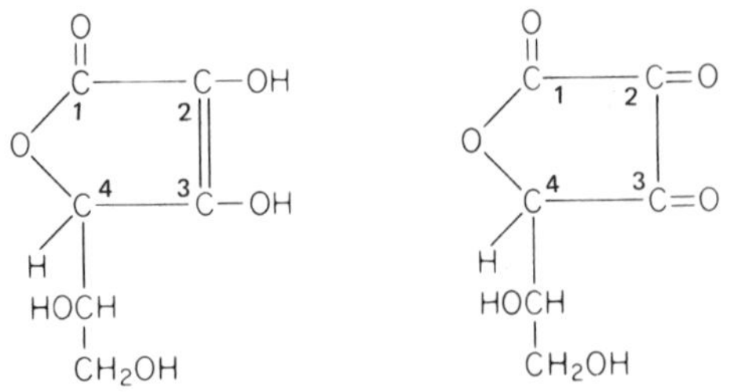

L-ascorbic acid L-dehydroascorbic acid

In solution, the hydroxyl on C-3 readily ionizes ($pK_1 = 4.04$ at 25°C) and a solution of the free acid gives a pH of 2.5. The second hydroxyl is much more resistant to ionization ($pK_2 = 11.4$).

2. Stability

Ascorbic acid is highly sensitive to various modes of degradation. Factors which can influence the nature of the degradative mechanism include temperature, salt and sugar concentration, pH, oxygen, enzymes, metal catalysts, initial concentration of ascorbic acid, and the ratio of ascorbic acid to dehydroascorbic acid [10-19].

Since so many factors can influence the nature of ascorbic acid degradation, it is not feasible to construct clearly defined precursor-product relationships for any but the earliest products in the reaction pathway. Reaction mechanisms and pathways are based upon both kinetic and physical-chemical measurements, as well as on structure determination of isolated products. Many of these studies have been conducted in model systems at a pH less than 2 or in high concentrations of organic acids, and therefore may not duplicate the exact pattern of a particular ascorbic acid containing food product.

The scheme shown in Fig. 2 demonstrates the influence of oxygen and heavy metals on the route and products of the reaction [13,15-17,20,21].

When oxygen is present in the system, ascorbic acid is degraded primarily via its monoanion (HA^-) to dehydroascorbic acid (A), the exact pathway and overall rate being a function of the concentration of metal catalysts (M^{n+}) in the system. The rate of formation of (A) is approximately first-order with respect to $[HA^-]$, $[O_2]$, and $[M^{n+}]$. When the metal catalysts are Cu^{2+} and Fe^{3+}, the specific rate constants are several orders of magnitude higher than for spontaneous oxidation, and therefore even a few ppm of these metals may cause serious losses in food products. The uncatalyzed reaction is not proportional to oxygen concentration at low partial oxygen pressures, as shown in Table 3. Below a partial pressure of 0.40 atm, the rate seems to level off, indicating a different oxidative path. One possibility is direct oxidation by hydroperoxyl radicals ($HO_2\cdot$) or hydrogen

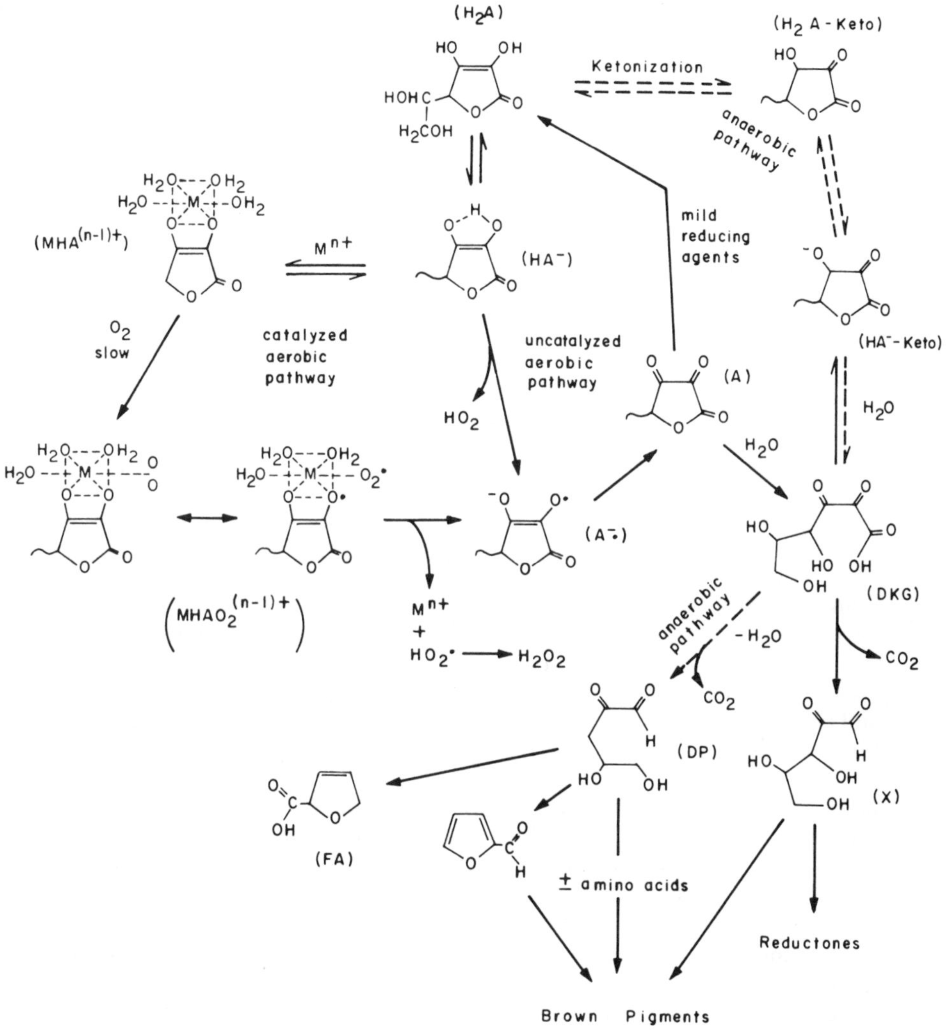

Fig. 2. Degradation of ascorbic acid.

peroxide. In contrast, the rate in the catalyzed pathway is proportional to oxygen concentration for partial pressures down to 0.19 atm. The postulated pathway involves formation of a metal anion complex, $MHA^{(n-1)+}$, which combines with oxygen to give a metal-oxygen-ligand complex, $MHAO_2^{(n-1)+}$. This latter complex has a resonance form of a diradical which rapidly decomposes to give the ascorbate radical anion ($A^{\cdot-}$), the

TABLE 3 Variation of Rate Constants (sec^{-1}) of Uncatalyzed Ascorbic Acid Oxidation with Oxygen Partial Pressure

Partial oxygen pressure (atm)	Specific rate constant for ascorbate anion x 10^4
1.00	5.87
0.81	4.68
0.62	3.53
0.40	2.75
0.19	2.01
0.10	1.93
0.05	1.91

Source: Khan and Martell [20].

original metal ion (M^{n+}) and ($HO_2\cdot$). The radical anion ($A^{\dot{-}}$) then rapidly reacts with O_2 to give dehydroascorbic acid (A). The oxygen dependence of the catalyzed and uncatalyzed reactions is a key to establishing this mechanism, in which oxidation takes place in the rate-determining step of $MHAO_2^{(n-1)+}$ formation. This oxygen dependence is of considerable importance in explaining the influence of sugars and other solutes on ascorbic acid stability, where at high solute concentrations there is a salting-out effect on dissolved oxygen.

In the uncatalyzed oxidative pathway, the ascorbate anion (HA^-) is subject to direct attack by molecular oxygen in a rate-limiting step, to give first the radical anions ($A^{\dot{-}}$) and ($HO_2\cdot$), followed rapidly by formation of (A) and H_2O_2. Thus, the catalyzed and uncatalyzed pathways have common intermediates, and are indistinguishable by product analysis. Since dehydroascorbic acid is readily reconverted to ascorbate by mild reduction, loss of vitamin activity comes only after hydrolysis of the lactone to form 2,3-diketogulonic acid (DKG).

The pH-rate profile for uncatalyzed oxidative degradation is an S-shaped curve which increases continuously through the pH corresponding to pK_1 of ascorbic acid, tending to flatten out above pH 6. This is taken as evidence that it is primarily the monoanion that participates in oxidation. In catalyzed oxidation the rate is inversely proportional to $[H^+]$, indicating that (H_2A) and (HA^-) compete for O_2. However, the specific rate constant for (HA^-) is 1.5-3 orders of magnitude greater than for (H_2A) [20]. Under anaerobic conditions the rate reaches a maximum at pH 4, declines to a minimum at pH 2, and then increases again with increasing acidity [13]. The characteristics of the reaction below pH 2 are of little significance in foods, but the maximum at pH 4 is of considerable practical significance and remains the object of much experimentation and speculation. The scheme for anaerobic degradation shown in Figure 2 is speculative. Following the suggestion of Kurata and Sakurai [15-17], ascorbic acid is shown to react via its keto tautomer (H_2A-keto). The tautomer is in equilibrium with its anion (HA^--keto) which undergoes delactonization to (DKG).

Although the anaerobic pathway could also contribute to ascorbic acid degradation in the presence of oxygen, even the uncatalyzed oxidative rate is very much greater than the anaerobic rate at ambient temperatures. Therefore, both pathways may be operative in the presence of oxygen with the oxidative pathways being dominant. In the absence of oxygen there is no added influence of metal catalysts. However, certain chelates of Cu^{2+} and Fe^{3+} are catalytic in a manner independent of oxygen concentration, with catalytic effectiveness being a function of metal chelate stability [21].

In Fig. 2 further degradation is shown beyond (DKG). Although these reactions are not of nutritional importance (nutritional value is already lost at this point), the decomposition of ascorbic acid is closely tied to nonenzymatic browning in some food products. Evidence accumulated thus far tends to indicate a major divergence of products formed, depending on whether or not decomposition was oxidative. Since the divergence appears to occur following (DKG) formation, it is somewhat paradoxical that the reactions themselves do not require molecular oxygen. However, in oxidative degradation a relatively large proportion of ascorbate is rapidly converted to (A), which in turn will influence the reaction chemistry via interactions not explicitly shown in the scheme. Xylosone (X) may be formed by simple decarboxylation of (DKG), whereas 3-deoxypentosone (DP) is formed by β-elimination at C-4 of (DKG) followed by decarboxylation. It may be the accumulation rate of (DKG) which influences its mode of decomposition or it may be a more specific interaction with (A). In either case, the reaction at this stage begins to assume the characteristics of other carbohydrate nonenzymatic browning reactions [22]. Xylosone is further degraded to reductones and ethylglyoxal, whereas (DP) is degraded to furfural (F) and 2,5-dihydrofuroic acid (E). Any or all of these compounds may combine with amino acids to contribute to the browning of foods [14].

The most critical study of factors influencing ascorbic acid degradation is that of Spanyar and Kevei [19], who examined the copper-catalyzed reaction with respect to oxygen concentration; Fe^{3+} catalysis; pH and temperature; ascorbate, dehydroascorbate, and isoascorbate concentrations; cysteine and glutathione; and polyphenols. The most interesting findings of this study were the interactions of cysteine, Cu^{2+}, and pH. If sufficiently high concentrations of cysteine are present, ascorbate is completely protected, even when it is still in molar excess. This protection may be related to interaction of cysteine with copper, since the pH-rate profile of ascorbate degradation was no longer related to the reciprocal of $[H^+]$, but showed a minimum in the vicinity of pH 4. Although the authors are unaware of any foodstuffs that behave in this manner, an insufficient number of cases have been examined, and the cysteine effect may prove to be significant.

In a recent study, volatile degradation products were isolated from a solution of L-dehydroascorbic acid in phosphate buffer solutions at pH 2, 4, 6, and 8, heated under reflux for 3 hr or held at 25° for 200 hr [23]. Fifteen products were identified; the five main degradation products were 3-hydroxy-2-pyrone, 2-furancarboxylic acid, 2-furaldehyde, acetic acid, and 2-acetyfuran. Their formation depended on pH and temperature; the presence of oxygen had no pronounced effect.

Ascorbic acid is readily oxidized by nitrous acid. Thus, addition of ascorbic acid to foods has been proposed to prevent nitrosamine formation in products containing sodium nitrite [24]. The amount of ascorbic acid required to prevent nitrosamine formation depends markedly on pH and oxygen concentration [25].

3. Assay

There are a variety of analytical procedures for detecting ascorbic acid, but none is entirely satisfactory because of lack of specificity and numerous interfering substances contained in most foodstuffs. Analysis usually involves oxidation by a redox dye such as 2,6-dichlorophenolindophenol [26]. This procedure does not take into account dehydroascorbic acid, which has approximately 80% of the vitamin activity of ascorbic acid. Therefore, redox procedures are often employed in conjunction with treatment of the sample extract with reductants such as H_2S, E_0' (pH 7) $H_2A \rightarrow A = -0.08$ V. An alternative approach utilizes the carbonyl properties of (A) to form bisphenylhydrazone from phenylhydrazine. This procedure obviously would be susceptible to errors introduced by the presence of similarly reactive carbonyls which have no vitamin activity.

4. Effect of Processing

Since ascorbic acid is soluble in water, it is readily lost via leaching from cut or bruised surfaces of foods. However, in processed foods the most significant losses result from chemical degradation. In foods which are particularly high in ascorbic acid, such as fruit products, loss is usually associated with nonenzymatic browning. Composition tables may be unreliable for estimating expected concentrations of ascorbic acid, since significant amounts of ascorbic acid are used in many types of foods as a processing aid [27].

In foods such as canned juices, the loss of ascorbic acid would tend to follow consecutive first-order reactions, that is, a rapid initial reaction which is oxygen-dependent and proceeds until the available oxygen is completely exhausted, followed by anaerobic degradation. In dehydrated citrus juices, degradation of ascorbic acid appears to be only a function of temperature and moisture content [28]. The influence of water on stability of ascorbic acid in a variety of foods is shown in Fig. 3 [29]. Although ascorbic acid appears to be degraded even at very low moisture contents, the rate becomes so slow that long storage can be used without excessive ascorbate loss. Storage stability data on a variety of food and beverage products are summarized in Table 4.

Although the stability of ascorbic acid generally increases with lower temperatures, certain investigations have indicated that there might be an accelerated loss upon freezing or in frozen storage. This has been shown to be unlikely for most practical food situations [31]; however, storage temperatures above -18°C will ultimately lead to significant losses [32]. In general, the largest losses for noncitrus foods will occur during heating. This is illustrated in Figure 4 for peas processed by a variety of techniques. It is apparent that the leaching loss during heating far exceeds losses during other process steps. This observation applies to most water-soluble nutrients.

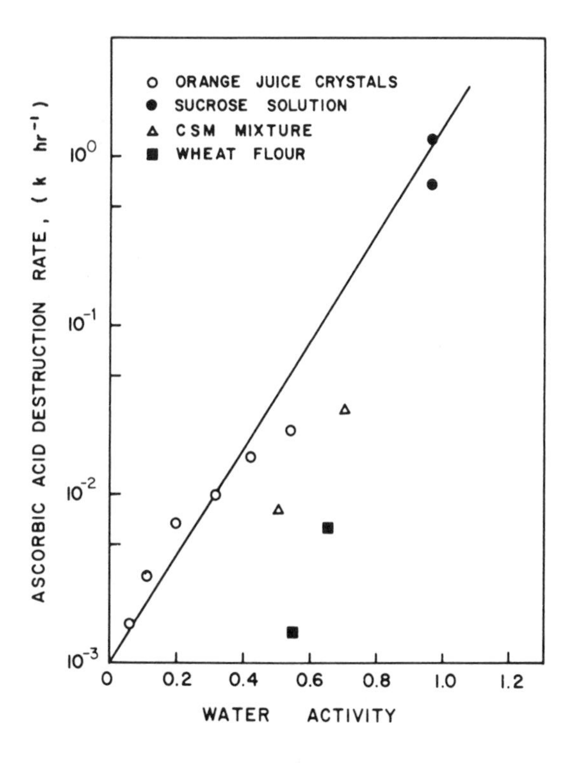

Fig. 3. Relation between water activity and rate of destruction of ascorbic acid (from Labuza [29]).

TABLE 4 Vitamin C Stability in Fortified Foods and Beverages After Storage at 23°C for 12 Months, Except as Noted

Product	No. of samples	Retention (%)	
		Mean	Range
Ready-to-eat cereal	4	71	60-87
Dry fruit drink mix	3	94	91-97
Cocoa powder	3	97	80-100
Dry whole milk, air pack	2	75	65-84
Dry whole milk, gas pack	1	93	-
Dry soy powder	1	81	-
Potato flakes[a]	3	85	73-92
Frozen peaches	1	80	-
Frozen apricots[b]	1	80	-
Apple juice	5	68	58-76
Cranberry juice	2	81	78-83
Grapefruit juice	5	81	73-86
Pineapple juice	2	78	74-82
Tomato juice	4	80	64-93
Vegetable juice	2	68	66-69
Grape drink	3	76	65-94
Orange drink	5	80	75-83
Carbonated beverage	3	60	54-64
Evaporated milk	4	75	70-82

Source: Bauernfeind and Pinkert [27] compiled by DeRitter [30].
[a]Stored for 6 months at 23°C.
[b]Thawed after storage in freezer for 5 months.

Another processing variable that can affect ascorbic acid losses is sulfur dioxide treatment. Fruit products treated with SO_2 have reduced ascorbic acid losses during processing, as well as during storage [35-36].

B. Thiamine

1. Chemistry

Thiamine, or vitamin B_1, consists of a substituted pyrimidine linked to a substituted thiazole by a methylene group. It is widely distributed throughout the plant and animal kingdom, since it plays a key role as a coenzyme in the intermediary metabolism of α-keto acids and carbohydrates. As a result, thiamine can exist in foods in a number of forms, including free thiamine, the pyrophosphoric acid ester (cocarboxylase), and bound to the

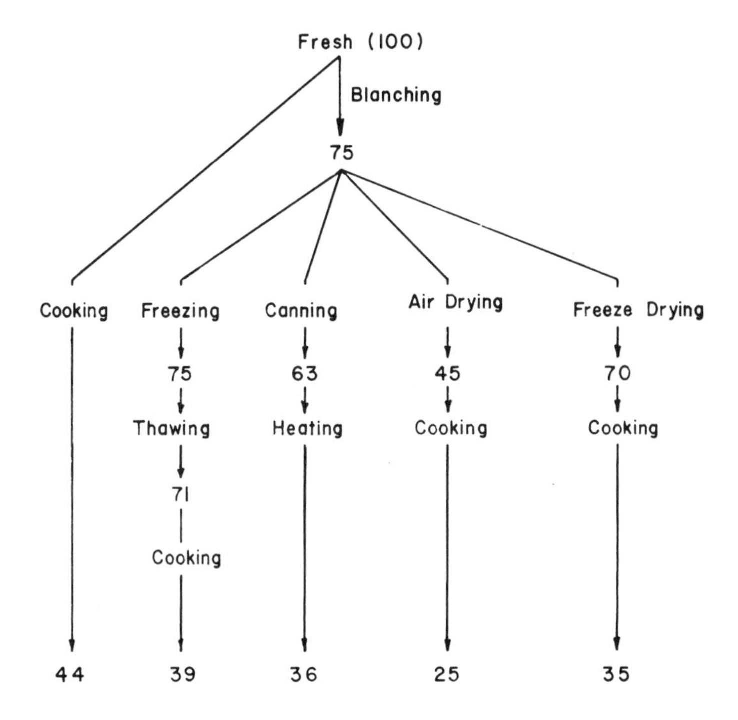

Fig. 4 Retention of ascorbic acid in processed peas (after Bender [33] and Mapson [34]).

respective apoenzyme. The details of these various structures are shown in Fig. 5.

Since thiamine contains a quaternary nitrogen, it is a strong base and will be completely ionized over the entire range of pH normally encountered in foods. In addition, the amino group on the pyrimidine ring will be ionized, the extent dependent upon pH ($pK_a = 4.8$). The coenzyme role of thiamine is elaborated through position 2 of the thiazole ring, which in its ionized form is a strong nucleophile. Studies on deuterium exchange at this position have shown a half-life at room temperature of 2 min at pH 5. At pH 7 the exchange reaction is too fast to follow by ordinary techniques [37].

2. Assay

Thiamine is characterized by strong UV absorption bands with pH-dependent absorption maxima (Table 5). These spectra are useful for analysis under the most limited circumstances only, since many thiamine

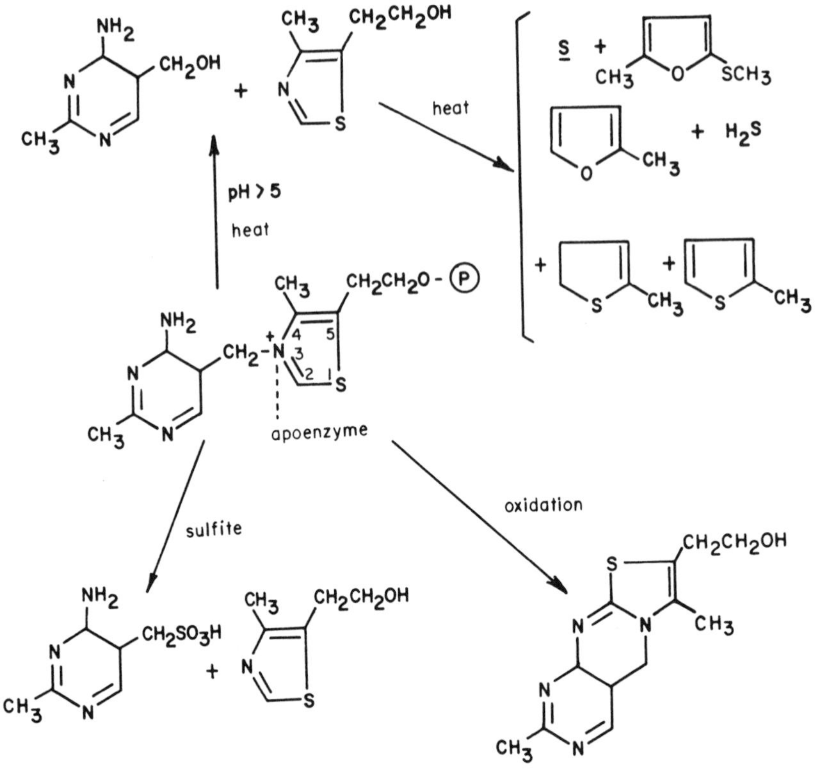

Fig. 5. Degradation of thiamine. The circled P is pyrophosphate.

degradation products are also UV absorbing. The method of choice for analysis is usually the thiochrome procedure, which involves treatment of thiamine with a strong oxidizing agent (e.g., ferricyanide or hydrogen peroxide) to effect conversion of strongly fluorescent thiochrome (see Fig. 5). Adaptation of this procedure to foods requires enzymic hydrolysis of combined forms of the vitamin to free thiamine, and a chromatographic clean-up prior to conversion to thiochrome [26].

3. Stability

Thiamine is one of the least stable of all the vitamins. Its stability depends on pH, temperature, ionic strength, buffer type, and other reacting species [38-42]. The typical degradative reaction appears to involve a

TABLE 5 Molar Absorptivities of Thiamine Solutions

System	λ_{max} (nm)	
0.005N HCl	247	14,500
pH 7.0	231	12,000
	267	8,600

nucleophilic displacement at the methylene carbon joining the two ring systems. Therefore, strong nucleophiles, such as HSO_3^- (sulfite) readily cause destruction of the vitamin. The similarity of the chemistry of degradation by sulfite and at alkaline pH is shown in Fig. 5. Both reactions yield 5-(β-hydroxyethyl)-4-methylthiazole and a corresponding substituted pyrimidine. In the case of sulfite, this is a 2-methyl-5-sulfomethylpyrimidine, whereas in the case of alkali it would be the corresponding hydroxymethylpyrimidine. The chemistry of sulfite cleavage has been extensively studied [43] since this reaction is of particular significance in dehydrated vegetables and fruits treated with sulfite to inhibit browning.

Thiamine is also inactivated by nitrite, possibly via reaction with the amino group on the pyrimidine ring. It was noticed very early [44] that this reaction is mitigated in meat products as compared to buffer solutions, implying a protective effect of protein. Casein and soluble starch also exhibit a protective effect against destruction of thiamine by sulfite [43]. In the latter case the protective effect was shown to be unrelated to binding of reactants, oxidation of reactive sulfhydryl groups, or competitive oxidation of sulfite by protein. Although the mechanisms of the protective effect are still unclear, it is probable that the effect itself extends to other degradative mechanisms, and may be another important source of discrepancies in the literature.

Since thiamine can exist in multiple forms, the overall stability of vitamin activity will depend on the relative concentrations of the various forms. Within given animal species the ratios will also depend on the nutritional status prior to death, with variations from one type of muscle to another, and dependence on postharvest or postmortem physiological stresses in both plants and animals. In the relatively few studies that have been conducted, the enzyme-bound forms (e.g., cocarboxylase) have been found to be less stable than the free vitamin. Farrer [45] suggested that the relative concentrations of the various forms of the vitamin may account for some discrepancies in the literature. Extensive losses of thiamine occur in cereals as a result of cooking or baking, and in meats, vegetables, and fruits, as a result of various processing operations and storage. The literature up through 1953 has been analyzed and summarized by Farrer [45] and only typical examples will be used for didactic purposes. The stability of thiamine is so strongly influenced by the nature and state of the system that it is difficult to extrapolate between systems, and numerous unexplainable differences exist in the literature (see Ref. 45 for examples).

Temperature is an important factor in thiamine stability. Table 6 shows differences in thiamine retention among various foods held at two different storage temperatures. Rate constants for thiamine degradation in various foods, including peas, carrots, cabbage, potatoes, and pork, range from 0.0020 to 0.0027 per min at 100°C, but these values cannot be extrapolated to other temperatures unless the Arrhenius activation energy is known for the particular system. Activation energies summarized by Farrer [45] vary by a factor of more than two, depending on the system and the reaction conditions. In phosphate buffer at pH 6.8, Goldblith and Tannenbaum [47] found the activation energy to be 22 kcal/mole for both conventional and microwave heating, which is similar to the value in pureed meats and vegetables found by Feliciotti and Esselen [48]. Leichter and Joslyn [49] found that sulfite lowered the activation energy to 13.6 kcal/mole. Since it is also possible that catalytic concentrations [50] of metals such as copper can accelerate the rate of degradation at a given pH, differences in activation energies could conceivably be ascribed to small compositional differences in the system. However, Feliciotti and Esselen [48] found similar values for activation energies in a wide variety of meats and vegetables, irrespective of sample pH.

TABLE 6 Thiamine Retention in Stored Foods

System	Retention after 12 months storage	
	38°C	1.5°C
Apricots	35	72
Green beans	8	76
Lima beans	48	92
Tomato juice	60	100
Peas	68	100
Orange juice	78	100

Source: Freed et al. [46].

4. Effect of Processing

The results of several recent studies of thiamine degradation in a variety of foods are summarized in Table 7.

Thermal destruction of thiamine leads to formation of a characteristic odor, which is involved in the development of "meaty" flavors in cooked foods. Some of the probable reactions [38, 39, 55-57] are summarized in Fig. 5. The reaction leading to the formation of pyrimidine and thiazole rings has already been described. Significant secondary products are then thought to arise from the thiazole ring, including elemental sulfur, hydrogen sulfide, a furan, a thiophene, and a dihydrothiophene. Reactions leading to these products are unclear, but extensive degradation and rearrangement of the thiazole ring must be involved.

TABLE 7 Recent Studies on Thiamine Degradation in Foods

Product	Treatment	Retention (%)	Reference
Cereals	Extrusion cooking	48-90	Beetner et al. [51]
Potatoes	Soaked in water for 16 hr, then fried	55-60	Oguntona and Bender [52]
	Soaked in sulfite for 16 hr, then fried	19-24	Oguntona and Bender [52]
Soybeans	Soaked in water and then boiled in water or bicarbonate solution	23-52	Perry et al. [53]
Mashed Potatoes	Various heating treatments	82-97	Ang et al. [54]
Vegetables	Various heating treatments	80-95	Ang et al. [54]
Meat items	Various heating treatments	83-94	Ang et al. [54]
Frozen fried fish	Various heating treatments	77-100	Ang et al. [54]

As previously indicated, the rate of thiamine degradation is extremely sensitive to pH. The pH-rate profiles for thiamine and cocarboxylase at elevated temperature are shown in Fig. 6. It is apparent that either the starch and/or protein components of cereal products exert a protective effect over the pH range examined. Cocarboxylase is more sensitive than thiamine, but the difference in sensitivity is a function of pH, disappearing completely at pH above 7.5. Since both the amino group on the pyrimidine ring and the 2 position on the thiazole ring are strongly influenced by pH in the region of interest for thiamine stability, either function could be implicated in the degradative reactions. However, based upon the nature of the secondary products, it appears that the thiazole ring is the most likely site for the reaction.

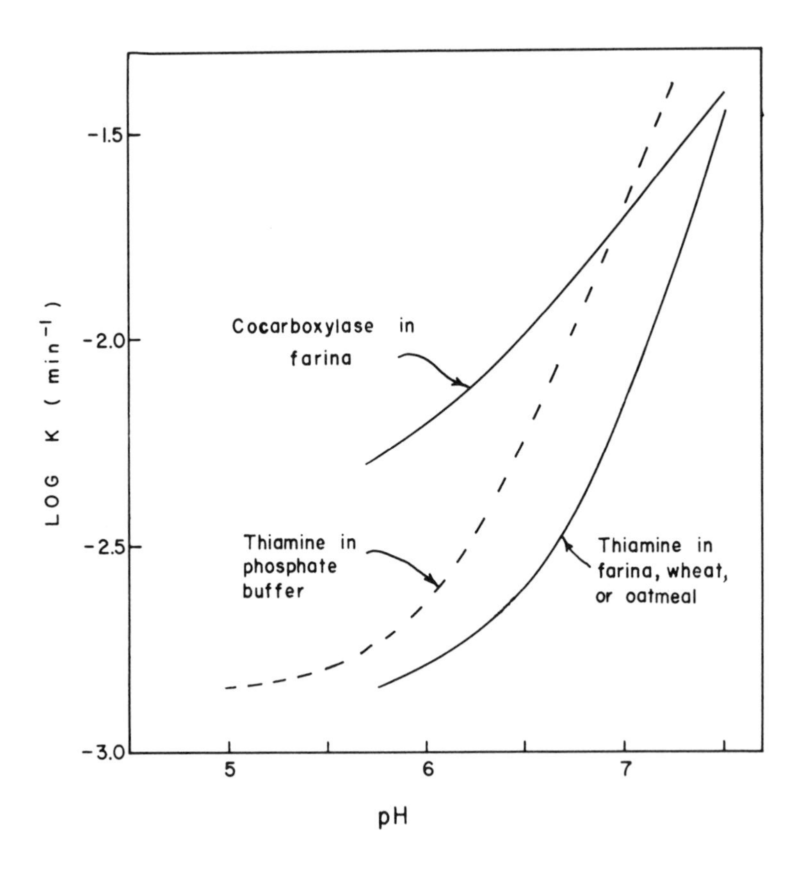

Fig. 6. The pH stability of thiamine and cocarboxylase (from Ref. 45).

As is true of other water-soluble vitamins, thiamine is extensively lost by leaching during cooking operations (see Table 2) and from cereals during milling (see Fig. 1). Little information is available on thiamine stability in dehydrated foods. Studies in dehydrated corn-soy-milk (CSM) indicate that degradation is influenced strongly by moisture content [58]. For example, storage at 100°F for 182 days caused no loss when the system was maintained below 10% moisture content, but extensive loss occurred at 13% moisture. Because of multiple possibilities of physical separation and chemical degradation, substantial losses of thiamine can occur in many foods unless great care is taken during all handling, storage, and processing operations.

Extracts from various fish and crustacea have been found to destroy thiamine [59-61] and an enzyme with antithiamine activity was proposed. Recently, however, the antithiamine factor from carp viscera was shown to be thermostable and probably not an enzyme. The antithiamine factor was identified as hemin or a related compound [62, 63]. Similarly, various heme proteins from tuna, pork, and beef have been shown to have antithiamine activity [64].

C. Riboflavin

1. Structure

Riboflavin (vitamin B_2) is an isoalloxazine derivative with a ribitol side chain. It exists in nature primarily as the phosphoric acid ester in the form of two coenzymes, flavin mononucleotide (FMN) and flavine-adenine dinucleotide (FAD). The enzymes associated with this vitamin are called flavoproteins and typically act as hydrogen-transfer agents in the oxidation of compounds such as amino acids and reduced pyridine nucleotides.

2. Stability

Riboflavin is thermostable and unaffected by atmospheric oxygen. It is stable in strongly acid solution but unstable in the presence of alkali. It decomposes readily if exposed to light. In alkaline solution, irradiation causes the photochemical cleavage of ribitol, shown below, yielding lumiflavin [65, 66], whereas in acid or neutral solution, irradiation leads to production of the blue fluorescent substance lumichrome together with varying amounts of lumiflavin [67].

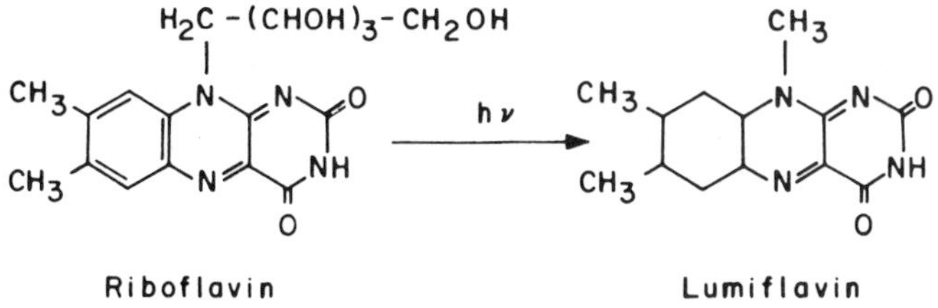

Riboflavin → Lumiflavin

Lumiflavin is apparently a stronger oxidizing agent than riboflavin and can catalyze destruction of a number of other vitamins, particularly ascorbic acid. Some years ago, when milk was sold in transparent bottles, the reaction sequence described above was a significant problem leading not only to a loss of nutritional value, but also to an acceptability problem known as "sunlight off-flavor." With the advent of paper or plastic milk container, this problem disappeared.

Riboflavin is quite stable in food under most processing or cooking conditions. In a recent study of the effects of various heating methods on vitamin retention in six fresh or frozen prepared food products, retention of riboflavin was always higher than 90% [54]. Riboflavin retention in peas and lima beans subject to blanching and processing was in all cases higher than 70% [68].

3. Assay

Riboflavin is usually assayed fluorometrically by measurement of the characteristic yellowish-green fluorescence. It can also be estimated microbiologically using *Lactobacillus casei* [26].

D. Nicotinic Acid

1. Structure

Nicotinic acid is pyridine β-carboxylic acid and nicotinamide is the corresponding amide. The acid and its amide are often given the collective

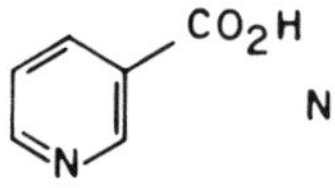

CO_2H Nicotinic Acid

name niacin. In living systems, nicotinamide is a constituent of the hydrogen-carrying coenzymes nicotinamide adenine dinucleotide (NAD) and nicotinamide adenine dinucleotide phosphate (NADP). Nicotinic acid is one of the most stable vitamins, being relatively insensitive to heat, light, air, acid, or alkali.

2. Assay

Niacin is assayed by first hydrolyzing the food with sulfuric acid to liberate nicotinic acid from combined forms (as coenzyme). The pyridine ring of the nicotinic acid is opened with cyanogen bromide and the fission product coupled with sulfanilic acid to yield a yellow dye with an absorption maximum at 470 nm [26].

3. Stability

Niacin is generally stable in foods, but losses occur in vegetables via trimming, leaching, etc., which parallel losses of other water-soluble vitamins [69,70]. Considerable losses of niacin occur during postmortem storage of pork and beef [71,72]. Roasting of meat leads to no losses, but drippings contain up to about 26% of the original niacin [73]. There appears to be no loss of niacin in milk processing [74].

E. Vitamin B_6

1. Structure

There are three closely related compounds which have vitamin B_6 activity: pyridoxal (I), pyridoxine or pyridoxol (II), and pyridoxamine (III).

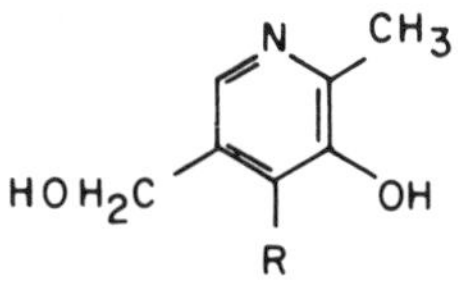

	R
I	$-CHO$
II	$-CH_2OH$
III	$-CH_2NH_2$

These compounds are widely distributed in the plant and animal kingdoms in the form of their phosphates. Pyridoxal phosphate is a coenzyme for many enzymatic transformations of amino acids (e.g., transamination, racemization, and decarboxylation reactions). It exerts its action as a coenzyme via a carbonyl-amine condensation with the amino acid yielding a Schiff base which is stabilized by chelation to a metal ion (IV).

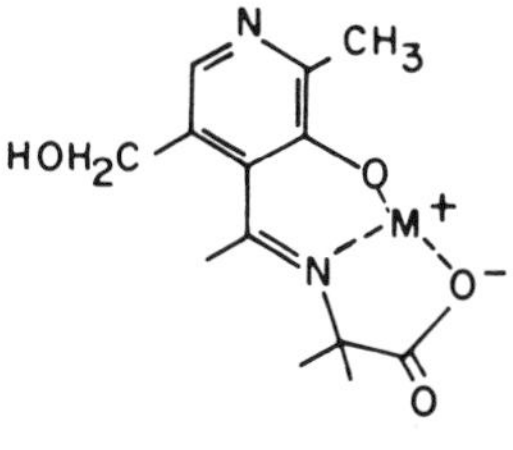

IV

2. Stability

The three forms of vitamin B_6 are quite stable to heat but are decomposed by alkali. The pyridoxol form of the complex is most stable and is used for fortification of foods. The compounds are converted by UV irradiation in the presence of oxygen to biologically inactive products such as 4-pyridoxic acid [75]. This reaction is probably of little importance in foods except milk [74].

When a solution of pyridoxal is heated with glutamic acid, a mixture including pyridoxamine and α-ketoglutaric acid results. In fact, heating a mixture of amino acids, pyridoxal, and polyvalent metal ions at 100°C, at any of a wide range of pH values, leads to the same products that would normally result from the conversion involving the holoenzyme [76].

When cysteine and pyridoxal are allowed to react under conditions similar to those encountered in the sterilization process, the reaction products have no vitamin B_6 activity for rats and approximately 20% activity for *Saccharomyces carlsbergensis*. A product of this reaction appears to be bis-4-pyridoxyldisulfide [77], possibly arising via a thiazolidine (V). A

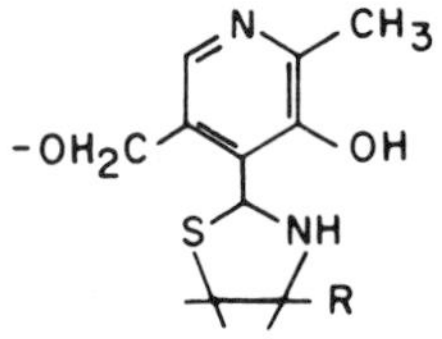

V

similar reaction sequence is possible by direct reaction of pyridoxal with sulfhydryl groups of proteins [78]. Since the main result of the interaction of B_6 with amino groups appears to be the interconversion between pyridoxal and pyridoxamine, both with full vitamin B_6 activity, the cysteine reaction may be the key to the stability of this vitamin in heat processed foods.

3. Assay

Vitamin B_6 is usually assayed microbiologically using *S. carlsbergensis* as test organism [26]. The individual compounds may be separated by ion-exchange chromatography prior to assay if required.

4. Effect of Processing

Information on the distribution of vitamin B_6 in its various forms in foods has become available only recently [79,80]. Although many foods have not been systematically studied for vitamin B_6 destruction during processing, it is nevertheless apparent that both the quantity and form of the vitamin are influenced by the processes of heating, concentration, and dehydration.

In view of the chemistry described above, it is not surprising to find an entirely different distribution of vitamin B_6 forms in fresh and processed foods. By recalculating some of the data of Polansky and Toepfer [80], it is evident that pyridoxal increases and pyridoxamine decreases in the dehydration of eggs. In fresh milk the main form is pyridoxal; in dried milk pyridoxal is still dominant, but there is more pyridoxamine than in fresh fluid milk, whereas in evaporated milk pyridoxamine is the main form. In raw pork loin the dominant form is pyridoxal; in fully cooked ham it is pyridoxamine.

A great deal of attention has been paid to the stability of vitamin B_6 in heat-processed milk. When pyridoxol is added to milk, it is stable during sterilization. However, the dominant natural form of vitamin B_6 in milk is pyridoxal. Studies have shown that sterilized liquid milk and formula milk contain less than half their original vitamin B_6 activity, the decrease continuing 7-10 days after processing [81, 82]. In addition, feeding tests with rats indicated that the decrease was even more substantial than that found using the test organism, S. carlsbergensis.

In another study [83], both rapid pasteurization (2-3 sec at 92°C) and boiling (2-3 min) gave only 3% loss of vitamin B_6. Sterilization of milk in bottles (13-15 min at 119-120°C) reduced the vitamin by 84%. Very rapid sterilization of milk by injection of steam at 143° for 3-4 sec into preheated milk caused negligible loss of vitamin B_6. Vitamin B_6 is believed to react with substances released from milk proteins upon heating, possibly cysteine [84, 85].

An extensive series of edible products has been analyzed for vitamin B_6 losses [86]. As a result of canning, large losses of the vitamin occurred in vegetables, varying from 57 to 77%. Frozen vegetables showed losses of 37 to 56%. Canning of seafood and meats led to losses of about 45% of vitamin B_6. Average loss of the vitamin from freezing fruits and fruit juices was 15% and from canning 38%. Processed and refined grains lost 51-94%, and processed meats lost 50-75% of vitamin B_6. Similar values have been reported by other workers [74, 83, 87].

The sensitivity of vitamin B_6 to conditions of processing led in 1952 to a situation which exemplifies the need for concern about the nutrient content of processed foods [88]. An infant food, available in liquid and dry form, was identified as the cause of at least 50-60 cases of convulsive seizures. The seizures occurred after consumption of the liquid form of the food and were attributed to instability of pyridoxal in the presence of milk proteins. The problem was solved by fortifying the product with the more stable pyridoxol.

F. Folic Acid

1. Structure

The structure of folic acid is shown in Fig. 7. It consists of a 2-amino-4-hydroxypteridine linked to p-amino benzoic acid, which in turn is coupled to glutamic acid (PABG). In biological systems, folic acid can exist in a wide variety of different forms. The pteridine ring may be reduced to yield di- or tetrahydrofolates; five different one-carbon substituents can

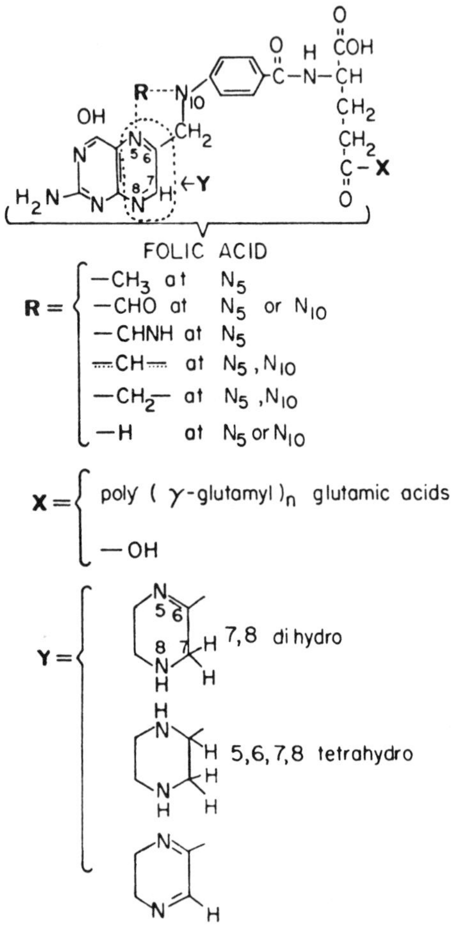

FOLIC ACID

R = $-CH_3$ at N_5; $-CHO$ at N_5 or N_{10}; $-CHNH$ at N_5; $=CH-$ at N_5, N_{10}; $-CH_2-$ at N_5, N_{10}; $-H$ at N_5 or N_{10}

X = poly (γ-glutamyl)$_n$ glutamic acids; $-OH$

Y = 7,8 di hydro; 5,6,7,8 tetrahydro

Fig. 7. Schematic representations of the theoretical structures of the folates (from Ref. 89).

be present at the N-5 and/or N-10 position; the glutamic acid residue can be extended to a poly-γ-glutamyl side chain of varied length. Assuming that the polyglutamyl side chain contains no more than six residues, the theoretical number of possible folic acid compounds exceeds 140, of which about 30 have been isolated and characterized.

Tetrahydrofolic acid is involved in the biological transfer of one carbon fragment, and its chemistry, function and biosynthesis have been reviewed in detail [90].

2. Stability

Folic acid is stable to alkali under anaerobic conditions. Alkaline hydrolysis under aerobic conditions, however, cleaves the sidechain to yield PABG and pterin-6-carboxylic acid [91]. Acid hydrolysis under aerobic conditions yields 6-methylpterin. Polyglutamate derivatives of folic acid are hydrolyzed by alkali in the absence of air to yield folic and glutamic acids [91]. Folic acid solutions are decomposed by sunlight to yield PABG and pterin-6-carboxaldehyde [92,93]. Irradiation of the 6-carboxaldehyde in turn yields the 6-carboxylic acid which is then decarboxylated to yield pterin [94]. These reactions are catalyzed by riboflavin and FMN [95].

Interaction of folic acid with two chemicals involved in food processing, sulfite and nitrite, has received some attention. Treatment with sulfurous acid leads to side-chain cleavage with production of reduced pterin-6-carboxyaldehyde and PABG [96]. Nitrous acid reacts with folic acid to yield the N-10 nitroso derivative in the cold [97]. At higher temperatures the 2-amino group also reacts to yield the 2-hydroxy-10-nitroso derivative [98]. We have recently shown that N^{10}-nitroso folic acid possesses some activity as a carcinogen in mice and hence may represent a hazard to man [99].

Di- and tetrahydrofolic acid (FH_2 and FH_4, respectively) are readily oxidized in air. In neutral solution, FH_4 is rapidly oxidized with formation of PABG and several pterins including xanthopterin, 6-methylpterin, and pterin itself [100-102], in addition to FH_2 and folic acid [102-104]. Under acid conditions, quantitative cleavage to PABG is observed [103]. Air oxidation of FH_4 is substantially reduced in the presence of thiols [105], cysteine [106], or ascorbate [107]. Dihydrofolic acid, FH_2, is somewhat more stable than FH_4, but is also subject to oxidative degradation. FH_2 is oxidized more rapidly in acid than in basic solution and the products are PABG and 7,8-dihydropterin-6-carboxaldehyde [108,109]. Again, reducing agents such as thiols or ascorbate retard oxidation.

The compound 5, 10-methylene-FH_4 is much more stable to oxidation than FH_4, particularly at high pH [107,110]. Stability is decreased by amines or ammonium salts [107]; 10-formyl-FH_4 is as unstable as FH_4 toward oxygen and other oxidizing agents. On the other hand, 5-formyl-FH_4

is stable to oxygen in neutral or mildly alkaline solution and reacts only slowly with iodine or dichromate in weakly acidic solution. Furthermore, 5-formyl-FH_4 shows considerable stability to alkaline hydrolysis, whereas 10-formyl-FH_4 readily loses its formyl group in alkali [111-113]. Under acidic conditions, both 5- and 10-formyl-FH_4 lose a molecule of water to form 5,10-methenyl-FH_4; 5-methyl-FH_4 is readily oxidized at alkaline pH by air or other oxidizing agents [114-117]. The structure of the oxidation product is probably 5,6-dihydro-5-methylfolate [116].

3. Assay and Distribution

Investigation of the distribution of folic acid derivatives in nature has only recently become feasible with the ready synthesis of polyglutamate derivatives [118,119]. Stokstad and his group have shown that more than 90% of cabbage folate is in the form of polyglutamates containing more than five glutamic acid residues, mainly as the 5-methyl derivatives [120].

The same workers have shown that soybeans contain mainly monoglutamates (52%), with some diglutamates (16%); pentaglutamates represented the major portion of the remaining polyglutamates [121]. Of the total folate activity in soybeans, 65-70% was in the 5-formyl-FH_4 form. Cow's milk contained about 60% monoglutamates with the remainder ranging from di- to hepta-conjugates. In this case, however, 90-95% of the total folate activity was in the 5-methyl-FH_4 form [121].

It has been shown recently that the rate of intestinal absorption of folates is inversely proportional to the length of the γ-glutamyl side chain [122]. Thus, before a food can be categorized as a good source of the vitamin, an estimate of chain lengths will be necessary. This type of information is not currently available for most foods.

Folic acid content in foods is usually assayed by microbiological methods, after the various glutamate conjugates are enzymatically cleaved by the enzyme conjugase to give free folic acids [26]. The free folic acid content for individual foods has been determined using these methods [123]. In typical American diets the total folate content has been found to be about 180 μg/day [124]. These values are significantly lower than the RDA (see Chapter 1), and folate deficiency is thought by some to be a significant public health problem [125].

4. Nutritional Stability

The 'nutritional' stabilities of several dietary tetrahydrofolate derivatives in their monoglutamyl forms were recently studied in solution and compared to folic acid itself [126]. Nutritional stability was determined by the ability of folate solutions to support growth of *Lactobacillus casei*. Tetrahydrofolate showed 67% retention of activity when held at 121°C for 15 min in the presence of 1% ascorbate. When the ascorbate concentration was reduced to 0.05%, no activity was left after similar treatment. Complete protection for 5-methyl-FH_4 was afforded by 0.2% ascorbate under these conditions. In the absence of ascorbate, 5-methyl-FH_4 showed its greatest nutritional stability in alkaline conditions, with a half-life of 330 hr in 0.1 M TRIS, pH 9, but only 25 hr in 0.1 M HCl (pH 1.0) at room temperature. The most stable reduced folate proved to be 5-formyl-FH_4 with a half-life of about 30 days at room temperature in the pH range of 4 to 10. With a half-life of about 100 hr at pH 7.0, 10-formyl-FH_4 was more stable than expected, though the stability was markedly dependent on the chemical nature of the buffer species. The unexpectedly high stability of 10-formyl-FH_4, however, was probably due to conversion to the more stable oxidation product, 10-formylfolic acid, which was nutritionally active in the assay system used. Folic acid itself was extremely stable with a half-life longer than 700 hr in 0.05 M citrate-phosphate buffer (pH 7.0) at room temperature. Folic acid was much less stable in the absence of citrate, however, having a half-life of only 48 hr under similar conditions. No mechanistic interpretation of this result was given and in none of the studies were degradation products identified. The stability of folic acid polyglutamate derivatives was not examined.

In another study, pure folic acid in the solid state under normal storage conditions (20°C and 65% relative humidity) had a decomposition rate of 1% per year [127].

5. Effect of Processing

The extent and mechanism of loss of folic acid derivatives in processed foods is not yet clear. Studies of processed and stored milk indicate that a primary inactivation process is oxidative [128-130]. The destruction of folate parallels that of ascorbate, and added ascorbate can stabilize folate. Both vitamins are greatly stabilized by deoxygenation of the milk, but both inevitably decline during 14 days of storage at room temperature (15-19°C).

Rapid pasteurization of milk (2-3 sec at 92°) caused about 12% loss of total folate; after boiling the milk for 2-3 min a loss of 17% total folate was observed [131]. Sterilization of milk in bottles (13-15 min at 119-120°C) produced the greatest loss in folate (39%). Rapid sterilization of preheated milk by injection of steam at 143° for 3-4 sec caused only 7% loss of total folates.

In a study of folic acid in garbanzo beans [132], washing and soaking resulted in about 95% retention of folates. In water-blanched beans, free folic acid retention decreased from 75% to 45% as the blanching time lengthened from 5 to 20 min. Total folates decreased from 77 to 54% under the same conditions. Steam blanching improved folic acid retention to some extent. Folic acid in garbanzo beans was shown to be quite stable toward heat processing. Retorting at 118°C for 30 min resulted in about 60% retention of free folic acid and about 70% retention of total folates based on the original dry bean value. There was no significant decrease in folate content when the processing time was lengthened from 30 to 53 min.

In a recent study, the fate of folate polyglutamates in meat during storage and processing was investigated, using chicken liver as a model [133]. In the intact tissue, only slight degradation of polyglutamates by endogenous conjugase was found after 48 hr at 4°C, and complete degradation took 120 hr. Homogenized tissue, however, showed complete degradation to folate monoglutamates and a small amount of diglutamate after 48 hr storage. If at any stage prior to or during storage the liver was heated to over 100°C, the polyglutamates were stabilized due to inactivation of the conjugase.

Table 8 summarizes several other studies of folate losses in foods subjected to various processes.

G. Vitamin B_{12}

1. Structure

Vitamin B_{12} is a red crystalline substance and is chemically by far the most complex of the vitamins. Its structure, shown in Fig. 8, has two characteristic components. In a nucleotide-like structure, 5,6-dimethylbenzimidazole is bound to D-ribose via an α-glycosidic bond. The ribose contains a phosphate group at the 3'-position. The central ring structure is a "corrin" ring system which resembles the porphyrins. Coordinated to the four inner nitrogen atoms of the corrin ring is a cobalt atom. (Vitamin B_{12} is also known as cobalamin.) In the form that is usually isolated, the sixth coordination position of the cobalt II atom is occupied by cyanide (cyanocobalamin). In the active coenzyme, the sixth coordination position is filled by 5-deoxyadenosine attached to the cobalt via its methylene group.

Vitamin B_{12} is found in animal tissues but is almost entirely absent from higher plants. It is unique among the vitamins in being synthesized exclusively by microorganisms. Vitamin B_{12} coenzyme is required for the action of several enzymes, including methylmalonylmutase and dioldehydrase, and, along with folic acid, for the formation of methionine from homocysteine.

2. Stability

Aqueous solutions of cyanocobalamin are stable at room temperature if they are not exposed to ultraviolet or intense visible light. The region of optimal stability is pH 4-6, and in this region only small losses are found even after autoclaving [74]. Heating in alkaline solution can quantitatively destroy vitamin B_{12}. Reducing agents such as thiol compounds at low concentrations can protect the vitamin, but in larger amounts they can cause destruction [143,144]. Ascorbic acid or sulfite can also destroy it. A combination of thiamine and nicotinic acid is slowly destructive to vitamin B_{12} in solution, though alone neither is harmful [145]. Iron protects the vitamin from this combination by combining with hydrogen sulfide, which is the destructive agent [146].

TABLE 8 Reported Observations of Folate Losses in Foods Subjected to Various Processes

Food product	Process method	Loss of folate activity (%)	Reference
Eggs	Frying Boiling Scrambling	18-24	Hanning and Mitts [135]
Sauerkraut	Fermentation	None	Cheldin et al. [136]
Liver	Cooking	None	Cheldin et al. [136]
Halibut	Cooking	46	Cheldin et al. [136]
Cauliflower	Boiling	69	Cheldin et al. [136]
Carrots	Boiling	79	Cheldin et al. [136]
Meats	γ-Irradiation	None	Alexander et al. [137]
Grapefruit juice	Canning and storage	Negligible	Krehl and Cowgill [138]
Tomato juice			
Yugoslavia	Canning	70	Suckewer et al. [139]
USA	Canning	50	Suckewer et al. [139]
	Storage in dark (1 year)	7	Suckewer et al. [139].
	Storage in light (1 year)	30	Suckewer et al. [139]
Maize	Refining	66	Metz et al. [140]
Flour	Milling	20-80	Schroeder [86]
Meat/ vegetable stew	Canning and storage (1 1/2 years)	Negligible	Hellendoorn et al. [141]
	Canning and storage (3 years)	Negligible	Hellendoorn et al. [141]
	Canning and storage (5 years)	Negligible	Hellendoorn et al. [141]

Source: Malin [134].

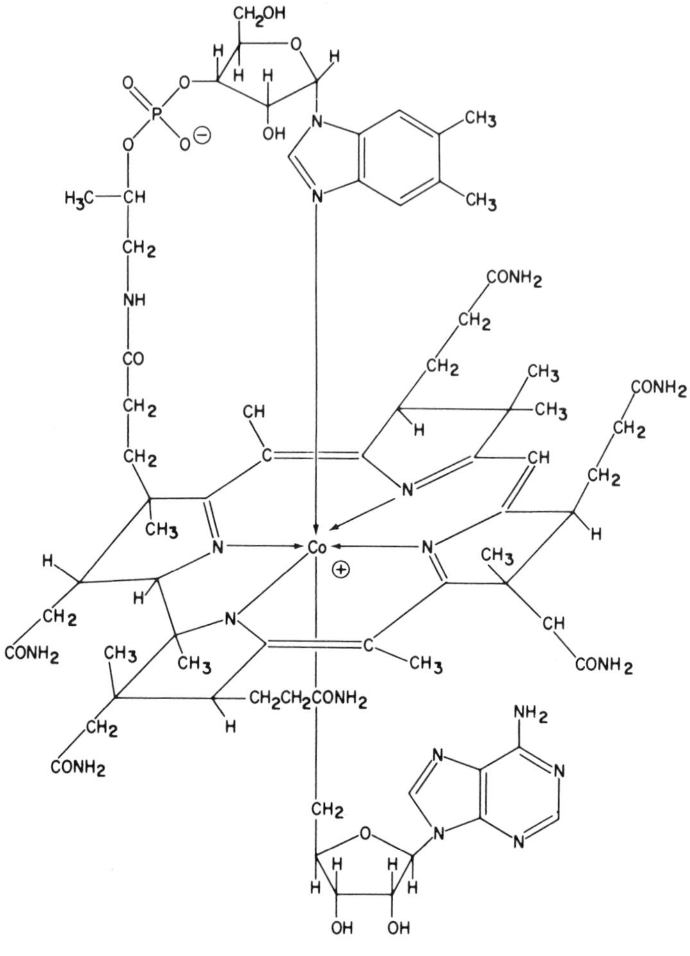

Fig. 8. Structure of vitamin B_{12} (from Whitaker [142]).

Ferric salts stabilize vitamin B_{12} in solution, but ferrous salts can cause rapid destruction [147].

3. Assay

Vitamin B_{12} is analyzed microbiologically using Lactobacillus leichmannii as tester organism [26].

Prior to absorption, free vitamin B_{12} is bound to a mucoprotein in gastric juice, called the intrinsic factor. Protein-bound vitamin B_{12} is readily liberated by digestive enzymes.

4. Effect of Processing

Vitamin B_{12} is not destroyed to any appreciable extent by cooking, unless boiled in alkaline solution. In liver, 8% of the vitamin is lost by boiling at 100°C for 5 min, while broiling muscle meat at 170°C for 45 min results in 30% loss [148]. In normal oven heating of frozen convenience dinners prior to serving, the retention of vitamin B_{12} ranged from 79 to 100% in products containing fish, fried chicken, turkey, and beef [149]. Studies of vitamin B_{12} stability during various heat treatments in the processing of milk are summarized in Table 9.

H. Pantothenic Acid

1. Structure, Stability, and Distribution

Pantothenic acid, D(+)-N-(2,4-dihydroxy-3,3-dimethyl-butyryl)-β-alanine (also known as vitamin B_5) is most stable in the pH range of 4 to 7. It is susceptible to both acid and base hydrolysis. Alkaline hydrolysis yields β-alanine and pantoic acid, whereas acid hydrolysis yields the γ-lactone of pantoic acid [150-152].

$$HOH_2C-C(CH_3)_2-CH(OH)-C(=O)-NH-CH_2CH_2CO_2H$$

Pantothenic Acid

TABLE 9 Losses of Vitamin B_{12} During the Heat Treatment of Milk

Treatment	Loss (%)
Pasteurization for 2-3 sec[a]	7
Boiling for 2-5 min[a]	30
Sterilization at 120°C for 13 min[a]	77
Sterilization at 143°C for 3-4 sec (steam injection)[a]	10
Evaporation[b]	70-90
Spray drying[b]	20-35

[a]Karlin et al. [131].
[b]Borenstein [74].

Pantothenic acid is widely distributed in biological material and occurs in foods primarily as a moiety of coenzyme A. The microbial growth factor pantetheine is pantothenylaminoethanethiol.

2. Assay

Because of the low concentration of pantothenic acid in nature, microbiological methods are generally used to assay for this vitamin. *Saccharomyces carlsbergensis* (detects 10 ng) [153] and *Lactobacillus plantarum* (detects 1 ng) are common test organisms [26].

3. Effect of Processing

In a recent study, 507 edible products were analyzed for pantothenic acid content [86]. In canned foods of animal origin, losses ranged from 20 to 35%, in vegetable foods, from 46 to 78%. Losses were also large in frozen foods, ranging from 21 to 70% for animal products to 37 to 57% in vegetable foods. Losses of pantothenic acid from fruits and fruit juices via freezing and canning were 7 and 50%, respectively. Processed and refined grains lost 37 to 74% of pantothenic acid, and processed meats lost 50 to 75%.

Losses of pantothenic acid are usually less than 10% during milk processing. Cheeses generally have lower levels than fresh milk [74].

I. Biotin

1. Structure and Distribution

Biotin, shown below, consists of two fused five-membered rings formed from urea and a thiophen ring. The structure contains three asymmetric centers and in addition the rings may be cis or trans fused. Of the eight possible stereoisomers, only one, the cis fused (+) biotin, is found naturally and has vitamin activity.

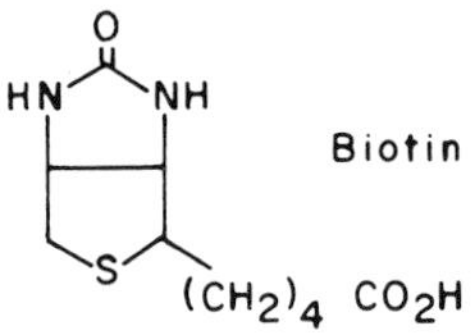

O
HN NH
S $(CH_2)_4\ CO_2H$

Biotin

Biotin is widely distributed throughout the animal and plant kingdoms. It is a coenzyme for carboxylation and transcarboxylation reactions. The coenzyme is covalently bound to biotin-dependent enzymes through an amide bond between the ϵ-amino group of a lysine residue on the enzyme and the carboxyl group of the valeric acid side chain of biotin.

2. Assay

Microbiological methods using Allescheria boydii (detects 0.5 ng) or Lactobacillus arabinosus (detects 0.05 ng) are used for the quantitative assay of biotin [153].

3. Stability

Pure biotin is reportedly stable to heat, light, air, and moderately acid and neutral solutions (optimum 5-8). Alkaline solutions are reasonably stable up to a pH of about 9 [154].

Oxidation of biotin with permanganate or hydrogen peroxide in acetic acid yields the sulfone [155,156]. Nitrous acid destroys the biological activity of biotin, presumably by formation of the nitrosourea derivative. Formaldehyde also inactivates the vitamin.

The limited data available suggest that biotin is very stable during commercial and domestic preparation of foodstuffs.

Biotin in egg white is strongly bound to the protein complex avidin. Avidin is denatured by heat, however, so that its antagonistic action is destroyed when eggs are cooked. Deficiencies of biotin are unlikely in man due to the extensive amount of synthesis by bacteria that occurs in the intestinal tract.

IV. FAT-SOLUBLE VITAMINS

A. Vitamin A

1. Structure

Vitamin A activity is contained in a series of C-20 and C-40 unsaturated hydrocarbons which are widely distributed in the plant and animal kingdoms. The structure of the vitamin is shown below, and it may occur as the free alcohol, esterified to fatty acids, or as the aldehyde or acid. In animals the vitamin is found in highest concentration in the liver where it is stored, existing generally as the free alcohol or in esterified form. In plants and fungi, vitamin A activity is contained in a number of carotenoids which are metabolically converted to vitamin A after absorption by the ingesting animal. The structures and provitamin A activities of a number of commonly occurring carotenoids are given in Table 10, the most potent provitamin being β-carotene, which yields two equivalents of vitamin A.

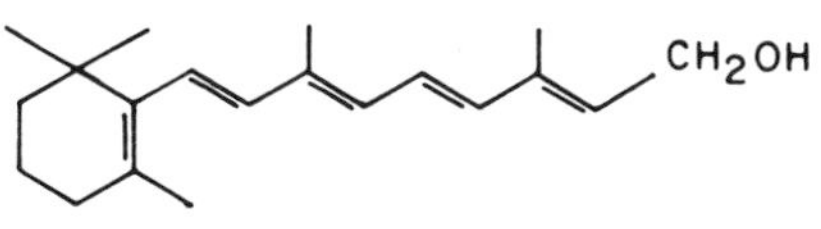

CH_2OH

Vitamin A Alcohol

TABLE 10 Carotenoids: Structure and Provitamin A Activity

Compound	Structure	Relative activity
β- Carotene (widely distributed)	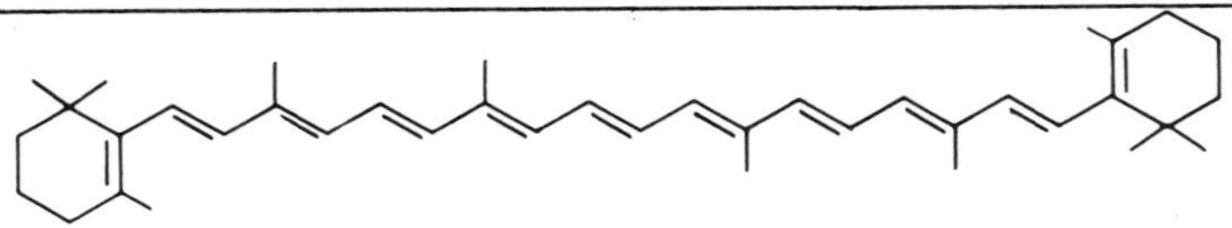	++++
β-apo-8'-carotenal	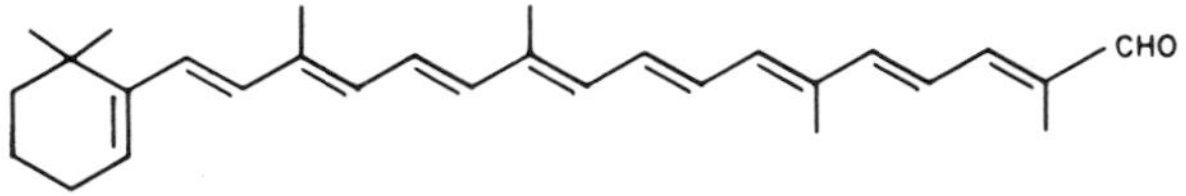 CHO	+++
Cryptoxanthin (orange)	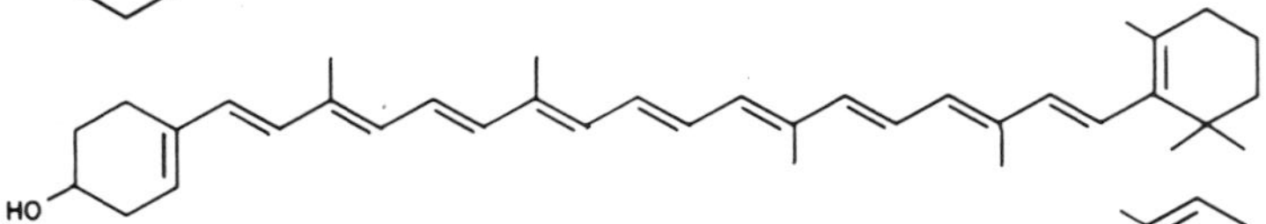 HO	+++
α-Carotene (widely distributed)	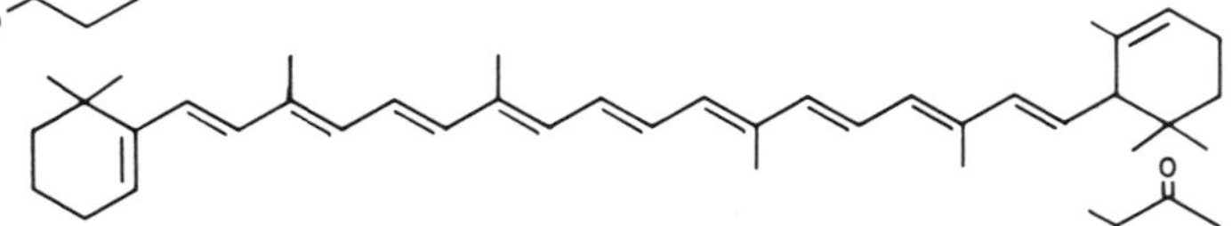	++
Echinenone (sea urchin)	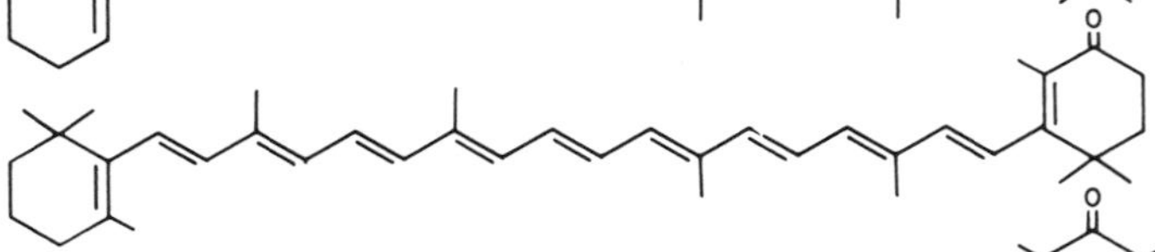 O	++
Astacene (crustacea)	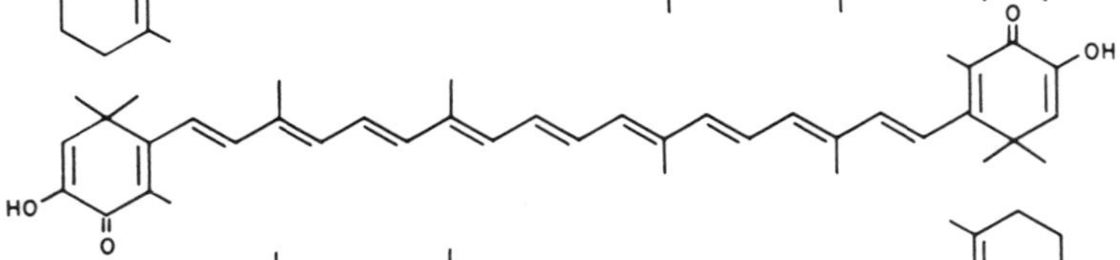 HO, O, O, OH	0
Lycopene (tomato)	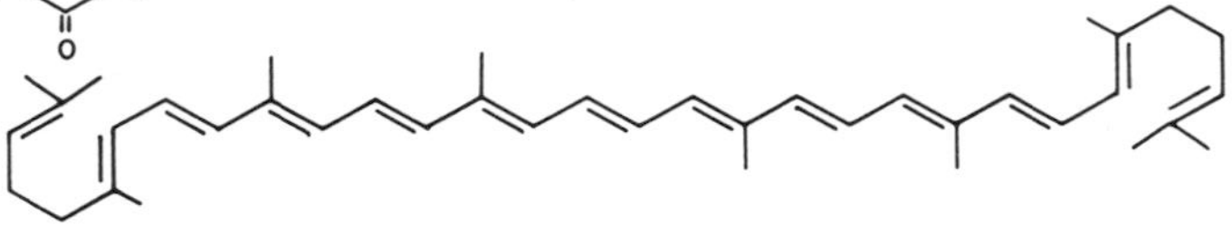	0

Since the carotenoids are predominantly hydrocarbon in nature, they are fat soluble and water insoluble, and are naturally associated with lipid-like structures. Carotene-protein complexes also occur [157] and may be associated with an aqueous phase. The highly unsaturated carotenoid systems give rise to a series of complex UV and visible spectra (300-500 nm) which account for their strong orange-yellow pigmentation. Extensive listings of carotenoids, their spectra, and provitamin A activities are given by Deuel [158] and Zechmeister [159]. A review of the chemistry of carotenoids in foods is given by Borenstein and Bunnell [160].

2. Assay

It is now generally accepted that the vitamin A activity of a food is best determined by chromatographic separation of the carotenoids, followed by summation of the activities in the various geometric and stereoisomers. The early analytical procedures, including the Carr-Price reaction with antimony trichloride, gave too high values, and especially neglected the possibility of cis-trans isomerization following food processing.

3. Stability

The destruction of provitamins A in processed and stored foods can follow a variety of pathways depending on reaction conditions, and is summarized in Fig. 9. In the absence of oxygen there are a number of possible thermal transformations, particularly cis-trans isomerization. This has been shown to occur in both cooked and canned vegetables [161, 162]. Overall losses may vary from 5 to 40%, depending on temperature, time, and the nature of the carotenoids. At higher temperatures, β-carotene can fragment to yield a series of aromatic hydrocarbons, the most prominent being ionene [163,164].

If oxygen is present, extensive losses of carotenoids will occur, stimulated by light, enzymes, and cooxidation with lipid hydroperoxides. Chemical oxidation of β-carotene appears to yield primarily the 5,6-epoxide which may later isomerize to mutachrome, i.e., the 5,8-epoxide. Light-catalyzed oxidation yields primarily mutachrome [165]. Isomerization of the 5,6-epoxide to mutachrome has been studied in orange juice [166]. Further fragmentation of the primary oxidation products yields a complex of compounds [167] similar to those obtained following oxidation of fatty acids. Oxidation of vitamin A results in complete loss of vitamin activity.

Dehydrated foods are most susceptible to loss of vitamin A and provitamin A activity on storage [168] because of their propensity to undergo oxidation [169]. Overall losses of vitamin A resulting from a number of types of dehydration processes are shown in Table 11.

As is the case for lipid oxidation, the rate of vitmain A loss is a function of enzymes, water activity, storage atmosphere, and temperature [171]. Therefore, the expected loss will depend on the severity of drying conditions

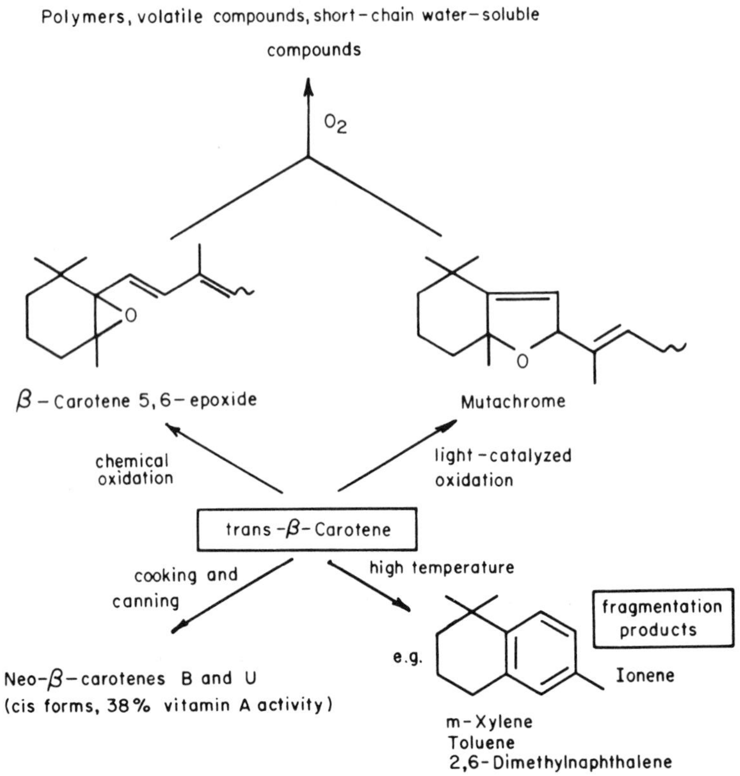

Polymers, volatile compounds, short-chain water-soluble compounds
O_2
β-Carotene 5,6-epoxide
Mutachrome
chemical oxidation
light-catalyzed oxidation
trans-β-Carotene
cooking and canning
high temperature
fragmentation products
e.g.
Ionene
Neo-β-carotenes B and U (cis forms, 38% vitamin A activity)
m-Xylene
Toluene
2,6-Dimethylnaphthalene

Fig. 9. Degradation of β-carotene.

TABLE 11 Concentration of β-Carotene in Cooked Dehydrated Carrots

Sample	Range of concentration (ppm solids)
Fresh	980-1860
Explosive puff dried	805-1060
Vacuum freeze dried	870-1125
Conventional air dried	636-987

Source: Dellamonica and McDowell [170].

(see Table 11) and the degree of protection during storage. The influences of water and oxygen concentrations are shown in Table 12. There is a measurable rate of loss even at very low oxygen partial pressures. Although the rate of loss in shrimp decreases with increasing moisture, other foods (corn) show more complex behavior [172]. In general, the stability of carotenoids would be expected to parallel that of unsaturated fatty acids in a given food; the complex interactions of carotenoid stability with water activity and other constituents of food are beyond the scope of this chapter [173].

B. Vitamin K

Vitamin K activity is found in a number of fat-soluble naphthaquinone derivatives and occurs primarily in green plants. Aside from the fact that these compounds are photoreactive, little is known of their chemical behavior in foods.

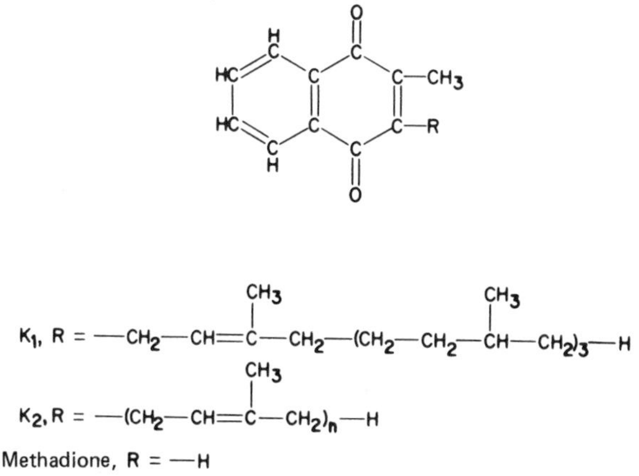

K_1, R = $-CH_2-CH=C(CH_3)-CH_2-(CH_2-CH_2-CH(CH_3)-CH_2)_3-H$

K_2, R = $-(CH_2-CH=C(CH_3)-CH_2)_n-H$

Methadione, R = —H

Vitamin K

TABLE 12 Carotenoid Destruction Rates

Product	H_2O (%)	Gas	k (hr^{-1})
Carrot flakes, 20°C	5	Air	6.1×10^{-4}
		2% O_2	1.0×10^{-3}
Shrimp, 37°C	<0.5	Air	1.1×10^{-3}
	3	Air	0.9×10^{-3}
	5	Air	0.6×10^{-3}
	8	Air	0.1×10^{-3}
E_a = 19 K_{cal}/mole			

Source: Labuza [29].

C. Vitamin D

Vitamin D activity is found in certain sterol derivatives, the structure of one member of this group being shown below. These compounds are fat soluble and are sensitive to oxygen and light. A significant fraction of milk sold in the United States is fortified with vitamin D and stability in foods does not appear to be a significant problem.

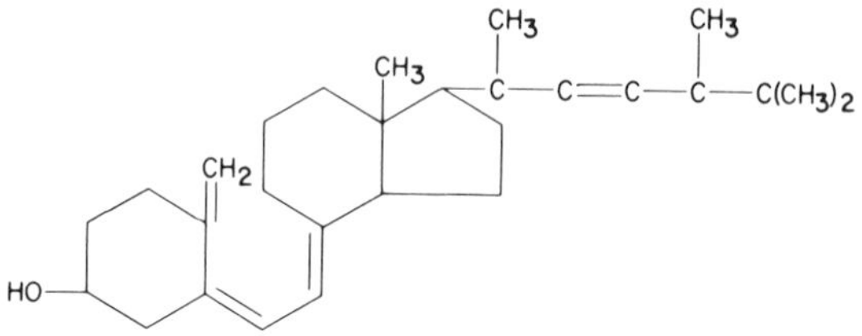

Vitamin D

D. Vitamin E

1. Structure

This fat-soluble vitamin is found in nature in the form of a number of tocopherols and tocotrienols. The most active compound in the group is α-tocopherol. The tocopherols are polyisoprenoid derivatives (chromanol nucleus) and all have a saturated C-16 side chain (phytyl); centers of asymmetry at the 2, 4', and 8' positions; and variable methyl substitution at R_1, R_2, and R_3. The naturally occurring stereoisomer, d-α-tocopherol, is defined as the 2D, 4'D, 8'D configuration, and various diastereoisomers are possible via chemical synthesis.

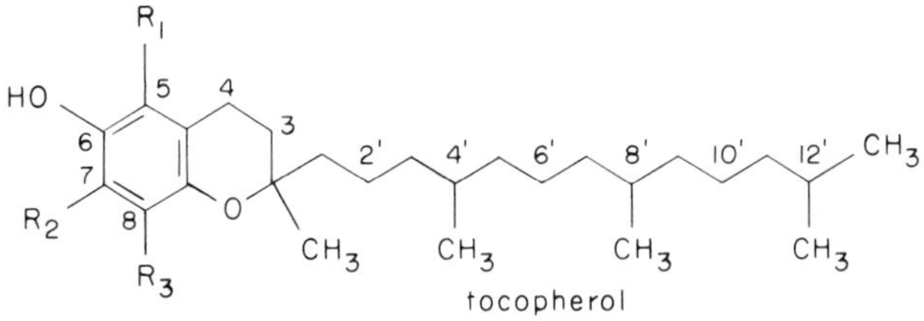

tocopherol

	R_1	R_2	R_3
α	CH_3	CH_3	CH_3
β	CH_3	H	CH_3
γ	H	CH_3	CH_3
δ	H	H	CH_3
tocol	H	H	H

2. Assay and Distribution

Vitamin E has multiple functions in human or animal bodies, but its exact biochemical function has still not been defined. Bioassays can be performed on the basis of a variety of biological responses. Ames [174] has categorized the physiological processes involved in bioassay as follows:

a. Bioassays involving a biological function such as the prevention of fetal resorption in rats or encephalomalacia in chicks
b. Measurement of a physiological parameter, such as prevention of erythrocyte hemolysis
c. Measurement of vitamin E levels in vivo such as liver storage and plasma responses

Since eight tocopherols are known to exist in nature, a simple colorimetric method has not proven feasible for food analysis. The analytical procedure involves extraction, saponification, and chromatography of the nonsaponifiable fraction by thin-layer (TLC) or gas-liqiud chromatography (GLC). The state of the art as of 1971 is reviewed by Bunnell [175].

Vitamin E activity is widely distributed among food groups, including seeds and seed oils, vegetable oil, grains, fruits, vegetables, and animal products [176-178]. Although α-tocopherol is most important insofar as biological potency is concerned, the other naturally occurring isomers exist in significant concentrations and make an important contribution in both vitamin and antioxidant activity, as seen in Table 13. Similar data

TABLE 13 Tocopherol Content of Cereal Grains

Product	Tocopherols, mg/100g			
	α	β	γ	α-Tocotrienol
Whole yellow corn	1.5	-	5.1	0.5
Yellow corn meal	0.4	-	0.9	-
Whole wheat	0.9	2.1	-	0.1
Wheat flour[a]	0.1	1.2	-	0
Whole oats	1.5	-	0.05	0.3
Oat meal	1.3	-	0.2	0.5
Whole rice	0.4	-	0.4	-
Milled rice	0.1	-	0.3	-

Source: Herting and Drury [179].
[a]74% extraction.

exist for vegetable oils [180]. The total tocopherol and α-tocopherol content of typical American diets has been analyzed [181]. The estimated range of daily intake was from 2.6 to 15.4 mg, with the most substantial contribution coming from food products containing, or processed with, vegetable oils or vegetable oil derived margarine or shortening. A more recent analysis [182] came to similar conclusions (4.4 to 12.7 mg/day).

3. Stability

The influence of processing and storage has been reviewed by Harris [183]. The loss of vitamin E activity can occur either via mechanical loss or by oxidation. An example of mechanical loss by degermination of grain is seen in Figure 1. Thus, any process causing separation or removal of the lipid fraction, or manufacturing procedures which involve refining or hydrogenation, are liable to cause loss of this vitamin.

Loss by oxidation usually accompanies lipid oxidation or may be caused by the use of food processing chemicals such as benzoyl peroxide or hydrogen peroxide. Dehydrated foods are particularly susceptible for the reasons discussed under vitamin A. The decomposition products of oxidized tocopherols include dimers, trimers, dihydroxy compounds, and quinones [184]. A mechanism for forming some of these compounds has been suggested [185], and the effects of pH, water content, and temperature have been investigated for the case of tocopherol oxidation with methyl linoleate hydroperoxides.

In the absence of oxygen, tocopherol can form addition compounds with linoleate hydroperoxide [186]. The initial oxidation product is apparently a semiquinone, which is further oxidized to tocopherolquinone or reacts anaerobically with an alkoxy radical to form the addition compound.

REFERENCES

1. R. S. Harris and E. Karmas, in Nutritional Evaluation of Food Processing, Avi Publishing Co., Westport, Conn., 1975.
2. W. Malewski and P. Markakis, J. Food Sci. 36:537 (1971).
3. R. F. Cain, Food Technol. 21:998 (1967).
4. W. R. Aykroyd and J. Doughty, Wheat in Human Nutrition, Food and Agricultural Organization of the U.N., Rome (1970).
5. F. A. Lee, Adv. Food Res. 8:63 (1958).
6. B. E. Proctor and S. A. Goldblith, Food Technol. 2:95 (1948).
7. M. H. Thomas, S. Brenner, A. Eaton, and V. Craig, J. Am. Dietet. Assoc. 25:39 (1949).
8. L. E. Clifcorn, Adv. Food Res. 1:39 (1948).
9. H. Dahn, L. Loewe, and C. A. Bunton, Helv. Chim. Acta 43:320 (1960).

10. K. M. Clegg and A. D. Morton, J. Sci. Food Agr. 16:191 (1965).
11. K. Clegg, J. Sci. Food Agr. 17:546 (1966).
12. P. Finholt, R. B. Paulssen, and T. Higuchi, J. Pharm. Sci. 52:948 (1963).
13. F. E. Huelin, I. M. Coggiola, G. S. Sidhu, and B. H. Kennett, J. Sci. Food Agric. 22:540 (1971).
14. M. A. Joslyn, Food Res. 22:1 (1957).
15. T. Kurata, H. Wakabayashi, and Y. Sakurai, Agr. Biol. Chem. 31:101 (1967).
16. T. Kurata and Y. Sakurai, Agr. Biol. Chem. 31:170 (1967).
17. T. Kurata and Y. Sakurai, Agr. Biol. Chem. 31:177 (1967).
18. V. Kyzlink and D. Curda, Z. Lebensm. Unters. Forsch. 143:263 (1970).
19. P. Spanyar and E. Kevei, Z. Lebensm. Unters. Forsch. 120:1 (1963).
20. M. M. T. Khan and A. E. Martell, J. Am. Chem. Soc. 89:4176 (1967).
21. M. M. T. Khan and A. E. Martell, J. Am. Chem. Soc. 89:7104 (1967).
22. J. E. Hodge, and E. M. Osman, in: Food Chemistry (O. R. Fennema, ed.), Dekker, New York, 1976.
23. J. Velisek, D. Jiri, V. Kubelka, Z. Zelinkova, and J. Pokorny, Z. Lebensm. Unters. Forsch. 162:285 (1976).
24. S. S. Mirvish, L. Wallcave, M. Eagan, and P. Shubik, Science 177:65 (1972).
25. M. C. Archer, S. R. Tannenbaum, T-Y. Fan, and M. Weisman, J. Natl. Cancer Inst. 54:1203 (1975).
26. W. Horowitz, ed., Official Methods of Analysis of the Association of Official Chemists, 10th ed., Association of Official Analytical Chemists, Washington, 1965.
27. J. C. Bauernfeind and D. M. Pinkert, Adv. Food Res. 18:219 (1970).
28. M. Karel and J. T. R. Nickerson, Food Technol. 18:104 (1964).
29. T. P. Labuza, Crit. Rev. Food Technol. 3:217 (1972).
30. E. DeRitter, Food Technol. 30:48 (1976).
31. L. U. Thompson and O. Fennema, J. Agr. Food Chem. 19:232 (1971).
32. J. L. Heid, in Nutritional Evaluation of Food Processing, (R. S. Harris and H. Von Loesecke, eds.), John Wiley and Sons, New York, 1960, pp. 146-148.
33. A. W. Bender, J. Food Technol. 1:261 (1966).
34. L. W. Mapson, Br. Med. Bull. 12:73 (1956).
35. H. R. Bolin and A. E. Stafford, J. Food Sci. 39:1034 (1974).
36. A. E. Bender, J. Sci. Food Agr. 9:754 (1958).
37. R. Breslow, J. Am. Chem. Soc. 80:3719 (1958).
38. B. K. Dwivedi and R. G. Arnold, J. Food Sci. 37:886 (1972).

39. B. K. Dwivedi, R. G. Arnold, and L. M. Libbey, J. Food Sci. 37:689 (1972).
40. B. K. Dwivedi and R. G. Arnold, J. Agr. Food Chem. 21:54 (1973).
41. E. A. Mulley, C. R. Stumbo, and W. M. Hunting, J. Food Sci. 40: 985 (1975).
42. E. A. Mulley, C. R. Stumbo, and W. M. Hunting, J. Food Sci. 40: 989 (1975).
43. J. Leichter and M. A. Joslyn, J. Agr. Food Chem. 17:1355 (1969).
44. B. W. Beadle, D. A. Greenwood, and H. R. Kraybill, J. Biol. Chem. 149:339 (1943).
45. K. T. H. Farrer, Adv. Food Res. 6:257 (1955).
46. M. Freed, S. Brenner, and V. O. Wodicka, Food Technol. 3:148 (1949).
47. S. A. Goldblith and S. R. Tannenbaum, Proc. Seventh Int. Cong. Nutr., 1966, Verlag, W. Germany.
48. E. Feliciotti and W. B. Esselen, Food Technol. 11:77 (1957).
49. J. Leichter and M. A. Joslyn, Biochem. J. 113:611 (1969).
50. R. G. Booth, Biochem. J. 37:518 (1943).
51. G. Beetner, T. Tsao, A. Frey, and J. Harper, J. Food Sci. 39:207 (1974).
52. T. E. Oguntona and A. E. Bender, J. Food Technol. 11:347 (1975).
53. A. K. Perry, C. R. Peters, and F. O. VanDuyne, J. Food Sci. 41: 1330 (1976).
54. C. Y. W. Ang, C. M. Chang, A. E. Frey, and G. E. Livingston, J. Food Sci. 40:997 (1975).
55. R. G. Arnold, L. M. Libbey, and R. C. Lindsay, J. Agr. Food Chem. 17:390 (1969).
56. B. K. Dwivedi and R. G. Arnold, J. Agr. Food Chem. 19:923 (1971).
57. T. D. Morfee and B. J. Liska, J. Dairy Sci. 54:1082 (1971).
58. G. N. Bookwalter, H. A. Moser, V. F. Pfeifer, and E. L. Griffin, Jr., Food Technol. 22:1581 (1968).
59. A. Fujita, Adv. Enzymol. 15:389 (1954).
60. R. S. Green, W. E. Carlson, and C. A. Evans, J. Nutr. 23:165 (1942).
61. W. H. Yudkin, Physiol. Rev. 29:389 (1949).
62. H. Kundig and J. C. Somogyi, Int. Z. Vitaminforsch. 37:476 (1967).
63. J. C. Somogyi, Nutr. Dieta 8:74 (1966).
64. M. A. Porzio, N. Tang, and D. M. Hilker, J. Agr. Food Chem. 21:308 (1973).
65. R. Kuhn, H. Rudy, and T. Wagner-Jauregg, Ber. 66:1950 (1933).
66. O. Warburg and W. Christian, Biochem. Z. 266:377 (1933).
67. P. Karrer, H. Salomon, K. Schöpp, E. Schlittler, and H. Fritzche, Helv. Chim. Acta 17:1010 (1934).
68. N. B. Guerrant and M. B. O'Hara, Food Technol. 7:473 (1953).

69. B. B. Cook, B. Gunning, and D. Uchimoto, J. Agr. Food Chem. 9:316 (1961).
70. L. B. Rockland, C. F. Miller, and D. M. Hahn, J. Food Sci. 42:25 (1977).
71. E. E. Rice, E. M. Squires, and J. F. Fried, Food Res. 13:195 (1948).
72. B. Meyer, J. Thomas, and R. Buckley, Food Technol. 14:190 (1960).
73. S. Cover, E. M. Pilsaver, R. M. Hayes, and W. H. Smith, J. Am. Dietet. Assoc. 25:949 (1949).
74. B. Borenstein, Vitamins and Amino Acids, in Handbook of Food Additives (T. E. Furia, ed.), CRC Press, Cleveland, 1968, pp. 107-137.
75. H. Reiber, Biochem. Biophys. Acta 279:310 (1972).
76. E. E. Snell, Vitamins Hormones 16:77 (1958).
77. G. Wendt and F. W. Bernhart, Arch. Biochem. Biophys. 88:270 (1960).
78. V. Srncova and J. Davidek, J. Food Sci. 37:310 (1972).
79. M. L. Orr, Pantothenic Acid, Vitamin B_6 and Vitamin B_{12} in Foods, Home Economics Res. Rept. No. 36, U.S. Dept. Agr., Washington, 1969.
80. M. M. Polansky and E. W. Toepfer, J. Agr. Food Chem. 17:1394 (1969).
81. J. B. Hassinen, G. T. Durbin, and F. W. Bernhart, J. Nutr. 53:249 (1954).
82. R. M. Tomarelli, E. R. Spence, and F. W. Bernhart, J. Agr. Food Chem. 3:338 (1955).
83. C. H. Lushbough, J. M. Weichman, and B. S. Schweigert, J. Nutr. 67:451 (1959).
84. F. Bergel and D. R. Harrap, J. Chem. Soc. 4051 (1961).
85. V. Buell and R. E. Hansen, J. Am. Chem. Soc. 82:6042 (1960).
86. H. A. Schroeder, Am. J. Clin. Nutr. 24:562 (1971).
87. I. R. Richardson, S. Wilkes, and S. J. Ritchey, J. Nutr. 73:363 (1961).
88. E. M. Nelson, Public Health Rep. 71:445 (1956).
89. C. M. Baugh and C. L. Krumdieck, Ann. N.Y. Acad. Sci. 186:7 (1971).
90. R. L. Blakley, The Biochemistry of Folic Acid and Related Pteridines, North Holland, London, 1969.
91. E. L. R. Stokstad, B. L. Hutchings, J. H. Mowat, J. H. Boothe, C. W. Waller, R. B. Angier, J. Semb, and Y. SubbaRow, J. Am. Chem. Soc. 70:5 (1948).
92. E. L. R. Stokstad, D. Fordham, and A. Grunigen, J. Biol. Chem. 167:877 (1947).
93. O. H. Lowry, O. A. Bessey, and E. J. Crawford, J. Biol. Chem. 180:389 (1949).

94. H. M. Rauen and H. Waldmann, Z. Physiol. Chem. 286:180 (1950).
95. D. Roberts, Biochim. Biophys. Acta. 54:572 (1961).
96. C. W. Waller, A. A. Goldman, R. B. Angier, J. H. Boothe, B. L. Hutchings, J. H. Mowat, and J. Semb, J. Am. Chem. Soc. 72:4630 (1950).
97. D. B. Cosulich and J. M. Smith, Jr., J. Am. Chem. Soc. 71:3574 (1949).
98. R. B. Angier, J. H. Boothe, J. H. Mowat, C. W. Waller, and J. Semb, J. Am. Chem. Soc. 74:408 (1952).
99. G. N. Wogan, S. Paglialunga, M. C. Archer, and S. R. Tannenbaum, Cancer Res. 35:1981 (1975).
100. R. L. Blakley, Biochem. J. 65:331 (1957).
101. M. Silverman, J. C. Keresztesy, G. J. Koval, and R. C. Gardiner, J. Biol. Chem. 266:83 (1957).
102. S. F. Zakrezewski, J. Biol. Chem. 241:2962 (1966).
103. S. F. Zakrezewski, J. Biol. Chem. 241:2957 (1966).
104. M. J. Osborn and F. M. Huennekens, J. Biol. Chem. 233:969 (1958).
105. R. L. Blakley, Biochem. J. 74:71 (1960).
106. J. C. Rabinowitz, in The Enzymes, Vol. 2, (P. D. Boyer, H. Lardy, and K. Myrback, eds.), Academic Press, New York, 1960, p. 185.
107. B. V. Ramasastri and R. L. Blakley, J. Biol. Chem. 239:106 (1964).
108. B. L. Hillcoat, P. F. Nixon, and R. L. Blakley, Anal. Biochem. 21:178 (1967).
109. J. M. Whiteley, J. Drais, J. Kirchner, and F. M. Huennekens, Arch. Biochem. Biophys. 126:956 (1968).
110. M. J. Osborn, P. T. Talbert, and F. M. Huennekens, J. Am. Chem. Soc. 82:4921 (1960).
111. D. B. Cosulich, B. Roth, J. M. Smith, Jr., M. E. Hultquist, and R. P. Parker, J. Am. Chem. Soc. 74:3252 (1952).
112. M. May, T. J. Bardos, F. L. Barger, M. Lansford, J. M. Ravel, G. L. Sutherland, and W. Shive, J. Am. Chem. Soc. 73:3067 (1951).
113. B. Roth, M. E. Hultquist, M. J. Fahrenbach, D. B. Cosulich, H. P. Broquist, J. A. Brockman, Jr., J. M. Smith, Jr., R. P. Parker, E. L. R. Stokstad, and T. H. Jukes, J. Am. Chem. Soc. 74:3247 (1952).
114. J. C. Keresztesy and K. O. Donaldson, Biochem. Biophys. Res. Commun. 5:286 (1961).
115. V. A. Gupta and F. M. Huennekens, Arch. Biochem. Biophys. 120:712 (1967).
116. K. O. Donaldson and J. C. Keresztesy, J. Biol. Chem. 237:3815 (1963).
117. A. R. Larrabee, S. Rosenthal, R. E. Cathou, and J. M. Buchanan, J. Am. Chem. Soc. 83:4094 (1961).
118. C. L. Krumdieck and C. M. Baugh, Biochemistry, 8:1568 (1969).

119. J. K. Coward, P. L. Chello, A. R. Cashmore, K. N. Parameswaran, L. M. DeAngelis, and J. R. Bertino, Biochemistry 14:1548 (1975).
120. C. Chan, Y. S. Shin, and E. L. R. Stokstad, Can. J. Biochem. 51:1617 (1973).
121. Y. S. Shin, E. S. Kim, J. E. Watson, and E. L. R. Stokstad, Can. J. Biochem. 53:338 (1975).
122. C. M. Baugh, C. L. Krumdieck, H. J. Baker, and C. E. Butterworth, J. Clin. Invest. 50:2009 (1971).
123. R. Santini, C. Brewster, and C. Butterworth, Jr., Am. J. Clin. Nutr. 14:205 (1964).
124. C. E. Butterworth, Jr., R. Santini, and W. B. Frommeyer, J. Clin. Invest. 42:1929 (1963).
125. Ten-State Nutrition Survey 1968-70. U.S. Department of Health, Education and Welfare, Health Services and Mental Health Administration. DHEW Publication No. (HSM) 72-8134. Center for Disease Control, Atlanta, Georgia.
126. J. D. O'Broin, I. J. Temperley, J. P. Brown, and J. M. Scott, Am. J. Clin. Nutr. 28:438 (1975).
127. F. Y. Tripet and U. W. Kesselring, Pharm. Acta Helv. 50:318 (1975).
128. H. Burton, J. E. Ford, J. G. Franklin, and J. W. G. Porter, Dairy Res. 34:193 (1967).
129. H. Burton, J. E. Ford, A. G. Perkin, J. W. G. Porter, K. J. Scott, S. Y. Thompson, J. Toothill, and J. D. Edwards-Webb, J. Dairy Res. 37:529 (1970).
130. J. E. Ford, J. W. G. Porter, S. Y. Thompson, J. Toothill, and J. Edwards-Webb, J. Dairy Res. 36:447 (1969).
131. R. Karlin, C. Hours, C. Vallier, R. Bertoye, N. Berry, and H. Morand, Int. Z. Vitamin Forsch. 39:359 (1969).
132. K. C. Lin, B. S. Luh, B. S. Schweigert, J. Food Sci. 40:562 (1975).
133. B. Reed, D. Weir, and J. Scott, Am. J. Clin. Nutr. 29:1393 (1976).
134. J. D. Malin, World Rev. Nutr. Dietetics 21:198 (1975).
135. F. Hanning and M. L. Mitts, J. Am. Dietet. Assoc. 25:226 (1949).
136. V. M. Cheldin, A. M. Woods, and R. J. Williams, J. Nutr. 26:477 (1943).
137. H. D. Alexander, E. J. Day, H. E. Sauberlich, and W. D. Salmon, Fed. Proc. 15:921 (1956).
138. W. A. Krehl and G. R. Cowgill, Food Res. 15:179 (1950).
139. A. Suchewer, J. Bartinsk, and B. Secomska, Higiene 21:619 (1970).
140. J. Metz, A. Lurie, and M. Konidaris, S. Afr. Med. J. 44:539 (1970).
141. E. W. Hellendoorn, A. P. Groot, C. P. van de Mijil Dekker, P. Slump, and J. J. Willems, J. Am. Dietet. Assoc. 58:434 (1971).
142. J. R. Whitaker, in Principles of Enzymology for the Food Sciences, Marcel Dekker, Inc., New York, 1972.
143. D. V. Frost, M. Lapidus, K. A. Plant, E. Scherfling, and H. H. Fricke, Science 116:3005 (1952).

144. C A. Lang and B. F. Chow, Proc. Soc. Expt. Biol. Med. 75:39 (1950).
145. M. Blitz, E. Eigen, and E. Gunsberg, J. Am. Pharm. Assoc. Sci. Ed. 43:651 (1954).
146. S. P. Sen, Chem. Ind. (London) 94 (1962).
147. K. G. Shenoy and G. B. Ramasarma, Arch. Biochem. Biophys. 55: 293 (1955).
148. R. M. Heyssel et al., Am. J. Clin. Nutr. 18:176 (1966).
149. E. DeRitter, M. Osadca, J. Scheiner, and J. Keating, J. Am. Dietet. Assoc. 64:391 (1974).
150. H. Weinstock, H. Mitchell, E. F. Pratt, and R. J. Williams, J. Am. Chem. Soc. 61:1421 (1939).
151. H. K. Mitchell, H. H. Weinstock, E. E. Snell, S. R. Stanberg, and R. J. Williams, J. Am. Chem. Soc. 62:1776 (1940).
152. R. J. Williams and R. T. Major, Science, 91:246 (1940).
153. R. Strohecker and H. M. Henning, in Vitamin Assay, Verlag-Chemie Darmstadt, 1965.
154. Merck Index, 8th ed., Merck and Co., N.J., 1968.
155. K. Hofmann, D. B. Melville, and V. deVigneaud, J. Biol. Chem. 141:207 (1941).
156. F. Kögel and T. J. de Man, Z. Physiol. Chem. 269:81 (1941).
157. M. Nishimura and K. Takamatsu, Nature, 180:699 (1957).
158. H. J. Deuel, in The Lipids, Interscience, New York, 1951.
159. L. Zechmeister, in Cis-Trans Isomeric Carotenoids, Vitamin A and Arylpolyenes, Academic Press, New York, 1962.
160. B. Borenstein and R. H. Bunnell, Adv. Food Res. 15:195 (1966).
161. J. P. Sweeney and A. C. Marsh, J. Am. Dietet. Assoc. 59:238 (1971).
162. K. G. Weckel, B. Santos, E. Hernan, L. Laferriere, and W. H. Gabelman, Food Technol. 16:91 (1962).
163. I. Mader, Science 144:533 (1964).
164. W. C. Day and J. G. Erdman, Science 141:808 (1963).
165. G. R. Seely and T. H. Meyer, Photochem. Photobiol. 13:27 (1971).
166. A. L. Curl and G. F. Bailey, J. Agr. Food Chem. 4:159 (1956).
167. W. M. Walter, Jr., A. E. Purcell, and W. Y. Cobb, J. Agr. Food Chem. 18:881 (1970).
168. G. MacKinney, A. Lukton, and L. Greenbaum, Food Technol. 12:164 (1958).
169. T. P. Labuza, S. R. Tannenbaum, and M. Karel, Food Technol. 24: 543 (1970).
170. E. S. Dellamonica and P. E. McDowell, Food Technol. 19:1957 (1965).
171. H-E. Chou and W. M. Breene, J. Food Sci. 37:66 (1972).
172. F. W. Quackenbush, Cereal Chem. 40:266 (1963).
173. T. P. Labuza, Crit. Rev. Food Technol. 2:355 (1971).
174. S. R. Ames, Lipids 6:281 (1971).

175. R. H. Bunnell, Lipids 6:245 (1971).
176. H. T. Slover, Lipids, 6:291 (1971).
177. K. C. Davis, Am. J. Clin. Nutr. 25:933 (1972).
178. K. C. Davis, J. Food Sci. 38:442 (1973).
179. D. C. Herting and E-J. E. Drury, J. Agr. Food Chem. 17:785 (1969).
180. T. Gutfinger and A. Letan, Lipids 9:658 (1974).
181. R. H. Bunnell, J. Keating, A. Quaresimo, and G. K. Parman, Am. J. Clin. Nutr. 17:1 (1965).
182. J. G. Bieri and R. P. Evarts, J. Am. Dietet. Assoc. 62:147 (1973).
183. R. S. Harris, Vitamins Hormones 20:603 (1962).
184. A. S. Csallany, M. Chiu, and H. H. Draper, Lipids 5:1 (1970).
185. E. H. Gruger, Jr. and A. L. Tappel, Lipids 5:326 (1970).
186. H. W. Gardner, K. Eskins, G. W. Grains, and G. F. Inglett, Lipids 7:324 (1972).

CHAPTER 4

LIPIDS

Lloyd A. Witting

Supelco, Inc.
Supelco Park
Bellefonte, Pennsylvania

Edward G. Perkins
Fred A. Kummerow

Department of Food Sciences
Burnsides Research Laboratory
University of Illinois
Urbana, Illinois

I. RAW MATERIALS

Before considering the subject of food processing, it seems desirable to consider three questions related to the raw materials to be processed. (a) What constituents essential to man are present in the raw materials? (b) What constituents undesirable in the diet of man are present in the actual or potential raw materials? and (c) What may happen to the lipids in the raw materials prior to processing?

A. Essential Nutrients

1. Calories

At 9 kcal/g, fats are our most concentrated source of calories and supply approximately 40% of the calories in a normal mixed diet. Approximately 90-95% of the dietary fat is ingested in the form of triglycerides. The fatty acids commonly occurring in these triglycerides are straight-chain molecules with an even number of carbon atoms and may contain from zero to six methylene interrupted cis double bonds, $CH_3(CH_2)_x$-$(CH{=}CHCH_2)_{0-6}(CH_2)_yCO_2H$. These fatty acids may be separated into four groups according to chain length. (a) Short-chain fatty acids, C_4 - C_8, tend to be rather rare except in milk fats; (b) medium chain-length fatty acids, C_{10} - C_{14}, occur in certain seed fats such as coconut oil; (c) long-chain fatty acids, C_{16} - C_{18}, are the predominant fatty acids in a great variety of lipids; and (d) very long saturated-chain fatty acids (greater than C_{20}) are frequently restricted to specialized uses such as wax ester formation, or appear in specific lipid classes such as sphingolipids or glycosphingolipids. Fatty acids containing 18-22 carbon atoms usually contain one or more cis double bonds.

For the fat to be utilized as a source of calories it must be digested and absorbed into the body. The first step in this process is the isolation of the fat from the food and organization into small globules, or micelles, in the digestive tract. This requires that the fat be not too high melting. Deuel has suggested that any fat melting at 50°C (122°F) or lower is fully digested [1-4]. Trilaurin (mp, 49°C) and trimyristin (mp, 56°C) have coefficients of digestibility of 97 and 77, respectively [5]. Tripalmitin (mp, 67°C) and tristearin (mp, 70°C) are poorly absorbed with coefficients of digestibility of 28 and 19, respectively [5]. The presence in the triglyceride of a fatty acid with one or more cis double bonds results [6] in a lowering of the melting point and increased absorption, as in the case of distearylmonoolein with a coefficient of digestibility of 61.

Digestion of fats requires bile acids [7] for emulsification and pancreatic lipases for hydrolysis [8]. Where the production or release of either is impaired, as in cystic fibrosis or biliary atresia, malabsorption and steatorhea result. Similarly, a decrease in absorptive area [9], as in surgical excision of a portion of the digestive tract, results in reduced fat absorption. It is possible to ameliorate these problems by feeding triglycerides of medium chain-length fatty acids [9,10], which are absorbed via the portal vein rather than via the lymphatic system [11]. Other specialized situations are known where this modification of the dietary fat may be desirable.

In the chronic alcoholic, there may be a pathological accumulation of triglyceride in the liver [12,13]. The output of triglyceride by the liver is directly related to fatty acid chain length and inversely related to the degree of unsaturation of the available fatty acids [14]. Replacement of the normal dietary fat with triglycerides of medium chain length fatty acids, such as those of coconut oil, has been reported to reduce fatty liver formation [15].

Serious questions have been raised regarding the caloric value of highly unsaturated fatty acids containing five or six double bonds [16,17]. Kaneda and Ishii [18] have shown that such fatty acids are as nutritious as oleic acid if they are rigorously protected from autoxidation. This is not always easy to accomplish, since such oxidation may take place in the digestive tract [19]. When it was necessary for experimental purposes [20] to feed pure synthetic arachidonate, for instance, this material had to be incorporated into a complex which required proteolytic digestion for release of the lipid in the gut.

2. Essential Fatty Acids

Polyunsaturated fatty acids are frequently described in terms of an omega (ω) nomenclature [21], where the position of the double bonds is designated with regard to the terminal methyl group. Use of this nomenclature is restricted to fatty acids with only methylene interrupted double bonds. The series of fatty acids 9, 12-octadecadienoic acid; 6, 9, 12-octadecatrienoic acid; 8,11,14-eicosatrienoic acid; 5,8,11,14-eicosatetraenoic acid are all members of the ω-6 family and are usually abbreviated as 18:2ω6, 18:3ω6, 20:3ω6, and 20:4ω6, respectively. Similarly, 9,12,15-octadecatrienoic acid; 6,9,12,15-octadecatrienoic acid; 8,11,14,17-eicosatetraenoic acid; 5,8,11,14,17-eicosapentaenoic acid; 7,10,13,16,19-docosapentaenoic acid, and 4,7,10,13,16,19-docosahexaenoic acid are members of the ω3 series, abbreviated 18:3ω3, 18:4ω3, 20:4ω3, 20:5ω3, 22:5ω3, and 22:6ω3, respectively. Except in certain rare or special situations, the double bonds in the naturally occurring fatty acids are all in the cis configuration.

Arachidonic acid, 20:4ω6, appears to be the true essential fatty acid, but linoleic acid, 18:2ω6, is the biological precursor available in quantity in the normal diet [22,23]. Plants of the _Liliaceae_ [24] and _Boraginaceae_ [25] families contain 18:3ω6, and if green algae ever becomes a popular foodstuff, a source of 16:2ω6 would be available [26]. The interrelationship between the ω6 and ω3 series of fatty acids in terms of essential fatty acid activity is relatively complex. Fish may have an absolute requirement for $\omega 3$ polyunsaturated fatty acids, PUFA, which is not satisfied by $\omega 6$ PUFA [27]. In man, the $\omega 3$ PUFA lower the requirement for $\omega 6$ PUFA but are not completely satisfactory for total replacement of $\omega 6$ PUFA [22, 23].

In the absence of dietary essential fatty acids, growth is retarded and dermal symptoms appear. Generally such observations are restricted to children [23], since it is extremely difficult to deplete the adult of essential fatty acids stored in the adipose tissue. The exceptions are prolonged malabsorption syndromes or intravenous feeding with fat-free preparations, particularly where there has been a gross loss of adipose tissue. For example, eight adult men maintained at constant weight were depleted of essential fatty acids for five years. The one-half depletion rate time of linoleate stored in the adipose tissue was found to be approximately 26 months [28].

One sensitive test of the potential biological activity of various positional and geometric isomers of the essential fatty acids is the manner in which they are enzymatically elongated and desaturated [29-31]. Privett has shown that trans-trans isomer is not utilized [29] and the cis-trans isomers of $18{:}2\omega 6$ are poorly utilized for $20{:}4\omega 6$ formation [32]. Of the various positional isomers of odd-chain-length fatty acids only those retaining the all cis methylene interrupted sequence corresponding to linoleic acid are utilized [33]. Fatty acids containing trans double bonds in various positions found in the flesh and milk of ruminants are produced by rumen organisms [34,35] or during catalytic hydrogenation. By several criteria, such trans fatty acids are treated as "unnatural" materials. They do not pass through the placenta [36,37] and despite the presence of unsaturation are esterified to the alpha rather than the beta position of glycerol [38,39].

3. Fat-Soluble Vitamins

a. Vitamin A, Retinol. Vitamin A is required in vision and appears to be involved in glycosaminoglycan synthesis [40]. Recent experiments suggest that in certain reactions retinol may function as a carrier molecule somewhat in the manner of the phosphorylated polyisoprenols [41,42]. The recommended daily allowance [43] of 4000-5000 I.U. or 800-1000 retinol equivalents recognizes the production of retinol in the cleavage of various carotenoids [44]. Retinol is stored in the liver [45] where vitamin E is required to protect this highly unsaturated material [46]. Evidence of vitamin A toxicity is seen at grossly excessive intakes [47].

b. Vitamin D, Cholecalciferol. DeLuca [48] has presented strong arguments that cholecalciferol and its biologically activated forms involved in calcium and phosphorus metabolism are hormones since, with adequate exposure of the individual to sunlight, a dietary precursor is not required. For infants and children a recommended daily allowance of 400 I.U. is advised [43]. The well-known deficiency sign is, of course, rickets. Excessive intakes of vitamin D are considered to be dangerous [49].

c. Vitamin E, Tocopherol. Various biochemical reactions involve free-radical intermediates. Leakage of free radicals, such as the hydroxyl radical at the flavin level of electron transport, may adventitiously initiate lipid peroxidation [50]. Lipid hydroperoxides are destroyed in vivo by peroxidases, such as the selenium-containing enzyme glutathione peroxidase [51, 52]. Lipid peroxidation proceeds via a cyclic chain reaction capable of producing a large number of product molecules per single free-radical initiation [53]. The hydroperoxides or their secondary oxidation products may damage other tissue constituents [28, 54]. α-Tocopherol reacts competitively in the cyclic chain reaction and terminates it by withdrawing free radicals from the system [55]. This has the effect of moderating the yield of hydroperoxide per adventitious enzymatic initiation and thus reduces potential tissue damage prior to enzymatic destruction of the hydroperoxides [56].

Since this vitamin is stored in the tissues, particularly the liver [57], overt pathological signs of deficiency are rarely observed. Signs of a nutritional muscular dystrophy have been observed in man in connection with a malabsorption syndrome [58]. The recommended daily allowance [43] is stated as 12-15 I.U. of vitamin E activity. γ-Tocopherol, which occurs at approximately 2.5 times the level of α-tocopherol in a normal mixed diet [59], is assumed to have about 10% of the biological activity of α-tocopherol [60]. Vitamin E is essentially nontoxic [61] with gram quantities being consumed daily for years by faddists [62]. Since intestinal absorption is inversely related to dosage [63], relatively little of this grossly excessive intake actually reaches the tissue.

d. Vitamin K. Phylloquinone of plant origin is the major source of vitamin K in the human diet [64], although menaquinones of bacterial origin are directly ingested in ruminant tissues such as beef liver [65]. In man, absorption of bacterial menaquinones produced within his own digestive tract satisfies a major portion of the requirement for this vitamin. The vitamin is well-known for its involvement in blood clotting [66].

Tissue storage is an important factor with all the fat-soluble vitamins and essential fatty acids. The requirements for these nutrients may safely be met on the average over a period of time without placing emphasis on daily intake [43]. Although man does not have the storage capacity of some animals, several well-known facts may be cited to emphasize the documented extremes. It is possible, for instance, to administer in one or two massive doses enough vitamin A to establish liver stores in the rat sufficient to last the remainder of the animal's lifespan [67]. Similarly, polar bear liver contains levels of vitamin A and D that are toxic to man [68].

B. Undesirable Constituents

"Undesirable" is used here in the sense that the constituent is, or is thought to be, a potential source of problems under certain conditions.

Obviously, almost anything may be thought to be undesirable by someone under some conditions. Based on different criteria saturated any polyunsaturated fatty acids may be either undesirable or desirable in foods.

1. Oxygenated Fatty Acids

Lipids contain at least four types of oxygenated fatty acids. Sphingolipids are the usual source of α-hydroxy fatty acids, whereas waxes of various types may contain α-hydroxy, ω-hydroxy [69], or (ω-1)-hydroxy fatty acids [70]. Such lipids are not of particular nutritional importance.

Another type corresponds to the equivalent of direct addition of either oxygen or water to a double bond in oleic, linoleic, or linolenic acid. Coronaric [71] and vernolic [72] acids, 9,10-epoxyoctadec-12-enoic acid and 12,13-epoxyoctadec-9-enoic acid, respectively, are examples of this type of oxygenated fatty acid. Vernolic acid has been identified as a major fatty acid of seed oils from a number of species representing the families Compositae, Dipsacaceae, Euphorbiaceae, Onagraceae, and Valerianaceae [73-75]. Ricinoleic and densipolic acids [76], 12-hydroxyoctadec-9-enoic acid and 12-hydroxyoctadec-9,15-dienoic acid, respectively, are also examples of this third type. Ricinoleic acid which makes up 90% of the fatty acids in castor oil, the seed oil of Ricinus communis, is an intestinal irritant producing catharsis in man, although well digested by rats, rabbits, sheep, and guinea pigs [77,78].

Plants contain an enzyme, lipoxidase, which acts on systems containing a cis,cis-1,4-diene to produce a trans,cis conjugated fatty acid hydroperoxide [79]. Such compounds may be converted enzymatically to keto [80] and hydroxyketo fatty acids [81] and ethers [82]. Products of this type such as coriolic acid, 13-hydroxyoctadeca-cis-9-trans-11-dienoic acid, are occasionally encountered in seed fats [83]. The presence of such oxygenated fatty acids, however, is usually associated with damage to plant or seed tissue [84,85].

Oxygenated fatty acids in general are undesirable constituents in raw materials for at least two reasons. Such materials appear to strongly promote the autoxidation of polyunsaturated fatty acids [86,87]. Several of these compounds have been shown to irritate or damage the stomach and intestinal mucosa [88,89].

2. Fatty Acids Containing Rings

Fatty acids containing a cyclopropane ring are characteristic of gram-negative bacteria [90] but are rarely seen in plants [91]. Sterculic acid, 9,10-methyleneoctadec-9-enoic acid, a cyclopropene compound, is the Halphen reactive material in cottonseed oil [92]. A similar fatty acid, malvalic acid, 8,9-methylenehaptadec-8-enoic acid, is unusual in having an odd chain length [93]. These fatty acids are toxic to nonruminants [94].

The residual amounts of this lipid in cottonseed meal is sufficient to have an adverse effect on poultry and to result in pink discoloration of the white in stored chicken eggs [95]. Addition of 1% cyclopropene fatty acid to the diet of the weanling rat results in growth retardation, enlargement of the liver and kidneys with fatty infiltration, focal degeneration of the tubular epithelium and shriveling of the hepatic parenchymal nuclei, and increased saturation of the tissue lipids [96]. Malvalic acid blocks the desaturation of stearic acid to oleic acid in several animal species [97,98]. The various biological effects are obviated by hydrogenation of the cyclopropene ring [96]. These cyclopropene fatty acids are also known to rapidly undergo thermal polymerization [99].

A series of cyclopentenoid fatty acids is also known to occur in a family of tropical shrubs and trees [100] and are of interest only in terms of their use in the treatment of Hansen's disease.

3. Long Chain Monoenoic Acids

Rape (Brassica campestris) seed oil is a major source of edible fat in Canada and is exported to Europe in very large quantities. High levels of rape seed oil fed to male weanling rats resulted in transitory fatty infiltration of the heart, adrenals, and skeletal muscle with residual necrotic and fibrotic cardiac lesions [101,102]. Initially all of these observations were attributed to the erucic acid, cis-13-docosenoic acid, content of the oil [103]. The level of erucic acid in commercial rape seed oil was subsequently reduced from 45-55% to approximately 0.5% by genetic selection [104]. However, subsequent studies with such oils suggested that other constituents in the lipid may also be involved [105].

Partially hydrogenated marine fish oils, which are also a source of long-chain monoenoic acids, and cetoleic acid, cis-11-docosenoic acid, from herring oil have been shown to produce effects similar to those attributed to erucic acid, whereas cis-11-eicosenoic has little effect [106,107].

4. Saturated Fatty Acids

In the early 1950s, when very little was known regarding the intermediary metabolism of lipids and its regulation, attention was focused on the interrelations between dietary lipid, serum cholesterol, and the incidence of atherosclerotic heart disease [108-110]. Unfortunately a large-scale national dietary heart survey was inconclusive in documenting direct correlations [111]. Hyperlipoproteinemias are not a homogeneous entity and have been resolved into at least five distinct subclassifications requiring different treatments [112]. Several general dietary modifications, however, have received wide attention.

It has been demonstrated that the level of serum cholesterol is mildly, approximately 15%, influenced by the balance of saturated to polyunsaturated fatty acids in the diet [113,114]. As judged from studies of the fatty acid composition of human adipose tissue, the level of linoleate in the dietary fat has doubled, from approximately 10 to 20%, between 1959 and

1974, with the prospect of still further increases [115]. There have been several interesting attempts made to increase the linoleate content of raw materials. Hydrogenation of dietary fatty acids by rumen microorganisms tends to limit efforts to modify the polyunsaturated fatty acid content of veal, beef, and milk. Some success has been achieved by encapsulation of the lipid with protein [116-119].

Naturally occurring fats and oils contain levels of tocopherols proportional to their content of polyunsaturated fatty acids [120]. Any normal mixed diet tends, therefore, to be inherently balanced. Modification of basic dietary raw materials to increase their polyunsaturated fatty acid content may disturb this balance with regard to vitamin E [115].

5. Polyunsaturated Fatty Acids

Oils containing large amounts of linolenic acid, such as linseed oil, are classed as "drying" oils and are suitable for use in oil-based paints. "Drying" is used in the sense that a thin layer of the oil will polymerize to a tough film in a reasonable period of time on exposure to air. The presence of approximately 2-8% linolenic acid in soybean oil greatly restricts the potential uses of this oil. Some progress has been reported in the development of a copper catalyst to hydrogenate selectively the linolenic acid in fats [121,122]. When whale oils were available in quantity, problems in product stability were related to the presence of polyunsaturated eicosa and docosanoic acids. It is extremely difficult to protect fats containing highly unsaturated fatty acids against autoxidation during processing. The oxidized material then promotes further oxidation in the product or breaks down to small molecules with undesirable flavors and odors [123-129].

6. Sterols

The normal human diet contains approximately 300-700 mg cholesterol per day and the normal adult synthesizes approximately 1.7-2.4 g of cholesterol de novo from acetate. In some animals, such as the rat, there is a homeostatic balance between dietary intake of cholesterol and biosynthesis [130] and catabolism [131] of this sterol. Since this homeostatic balance is poor in man [132], the circulating level of cholesterol may be slightly reduced by restricting cholesterol intake. Plant and animal sterols appear to be absorbed at the same intestinal sites. It has been suggested that plant sterols, by competing for these absorptive sites, impede the absorption of cholesterol [133]. This may be of some importance in selecting dietary items for individuals at high risk of atherosclerotic heart disease.

The foregoing discussion is intended to convey the impression that fats and oils are complex materials that cannot be considered merely as a source of calories. It is essential to be aware of the fatty acid composition of the fats in the raw materials as it relates to digestibility, product stability, possible outward physiological responses, and the economic ramifications of current fads.

C. Changes Occurring in Raw Materials Prior to Processing

The chemical changes taking place in lipids are discussed in detail in the next section. At this point, however, it is necessary to point out that various deteriorative reactions may occur prior to processing. Contamination by agricultural chemicals or the production of toxic materials by the action of molds or other organisms is considered in other chapters; only hydrolytic and oxidative rancidity will be considered here.

1. Hydrolytic Rancidity

The ester bond in lipids may be hydrolyzed with the production of free fatty acids and partial glycerides. In raw materials such as milk, which contains short-chain fatty acids including butyric, caproic, caprylic, and capric acids, the potential foodstuff may become inedible or organoleptically unacceptable because of the strong flavor and odor of these acids. Free fatty acids may also bind metal ions which act as catalysts in the autoxidation reaction and otherwise detract from product quality during subsequent processing. In the production of edible oils, the free fatty acids are removed during processing and therefore may be an important economic component of refining loss. The presence of free fatty acids obviously requires the concomitant presence of mono- and/or diglycerides. Since these compounds are emulsifying agents their presence will also affect product characteristics.

2. Oxidative Rancidity

Lipid oxidation is discussed in detail below. At this point it will merely be noted that oxidative reactions may occur in raw materials prior to processing [129]. In addition to the production of undesirable flavors and odors, the oxidized lipids may promote further oxidation during processing or detract from the stability of the finished product [123-127].

II. SPECIFIC REACTIONS OF FATS AND OILS

The principal reactive sites in lipids are the carboxyl groups and double bonds in the fatty acid chain. For the purpose of the present discussion the reactions occurring at the double bonds are of primary importance. Certain processing steps affect only the position and configuration of the double bonds without otherwise altering the molecule. However, most reactions involving oxidation, including heat or thermal oxidation, modify the double bond by insertion or formation of new functional groups or structures. It will also be necessary to distinguish between oxidations in the lipid phase and in aqueous emulsions.

A. Positional and/or Configurational Changes

In a previous section it was noted that positional and configurational isomers of unsaturated fatty acids may be of natural origin. In ruminants, the rumen microoganisms hydrogenate dietary polyunsaturated fatty acids with the production of various positional and trans isomers which are incorporated into milk and tissue lipids [34,35].

Liquid oils are converted to solid fats commercially by partial hydrogenation, usually in the presence of a nickel catalyst. At the catalyst surface both hydrogen addition and hydrogen abstraction occur [134]. The process proceeds in the direction of saturation but the residual double bonds are in somewhat random positions and may, depending on the selectivity of the catalyst, be largely in the trans configuration.

Hydrogenation is seldom carried to completion, since a fully hydrogenated fat would be a hard solid having a relatively high melting point and a low coefficient of digestibility. For practical purposes it may generally be stated that fatty acids tend to be hydrogenated in terms of their degree of unsaturation. That is, trienoic acids tend to be hydrogenated before dienoic acids, which tend to be hydrogenated before monoenoic acids. Reduction of all the unsaturated fatty acids in soybean oil to cis monoenoic acids would produce a liquid fat containing approximately 88% "oleic" acid. However, if the monoenoic acids are largely present in the trans configuration, such as positional isomers of elaidic acid, a solid fat may be produced, since an acyl dielaiden [135] has a melting point approximately 25-30°C higher than the corresponding acyl diolein [136].

Frequently it is necessary to produce a fat which has a wide "plastic range." Such a fat is soft and spreadable upon removal from refrigerated storage at 5°C and retains these properties, without melting, on standing at room temperature. Fats of this type are composed of liquid oil suspended in a solid fat matrix. Elaidic acid and other trans monoenoic acids contribute this solid matrix without exceeding the melting point at which the coefficient of digestibility begins to decrease rapidly. By careful control of the degree and selectivity of the hydrogenation, it is possible to adjust the properties of fats for various uses.

Oleomargarines containing lineoleic acid are best obtained by blending unhydrogenated triglycerides with a suitable matric fat, since the position and configuration of the double bonds in the dienoic acids in a partially hydrogenated oil is quite variable.

B. Oxidative Reactions

Autoxidation [53] of polyunsaturated fatty acids is usually initiated by the catalytic action of trace metals with abstraction of a proton and formation of a fatty acid free radical. This free radical may exist in a number of resonance forms but in the predominant structure the double bond moves into conjugation with the next adjacent double bond and is inverted to the trans configuration.

$$CH_3(CH_2)_4CH{=}CH{-}CH_2{-}CH{=}CH{-}(CH_2)_6CO_2R'$$

$$\downarrow$$

$$CH_3(CH_2)_4CH{-}CH{=}CH{-}CH{=}CH{-}(CH_2)_6CO_2R'$$

Autoxidation then proceeds via a chain-reaction process to produce fatty acid hydroperoxides.

$$RH \xrightarrow{r_i} R\cdot$$

$$R\cdot + O_2 \xrightarrow{k_2} RO_2\cdot$$

$$RO_2\cdot + RH \xrightarrow{k_3} ROOH + R\cdot$$

Chain branching or free-radical multiplication results in an autocatalytic reaction. The presence of even very small quantities of hydroperoxides in the lipid is thus a serious threat to the stability of the product.

$$ROOH \longrightarrow RO\cdot + \cdot OH$$

$$2\ ROOH \longrightarrow RO_2\cdot + RO\cdot + H_2O$$

Lipid antioxidants (AH) such as the tocopherols do not prevent lipid peroxidation: they react with hydroperoxides (ROOH) or destroy hydroperoxides [55]. Antioxidants minimize the yield of hydroperoxide per free-radical initiation by competing with the lipid-bound fatty acids (RH) for reaction with the peroxy free radical ($RO_2\cdot$) [61,137].

$$RO_2\cdot + AH \xrightarrow{k_4} ROOH + (A\cdot)$$

The antioxidant withdraws free radicals from the system by reactions such as quinone formation or dimerization [138,139]. Some other antioxidants do react directly with the peroxy free radical, but this is still a competitive reaction and hydroperoxide formation is minimized but not prevented [55].

$$2\ RO_2\cdot + HO\text{-}C_6H_2R'_2\text{-}CH_3 \longrightarrow O{=}C_6H_2R'_2(OOR)(CH_3) + ROOH$$

The conjugated diene formed in the original reaction is more reactive than the original fatty acid and secondary reactions rapidly become important [140-143]. Peroxide concentrates are best prepared at low temperatures. At room temperature, the reaction mixture in pure linoleate reaches a level of approximately 8% simple monohydroperoxide [144], at which stage secondary and primary reactions proceed at similar rates and there is no further net accumulation of the hydroperoxide. At moderate temperatures oxygen-bonded polymers, probably of a peroxide type R-O-O-R', are formed [145-148]. These polymers may contain a number and variety of functional groups and undergo rearrangements and chain scission on storage even at low temperatures.

In nonaqueous media chain scission may occur to produce a variety of products [149-167], including saturated, mono- and diunsaturated aldehydes, RCHO, R'CH=CHCHO, and R"CH=CHCH=CHCHO, respectively. Dialdehydes, OHCRCHO, and semialdehydes, $OHCR'CO_2H$, are also produced, although the latter may remain bound to the glyceride. The aldehydes may be oxidized to the corresponding acids, and the unsaturated aldehydes may undergo further chain scissions [168,169]. In aqueous media the aldehydes tend to contain hydroxyl groups [170] and in milk, lactones are a common product [171,172]. When compounds such as proteins, which contain free amino groups, are present, condensation products are formed.

Various tabulations of volatile products obtained under specified conditions have appeared. The techniques of gas chromatography and mass spectrometry have been particularly useful in this area [154-156]. In addition to the compounds noted above, which are present in homologous series, these lists include alcohols, esters, ketones, and aromatic as well as aliphatic hydrocarbons.

C. Thermal Reactions

Foodstuffs are not normally heated to high temperatures in the absence of air. However, an extensive literature exists on the anaerobic thermal polymerization of oils for other purposes. It is usually assumed that some of the same products appear in fats and oils heated in the presence of air at frying temperatures. The mixture of reaction products obtained during anaerobic heat treatment is less complex than that obtained during aerobic heat treatment, thus facilitating analysis and characterization. In considering thermal reactions, serious controversy arises as to what temperature range should reasonably be considered. Many studies have used temperatures of 250-350°C (482-662°F). It is somewhat arbitrarily stated that the "nonthermal" oxidation temperature range extends to 100°C, since polymerization by carbon-to-carbon bond formation is noted only at this or higher temperatures [173,174].

The nature of the thermal reaction products will vary with the degree of unsaturation of the fatty acids under investigation. Historically, Chinese tung oil, which contains approximately 75% eleostearic acid, octadeca-9,11,13-trienoic acid, was an extremely useful oil in the formulation of paints. The drying properties of this oil are, of course, related to the presence of preformed conjugated triene. Linseed oil, which contains approximately 55% linolenic acid, octadeca-9,12,15-trienoic acid, is not as good a drying oil and is therefore frequently "boiled" or partially polymerized, before incorporation into paints. The conjugation of linolenic acid and its thermal polymerization has been extensively studied.

Thermal dimers [175] are commercially produced also for other purposes. At room temperature these compounds are viscous liquids with good low-temperature characteristics and do not crystallize. They are easily soluble in hydrocarbons and are nonvolatile. Dimer acids and their esters have been used as additives to control corrosion, improve lubricating properties, or as crystallization inhibitors. As polyester derivatives, they are used in rubber-like products, urethane coatings and foams, alkyd resins, and epoxy coatings. As polyamide resins the dimer acids are used in metal coatings, textile fiber, glass fiber, plastic solders, and printing-ink formulations.

Under mild conditions (180°C) a cyclic monomer, a 1,2-dialkyl-3,5-cyclohexadiene, is produced from pure methyl eleostearate [176,177].

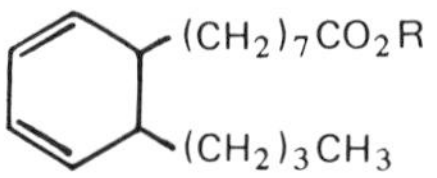

$(CH_2)_7CO_2R$
$(CH_2)_3CH_3$

The cyclic monomers from alkali-conjugated linseed oil [178-184] are 1-alkyl-2-alkenecarboxycyclohexenes with the side-chain double bond in various positions [185-187]. Since the double bonds are concentrated toward the terminal methyl group, the alkyl group is relatively short, containing approximately 1 to 4 carbons, whereas the saturated or unsaturated side chain terminating in the carboxyl group contains 8-11 carbons [188]. Similar compounds containing only one double bond have also been isolated in very low yield (0.6%) from linoleate heated at 200°C exposed to air [189,190]. Some excellent analytical work in this area has appeared in the German literature [191-198].

Given a source of conjugated diene, the Diels-Alder reaction provides a route to tetra-substituted cyclohexenes [199-202] and to trimers containing two tetra-substituted cyclohexene rings. Such compounds are indeed

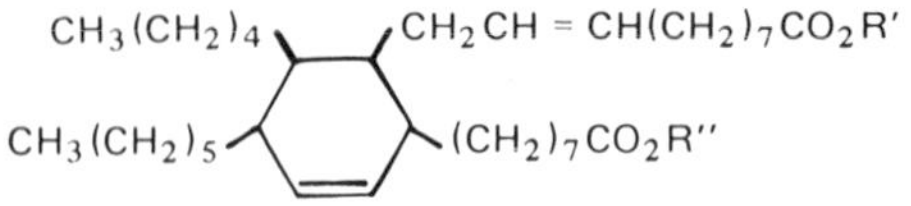

$CH_3(CH_2)_4$ $CH_2CH=CH(CH_2)_7CO_2R'$
$CH_3(CH_2)_5$ $(CH_2)_7CO_2R''$

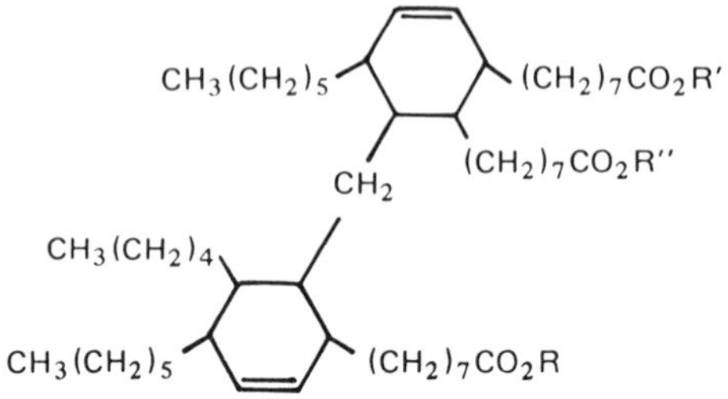

$CH_3(CH_2)_5$ $(CH_2)_7CO_2R'$
$(CH_2)_7CO_2R''$
CH_2
$CH_3(CH_2)_4$
$CH_3(CH_2)_5$ $(CH_2)_7CO_2R$

present in heated oils but the conjugation of the double bonds is a rate-limiting reaction requiring the presence of free radicals [203].

Early studies determined the tetra-substituted cyclohexene dimers by conversion to prehnitic acid [200,204]. Less than one-half as much prehnitic acid could be obtained from the thermal dimers of linoleate as from the thermal dimers of eleostearate. This amount was reported to form early in the heating period and not to increase thereafter [205]. More recent experiments, in which great stress was placed on attaining anaerobic conditions and definitive analyses were attained by the use of gas chromatography-mass spectrometry, were stated to disprove the occurrence of the Diels-Alder reaction in linoleate heated at 280°C (536°F) for 65 hr [203, 206].

The thermal dimers of linoleate have been reported to contain an average of 1.3 rings per dimer unit [207] and mono-, di-, and tricyclic dimers have been detected by mass spectrometry [203,206]. In an alternative polymerization process, free radicals are formed by hydrogen abstraction from the methylene group adjacent to a double bond. Acyclic dimeric radicals are formed by addition of the free radical to the double bond of a second fatty acid [203,206,208-210]. In a monoenoic acid, such as oleic acid, an intra-

$$R\text{-}CH_2CH{=}CH\text{-}R' \longrightarrow R\text{-}\dot{C}HCH{=}CH\text{-}R' \quad (X\cdot)$$

$$R\text{-}CH_2CH{=}CH\text{-}R' \xrightarrow{X\cdot} \begin{array}{l} R\text{-}CH_2CH\dot{C}H\text{-}R' \\ \quad\;\; | \\ R\text{-}CHCH{=}CH\text{-}R' \end{array} \longrightarrow \begin{array}{l} R\text{-}CH_2CH\text{—}CH\text{-}R' \\ \quad\;\; | \qquad\;\; | \\ R\text{-}CH\dot{C}HCH\text{-}R' \end{array}$$

molecular radical addition within the acyclic dimer free radical leads to a cyclic dimer free radical containing a cyclopentane ring. The reaction path itself becomes a cyclic chain reaction, since the cyclic dimer free radical abstracts an H from oleate.

Comparable reactions in linoleate are potentially more complicated, since the initial free radical has a number of resonance structures which include conjugated dienes. As in the monoenoic acids, intermolecular radical addition leads to an acyclic dimer radical which, via an intramolecular radical addition, forms a cyclic dimer radical. The cyclic dimer formed from linoleate has two exocyclic double bonds, thus allowing successive formation of bicyclic and tricyclic dimers. This topic has been reviewed in detail by Figge [211,212]. Free radicals may be lost from the system by radical disproportionation, but acyclic dimerization and ring formation do not result in a net loss of free radicals. If the reaction is catalyzed by addition of tert-butyl peroxide, thermal polymerization proceeds readily at 180°C (356°F) or even 134°C (273°F).

In heat-treated oils, two other groups of compounds also occur. These include the volatile products characteristic of the pyrolysis of organic compounds in general such as CO_2, CO, H_2O, olefins, ketones, and aldehydes, with acrolein arising from the glycerol portion of the lipid [213, 214]. Chain scission leaves a portion of the original fatty acid attached to the glycerol unit, thus producing a group of nonvolatile degradation products. Production of relatively large molecules via ketone formation, such as 18-ketopentatriacontane from stearate [215] or di-n-nonyl ketone from tricaprin [216], requires the presence of oxygen.

D. Thermal Oxidation

The nature and relative proportion of various products formed during thermal oxidation will be dependent on the temperature and the degree of aeration. To some investigators, thermal oxidation may mean heating an oil with its surface open to the atmosphere. One then has to specify carefully the ratio of surface area to total volume. Many investigations have involved bubbling air through the heated oil. The variables in this case are flow rate and degree of dispersion and some consideration should be given to the possible effects of blowing out somewhat volatile reaction products. As will be apparent below, it is also important to distinguish between continuous and discontinuous heating. For these reasons it is often difficult to compare the composition of various thermally oxidized oils produced by different investigators.

The observations from a study of thermally oxidized corn oil are fairly typical [217], and two reasonably distinct reaction phases were noted. In the first phase, oxygen was taken up and nonconjugated acids were converted to conjugated fatty acids. The carbonyl value increased markedly while the refractive index and viscosity increased only slightly. During the second phase, conjugated acids "disappeared" and the carbonyl value decreased while the refractive index and viscosity increased indicating polymer formation. The polymer content of cottonseed oil was found to increase regularly with heating time [218,219]. Intermittent heating and cooling increased the yield of altered fatty acids formed per hour of heating time [219]. This was attributed to the buildup of peroxides during the cooling period. Foaming of the oil is related to the formation of highly oxygenated polar polymers [220-222].

As in low-temperature oxidation, chain scission occurs with the production of numerous short-chain volatile compounds [223-229]. While the compounds produced are similar, a greater variety of aromatic compounds are noted in thermally oxidized oils. In addition to phenol, toluene, benzaldehyde, and benzoic acid, various aliphatic carbonyl compounds with an aromatic substituent, such as 6-phenyl-3-hexanone, have been characterized. It should be remembered that a portion of the fragmented fatty acid chain usually remains attached to the glycerol.

From the analytical viewpoint it is easier to work with either the oxygenated monomers or the less polar compounds in the nonvolatile fraction. Reported characterizations have therefore included a number of nonoxygenated compounds containing cyclohexane, cyclohexene, or aromatic rings [190,194,230-233]. The reaction mixture has frequently been subjected to thin-layer chromatography in solvent systems that move the components upward a distance inversely related to their polarity. Characterizations have tended to proceed from the front toward the origin. More recent fractionations have frequently utilized gel permeation [234-238]. Fatty acids in the polymers are linked by carbon-to-carbon bonds. Although reports of cyclic linkages predominate [239], there have been reports [240-242] of dimers which could not be aromatized and were, therefore, presumably noncyclic. In a study of the thermal oxidation of trilinolein, the level of cyclic dimer was approximately twice the level of acyclic dimer [227]. The hydroxyl and carbonyl content of the cyclic dimers is dependent on experimental conditions during the thermal oxidation.

When a heated and/or oxidized oil is converted to methyl or ethyl esters and treated with urea, simple straight-chain molecules form a clathrate and dimers, polymers, and compounds containing a nonterminal ring are excluded. The older literature frequently considered the material not forming a urea adduct to be polymeric [243], although cyclic monomers are included in this fraction. A fraction described as distillable, nonurea adduct forming material is also a concentrate of cyclic monomer. The nature of the materials formed in oxidized and thermally oxidized oils will be further considered in a later section dealing with their nutritional and physiological properties.

III. GENERAL EFFECTS OF PROCESSING ON LIPIDS

A. Processing of Foods by Methods Other than Frying

To a great extent the effect of various processing techniques on food lipids is easily predicted on the basis of tissue damage, exposure to oxygen, and the temperature range involved.

Simple rapid freezing of raw unblanched plant tissue raises the potential problem of lipoxidase action. The extent of enzymatic action will be a factor of tissue damage, storage temperature, and time. Blanching will inactivate some enzymes but reactivation during storage must be considered. Low-temperature storage reduces the rate of lipid autoxidation but does not prevent the reaction.

Dehydration or freeze-drying have the effect of exposing the food lipid to air while distributed as a thin film. Autoxidation proceeds most rapidly in the presence of a critical residual amount of water [244, 245]. Excess water protects the lipid from contact with air but a solution of metal ions catalyzes the autoxidation of the lipid. Again, storage temperature and time are important in determining the extent of oxidative damage. Since dehydration may remove some natural flavor and odor constituents, detection of oxidatively produced off-flavors may be facilitated.

As noted previously [170], the products of lipid autoxidation in an aqueous emulsion differ from those formed under somewhat anhydrous conditions, with hydroxylated intermediates leaving the lipid and being extracted into the aqueous phase. It is interesting to note that these compounds have been reported to inhibit glycolysis and respiration of tumor cells in vitro and to cause morphological changes which quickly lead to their death while not affecting normal cells. Myoglobin, hemoglobin, and other metalo-proteins are particularly effective catalysts of lipid autoxidation although histidine [246, 247] has been used in many model system studies.

Distribution of lipids over a large surface area also occurs in many baked products. Such products, however, are usually consumed soon after production. The bleaching of wheat flour with strong oxidizing agents also oxidizes the residual lipid which may combine with the bleaching agent. Some types of fermentations may deliberately produce compounds which would, in other products, be considered undesirable. Consider, for instance, the production of short-chain acids and carbonyl compounds in cheeses. Some of the fermented fish and soy products consumed in the Orient are quite rancid by occidental standards.

Ionizing radiation [248-251] has a tendency to destroy essential nutrients and to produce undesirable products. The compounds from gamma-irradiated fat resemble those from oxidized fat. Additional products arise from the interaction of lipid degradation products, such as unsaturated carbonyl compounds, with protein degradation products such as methyl mercaptan. Sterilizing dosages of radiation fail to inactivate many enzymes and a heating step prior to irradiation is required with many products.

Lipid antioxidants such as butylated hydroxytoluene or butylated hydroxyanisole and metal chelating agents are routinely added to many foods to stabilize the lipids. Where products are prepared for faddists, the exclusion of these materials shortens the shelf-life. Similarly, where the polyunsaturated fatty acid content of a foodstuff is increased, the prevention of autoxidation becomes a greater problem than usual.

The introduction of carcinogenic polycyclic hydrocarbons into the food during processing, such as may occur in charcoal broiling, is extraneous to this discussion since this is not a conventional commercial-scale process.

Generally speaking, the processing techniques noted above may in extreme cases adversely affect product quality and shelf-life and detract from vitamin content without greatly affecting the nutritional or safety aspects of the product. Some controversy exists, however, as to whether or not these statements are equally true regarding the techniques used in the preparation of bulk oils and fats and the processing of foods by frying. The processing of fats and oils as distinguished from the processing of complex foods is discussed in Sec. VI.

B. Frying

Frying operations may be divided into three classes. The skillet process [252] is usually ignored and attention is focused on discontinuous restaurant batch-type operations and continuous commercial processing. In both cases the fried food withdraws oil from the cooking unit which must be replenished periodically or continuously. The critical factors requiring discussion are (a) exclusion of air [253], (b) removal of volatiles, and (c) attainment of steady-state conditions. Factors (a) and (b) are related to the presence of water in the food. The hot oil is depicted as blanketed by a layer of steam which reduces contact with air, and volatile decomposition products are continuously removed by steam distillation. It should be noted, however, that water increases the extent of oxidation [254]. No mention is usually made of the nonvolatile portion of the scissioned fatty acid chain still attached to the glycerol moiety.

Maintenance of an effective steam blanket requires spraying water on the oil and the presence of a cover over the frying unit [255] which may consist of a continuous belt carrying the product through a long rectangular tray or trough of heated oil. Such a cover will favor condensation of the volatiles and their reentry into the oil. Some entrained and contained air will, of course, enter the oil with the food to be processed. Factors (a) and (b) are variable and related to the design and operation of the particular frying unit, but the steam blanket is normally not a significant protective factor. Examination of such units quickly reveals that volatiles are, to a great extent, purged out of the oil and oxidation does indeed take place. Significant oxygen uptake is associated with the first phase of thermal oxidation; however, the second phase, polymer formation, does not require a great amount of oxygen [217].

Factor (c), attainment of steady state, is most readily discussed in mathematical terms. Continuous addition of fresh oil at the rate of 8%/hr, for instance, results in a "complete" change of oil approximately twice a day. However, a certain number of molecules of the original fat placed in the frying unit will remain in the unit no matter how much fresh oil is added. Consideration of the addition of fresh oil at discreet intervals requires simple arithmetic; however, discussion of continuous addition requires the use of calculus. If one-third of the total oil is withdrawn by the food in time a, the original volume is reattained by the addition of this amount of fresh fat at the end of period a. A series is generated as shown below.

Average "age" in fryer after a hours = a

After 2a hr = (1/3 x a) + (2/3 x 2a)

After 3a hr = (1/2 x a) + (2/9 x 2a) + (4/9 x 3a)

After 4a hr (1/3 x a) + (2/9 x 2a) + (4/27 x 3a) + (8/27 x 4a)

After 5a hr = (1/3 x a) + (2/9 x 2a) + (4/27 x 3a) + (8/81 x 4a) + (16/81 x 5a)

After 8a hr = (1/3 x a) + (2/9 x 2a) + (4/27 x 3a) + (8/81 x 4a) + (16/243 x 5a) + (32/729 x 6a) + (64/2187 x 7a) + (128/2187 x 8a) = 2.88a

The origin of the terms in the series should be readily apparent. If all fat molecules are "equivalent," two-thirds of those present at the start of an interval should be present at the end of the interval. With the exception that the final term is always twice the expected term, this is the expansion of $1/3(2/3)^{n-1}na$. This series approaches a limiting value, and steady-state or equilibrium conditions are considered to have been attained. According to this popular description of the frying operation, the fat is effectively heated for only a short period of time before being removed from the unit by the food product, and serious damage cannot occur.

Oil used in the discontinuous batch-type fryer must eventually be discarded because of either high viscosity or excessive foaming. Stable foams are produced by about 9% oxidized polymer [220]. Hydroxyl groups are more effective in contributing to foam production than are carbonyl groups [221,222]. Viscosity, in turn, is closely related to thermal polymer content [256]. Discarded oils are frequently found to contain approximately 25% polymer [234,257,258]. Since the addition of methyl silicone [259] to frying oils to prevent foaming became a standard practice, higher levels of polymer, approximately 35%, are also encountered.

Such badly damaged oils are not usually seen in a well-managed continuous frying operation. An increase in the viscosity of the oil of 1 cs at 100°C was reported [256] to increase the uptake of oil by fry cakes from 10 to 15%. An increase of 2.5 cs substantially reduced product quality. These increases in viscosity appeared to be related to polymer levels and nonurea adduct forming materials of 2-4% and 5-7%, respectively. An increase of 3.0 cs resulted in a 20% reduction in heat transfer by the oil. Consider that in the United States approximately 250,000 tons of oil are used in frying potato chips each year and another 100,000 tons are used in frying doughnuts [242]. Economic considerations would seem to dictate careful preservation of oil quality. It should be noted, however, that the increased oil uptake by the fry cakes with increased oil viscosity suggests that all fat molecules may not indeed be "equivalent" in the frying unit.

IV. PROCESSING OF EDIBLE FATS AND OILS

A. Extraction

The relation of oil or fat quality to the severity of processing techniques has long been recognized and is apparent in the descriptions historically applied to the various grades of lard, tallow, and olive oil. Most vegetable oils are currently obtained by solvent extraction with volatile petroleum hydrocarbons. It has been suggested that such solvents contain undesirable traces of carcinogenic multi-ring aromatic hydrocarbons which are selectively transferred to the oil. Such materials have not been detected [260], but if this contamination did occur, these materials would undoubtedly be removed during the deodorization process which will be discussed below. In studies of the effect of the various processing procedures on the polymer content of the oil, it was reported that 1-2% of polymeric material is present in the raw oil [261].

The fire and explosion hazard associated with the solvent extraction process cannot be avoided by the use of halogenated hydrocarbons. When this was tried commercially, it was found that the solvent reacts with the sulfhydryl groups of the seed protein and attaches to the sulfur atom. This modified protein presents a very serious nutritional and safety hazard [262].

B. Refining

Oils are treated with water and then base, usually in a centrifugal apparatus, to remove plant gums, free fatty acids, and the bulk of the phospholipids. Commercial soybean lecithin is a by-product of this type of operation and the free fatty acids are a source of soap stock. Removal of the various associated materials markedly reduces the stability of the oil.

Fatty acids are also excellent starting materials for the preparation of various industrial chemicals. Frequently this includes a fractional distillation of the fatty acids in the starting material. The still residues or "foots" may be used in animal feeds. It has been found that such foots may occasionally contain levels of polychlorinated hydrocarbons which produce pericardial edema in chicks [263,264]. Only when pesticide residues are concentrated by a process of this type does the question of the safety of the material arise. Actually the chick is rather unusual in responding to these levels which do not bother numerous other species. Such foots are not edible by human standards, but they are fed to animals consumed by man.

C. Bleaching and Deodorization

Carotenes, chlorophyll, and other plant pigments are removed or reduced in concentration by treatment of the oil with a bleaching clay. This is not a simple adsorptive process and involves oxidation and free-radical production [265,266]. Small quantities of positional and geometric isomers of the fatty acids are formed as are various hydroxy, keto, and epoxy derivatives. This process improves the quality of the oil at the expense of a minor contribution to future instability.

Deodorization by vacuum steam distillation of volatile odoriferous components from the oil [267-269] may be performed as a batch, semi-continuous, or continuous process. At temperatures of 190-230°C (375-450°F) approximately 2-5 hr is needed for efficient batch deodorization. There has been a trend toward use of shorter processing times by raising the temperature to 230-250°C (445-480°F) and/or increasing the vacuum from 6 to 1 mm. A by-product of this process, the deodorizer condensate, is a commercial source of plant sterols and for practical purposes is the only source of commercial quantities of the natural tocopherols [270]. Crude corn oil is frequently stated to contain 1-2 mg tocopherols per g oil, whereas the oil on the grocery store shelf may contain only 0.5-0.7 mg/g. A substantial reduction in the level of an essential nutrient and natural antioxidant occurs in this process.

When the oil is hydrogenated, it is essential to perform the deodorization after hydrogenation [271,272]. Under somewhat anaerobic conditions, the high temperature in the deodorizer tends to decompose those oxidized materials which would later tend to decompose under aerobic conditions to initiate the cyclic chain reaction of autoxidation. The formation of small quantities of thermal polymers at this stage is clearly preferable to the formation of potentially larger quantities of oxidized material in the product at a later stage. Removal of the volatiles produces a bland oil. The various processing operations tending to remove or produce polymers somewhat balance each other, and the finished oil has been reported to contain slightly less polymer than does the raw oil [261].

When a stock source of bland unsaturated oil is needed as a reference standard in organoleptic testing, it has been found that despite elaborate precautions changes occur with time [273]. Only the regeneration of fatty acid esters from the urea adduct provided a reproducible stable standard [274].

D. Hydrogenation

Addition of hydrogen to the double bonds of the fatty acids converts oxidatively unstable liquid oils to stable solid fats. As noted previously, however, the process must be halted at an intermediate stage short of complete saturation. Furthermore, a catalyst is frequently chosen to selectively produce trans isomers. Recently data presented for one such catalyst suggested an essentially quantitative conversion of the linoleic and linolenic acids in soybean oil to trans-monoenoic acids [275]. In many ways these trans-positional isomers of oleic acid, because of their low melting point [135], are the ideal approach to the production of a solid, digestible matrix [276] for the support of lower melting fats or liquid oils.

E. Rearrangement

A simple blending of fats and oils does not produce a homogeneous product, since the triglycerides of the oil retain their characteristic fatty acids as do the triglycerides of the fat. Introduction of a suitable catalyst, such as sodium methoxide, results in a rapid ester interchange and produces triglycerides with a random fatty acid composition [277]. This reaction may also be conducted in the directed manner [278]. If the reaction vessel is provided with a low-temperature region, high melting point triglycerides may be selectively crystallized out and withdrawn from the system as they are formed. The catalyst may be destroyed by addition of a stoichiometric amount of water and the residual methanol removed. This process may have a slight adverse effect on product stability [279].

A variation of this reaction is used in the laboratory to produce methyl esters from lipids by simply adding a catalyst and a large excess of methanol. Conversely, triglycerides may be synthesized in the laboratory by the ester interchange between glyceryl triacetin and fatty acid methyl esters. Removal of the volatile methyl acetate under vacuum drives the reaction to completion.

V. NUTRITIONAL AND SAFETY ASPECTS OF PROCESSING

The literature on oxidized and heated fat must be viewed in historical perspective. During and for a period after World War II, edible fats and oils were in short supply and soybeans and soybean oil were relatively novel and minor items in the occidental diet. Deliberate thermal polymerization was seriously considered as an alternative to hydrogenation in some countries [280,281]. The best-known studies in this area are those with linseed oil by Crampton and co-workers [282-288]. They heated this oil to 275°C (550°F) for 12 hr under an atmosphere of carbon dioxide. When fed at the 20% level in the diet to rats, the animals lost weight and mortality rate was high. Toxicity was attributed to products formed from linolenic acid. Subsequent studies indicated that the thermal dimers and higher polymers had low caloric value due to poor intestinal absorption and their nonabsorption resulted in diarrhea. Subsequent studies, involving fractionation of the heated oil, attributed the toxicity to the cyclic monomers. A concentrate of these cyclic monomers, prepared by distillation of the material not forming an adduct with urea, produced weight loss but not death when fed at the 2.5% level in the diet to rats [188]. However, 0.5 ml of a comparable concentrate administered by stomach tube for three days to 50-g weanling rats resulted in 40% mortality [289]. This might be equivalent to approximately 10% of the food an animal of this size would eat per day.

In the late 1930s and early 1940s, a series of reports by Roffo appeared in Argentinian journals not generally available in the United States ascribing carcinogenic activity to heated fats [290-296]. The starting points for much of the work on thermally oxidized frying fats were these reports of outright toxicity and carcinogenicity in related materials. These topics will be discussed separately below.

A. Nutritional Properties of Thermally Oxidized Fats

When considering some of the original experiments in this area, the difficulty of publishing negative reports should be borne in mind. In general, conditions were selected for the production of oxidized, heated, or thermally oxidized materials such that if toxic materials similar to those studied by Crampton and co-workers [282-288] could be produced, they would probably be produced by the conditions used. In the usual scientific experiments, once conditions are established which will produce toxic materials, it then becomes feasible to isolate and characterize these materials and to develop techniques for qualitative and quantitative analysis. At this point it is possible to determine if significant quantities of such materials are produced under "practical" conditions.

In principal, such fats should perhaps have been produced in somewhat the following manner: A relatively highly unsaturated frying fat is used for an excessively prolonged period of time for the actual frying of foods at the highest temperature likely to be encountered in gross misuse which does not ruin the product or involve an actual fire hazard. Heating should be intermittent rather than continuous. Makeup fat is added batchwise after significant removal of fat by the product to increase the ratio of exposed surface to total volume. The equipment should not be covered, and an efficient draft should be provided to minimize steam blanketing of the surface.

Preparation of experimental fats by such a procedure is not economically feasible in most cases. Instead, fats and oils were heated under vacuum or with air blown through the hot oil. Little or no actual product was fried in the oil and therefore the addition of fresh oil was minimal or nonexistent. Fats or oils prepared under such conditions did not reach steady-state conditions and "damage" was usually proportional to the treatment time.

Needless to say, reputable scientists in the research laboratories of the major commercial producers of frying fats and oils hastened to point out that data derived from the feeding of such oils were not particularly relevant to the ingestion of fats and oils used in the actual frying of foods [297-304], which were shown to be harmless in their experiments. The scientific literature is composed to a large extent of progress reports on various approaches to specific problems and only to a very limited extent of final solutions to major problems. The professional has the training to read and evaluate such reports in their proper perspective, whereas the interested public is usually not aware of the differences between approaches to a problem, the problem itself, and benefits derived from these negative reports.

Early in the feeding studies with thermally oxidized fats and oils, confirmation of a rather basic general principal was obtained. In general, the toxicity or carcinogenicity of a material is inversely related to the protein level of the diet furnished to the experimental animal [305,306]. With a specific heated corn oil an almost total lack of growth was seen in young rats fed 10% casein; severe growth depression was noted with 20% casein and a relatively mild effect when the diets contained 30% or more casein [307]. In adult man, the recommended dietary allowance for protein is 56 g or 224 kcal or 8.3% of calories in a 2700 kcal diet. In a rat diet containing 10% fat by weight and usually 4-5% minerals, 30% casein corresponds to 28% of calories. For man this would require 756 kcal protein or 189 g protein/day or 2 1/4 lb of roast beef.

Industrially oriented investigators are notably prone to refer to the use of "normal" diets comparable to those eaten by man rather than the customary semisynthetic animal diets, and feed to test animals a mixture containing 5% casein, 21% nonfat dry milk, 43% ground whole wheat, 3% dried egg white, 3% dried defatted liver, and 0.5% L-lysine [289]. It should be realized that although such an unnatural high protein diet cannot actually prevent the growth depressing or toxic effects of a severely treated oil, it may make it difficult or impossible to detect accurately any deleterious effect that might otherwise be produced by a less drastically treated oil. The rat differs from man in its protein requirement but maximum weight gain and efficiency of food conversion are achieved with 18-20% protein [308].

Although frying oils are described as thermally oxidized, there seems to be agreement regarding the absence of nutritionally significant quantities of hydroperoxide per se in these oils. Air blown through fresh corn oil at the rate of 150 ml/min per kg oil at 120, 160, and 200°C (248, 320, and 392°F, respectively) for 24 hr resulted in peroxide numbers of 81, 6, and 2, respectively [309]. Essentially all reported analytical data on heated oils are characterized by low peroxide values. More than 20 years ago, Dubouloz [310-315] described the destruction of lipid peroxides in the intestinal lumen by a metalo-protein. In other tissues, enzymes such as glutathione peroxidase [51, 52] are now known to destroy lipid peroxides efficiently. Lymph cannulations have shown that oxidized lipids are absorbed but the absorbed material does not contain the hydroperoxide group [316-320]. Hydroxylated fatty acids, such as those prepared from epoxidized soybean oil, are also well tolerated by the rat [321]. In view of the well-known intolerance of the human gut to a simple hydroxy fatty acid, the relevance of much of these data to the practical problem is questionable.

The aromatic cyclic monomer from fish oil depressed growth at the 0.5% level and was lethal at the 2.15% level [322]. The partially saturated cyclic monomer from linseed oil is not as toxic, since growth depression but not death resulted when fed at the 2.5% level [188]. Some fractions of this type, nonurea adduct forming, appear to be toxic, but others fed at the same level are not. One report [289] appears to state that the level of this fraction does not increase to any great extent in grossly mistreated oils (1.7-2.3%) as compared to fresh oils (1.5-2.0%), but nevertheless may result in 20-40% mortality when fed as the isolated material.

The cyclic dimers and higher polymers of linseed oil are nontoxic by virtue of poor absorption. The growth depressing effects of polymers from other heated oils are due in part to poor absorption with as little as one-third of the isolated polymers of some laboratory treated oils being absorbed [323-325]. However, the polymers, nondistillable, nonurea adduct forming material, from corn oil and cottonseed oil, and an isolated dimer from soybean oil appear to be somewhat toxic. This may be due in part to the formation of a definite amount of acyclic polymers [240-242] in oils containing linoleic acid rather than linolenic acid as the predominant polyunsaturated fatty acid. Such a dimer has recently been chromatographically isolated and partially characterized by Japanese investigators [241].

The laboratory thermal oxidation of oils commonly used for frying purposes definitely produces toxic compounds [326-340]. Even the most sincere advocates of the absolute safety of commercially used frying oils agree that, qualitatively, essentially the same compounds are produced under actual conditions of use. The critical and as yet unresolved problem is the question of the quantitative significance of such compounds in commercially used oils. In the dog, even the use of a 29.4% protein diet containing 5% casein, 21% nonfat dry milk, 17% toasted soybean meal, 3% blood meal, 30% toasted ground whole wheat, and 2% liver powder would not mask the growth depressing properties of 15% commercially used oil [304].

Oils thermally oxidized in the laboratory produce enlarged livers, kidneys and adrenals in the rat [341, 342]. Commercially used oils have been reported [300] to produce mild enlargement of these organs even when fed in a diet containing 15% lactalbumin, 20.5% ground whole wheat, 15% soybean oil meal, 10% nonfat dry milk solids, 10% meat and bone meal, and 2% alfalfa meal. Laboratory prepared oils produced statistically significant increases in liver weight, whereas in a short-term, seven-day study commerically used oils did not. Liver enlargement in these circumstances is viewed in part as increased synthesis of the hepatic microsomal mixed-function oxidase system for the metabolism of toxic compounds. It has recently been shown that rats fed heated fat synthesized less hemoglobin and retained more protein in the liver to cope with the metabolic effects of the damaged fat [343]. Modern frying fats contain a minimum level of polyunsaturated fats [344] and therefore are less susceptible to heat damage. However, their nutritional impact is probably more dependent on the protein level of the diet than on the extent of the heat damage.

B. Thermally Oxidized Fats and Cancer

All efforts to reproduce the production of gastric cancers reported by Roffo by feeding heated and/or oxidized fats have failed [345]. Noncancerous lesions of the stomach were produced after 18-24 months in Ivy's laboratory using rats of Roffo's strain [346]. One of the critical points in Roffo's studies [290-296] was the extreme period of time, 27-30 months, needed to produce the gastric lesions. Most investigators were unable to maintain their rats for more than two years. The possible exception to these negative results is the study by Sugai and co-workers [347].

It was considered possible that addition of a low level of a powerful chemical carcinogen such as acetylaminofluorene, AAF, to the diet might sensitize the tissues of the rat to the point where the cocarcinogenic activity of a weak carcinogen could be detected. It was indeed finally possible to carry out an experiment wherein all the rats fed 0.005% AAF and fresh fat survived for 30 months without detectable tumors, whereas none of the rats fed this level of AAF and 2.5% of the nonurea adduct forming material from thermally oxidized corn oil were free of tumors.

At the time these experiments were conducted (1954-1961), it was discovered that the N-hydroxylation of AAF enhanced the carcinogenicity of this compound [348-349]. The drug-metabolizing capacity of the liver is now known to increase in response to the ingestion of oxidized fats [350-352]. In retrospect, this experiment is now open to the criticism that the inclusion of the thermally oxidized fat in the diet had the side effect [353, 354] of increasing the animal's capacity to metabolize the carcinogen AAF to another compound, N-hydroxyl-2-acetylaminofluorene, which is an even more potent carcinogen than its precursor.

C. Lipids and Atherosclerosis

The role that dietary fats play in the nutritional perspective, that is, their actual role in atherosclerosis is still not clear. Recommendations are being made and conclusions drawn without knowledge of all the factors involved. For example, the final report of the National Diet Heart Study [355] stated:

> Extensive evidence implicates diet as the key factor in the etiology of atherosclerosis and suggests that the disease can be prevented by changes in diet, particularly by lowering serum cholesterol level.

In addition, a recent policy statement of the American Medical Association Council on Food and Nutrition [356] and the Food and Nutrition Board of the National Academy of Sciences indicated how changes in diet could be accomplished:

> Generally such lowering can be achieved most practically by partial replacement of the dietary sources of saturated fat with sources of unsaturated fat, especially those rich in polyunsaturated fatty acids.

This joint statement concluded:

> There is abundant evidence that the risk of developing CHD [cardiovascular heart disease] is positively correlated with the cholesterol in the plasma.

However, such a nutritional perspective has not fully considered that dietary sources of saturated fat and cholesterol, that is, meat, eggs, and dairy products, serve as the major source of protein, vitamins, and minerals in the American diet.

Serum cholesterol levels are somewhat responsive to dietary cholesterol level, the relative balance of polyunsaturated and saturated fatty acids in the dietary fat, and to certain drugs such as niacin and clofibrate. Either the dietary [111] or the drug approach [357] may lower serum cholesterol levels 10%. However, essentially negative results in two long-term national studies [111, 357] have been obtained with such an approach.

It has been shown by Van Deenen and co-workers [358, 359] that the replacement of an unsaturated by a saturated fatty acid influences the physical characteristics of the phospholipid into which it is incorporated and maintains the physical properties of the phospholipid molecule between certain limits. Data on liquid crystals and synthetic membranes support the hypothesis that the properties of membranes are dependent on the physical characteristics of the fatty acid composition of the phospholipids [360-362]. Van Deenen and co-workers [363] recently incorporated elaidic instead of oleic acid into the phospholipids of A. laidlawii. They noted a difference in the energy contents of the phase transitions of the isolated lipids, which they believed may have significance to the liquid crystalline state, as cholesterol was shown to preferentially interact with lipids which are in the liquid crystalline state. As phospholipids are important components in the cell membranes that make up the myocardium and the intima, the fatty acid composition of these phospholipids may be important to the rate at which lipid infiltration can occur through the cell membrane.

VI. SUMMARY

Food lipids may suffer loss of nutritional value under conditions which do not differ drastically from normal processing conditions. However, fats and oils rapidly become unpalatable, by occidental standards, after relatively mild adverse treatment. The role of dietary fat in atherosclerosis is still under study.

REFERENCES

1. A. D. Holmes and H. J. Deuel, Jr., J. Biol. Chem. 41:227 (1920).
2. A. D. Holmes and H. J. Deuel, Jr., Am. J. Physiol. 54:479 (1921).
3. H. J. Deuel, Jr., R. M. Johnson, C. E. Calbert, J. Gardner, and B. Thomas, J. Nutrition, 38, 369 (1949).
4. H. J. Deuel, Jr., Progr. Chem. Fats Lipids 2, 99 (1954).
5. A. L. S. Cheng, M. G. Morehouse, and H. J. Deuel, Jr., J. Nutri. 37:237 (1949).
6. K. F. Mattil and J. W. Higgins, J. Nutri. 29:255 (1945).
7. R. W. Harkins, L. M. Hagerman, and H. P. Sarett, J. Nutri. 87:85 (1965).
8. S. A. Hashim, H. B. Roholt, and T. B. Van Italie, Clin. Res. 10:394 (1962).
9. R. B. Zurier, B. G. Campbell, S. A. Hashim, and T. B. Van Italie, N. Engl. J. Med. 274:490 (1966).
10. P. R. Holt, S. A. Hashim, and T. B. Van Italie, Am. J. Gastroenterol. 43:549 (1965).
11. B. Bloom, I. L. Chaikoff, and W. D. Reinhart, Am. J. Physiol. 166:451 (1951).
12. E. A. Porta, O. R. Koch, and W. S. Hartroft, Exp. Mol. Pathol. 12:104 (1970).
13. R. D. Hawkins and H. Kalant, Pharmacol. Rev. 24:67 (1972).

14. M. Kohout, B. Kohoutova, and M. Heimberg, Biochem. Biophys. Acta 210:177 (1970).
15. R. C. Theur, W. H. Martin, T. J. Friday, B. L. Zoumas, and H. B. Sarett, Am. J. Clin. Nutr. 25:175 (1972).
16. M. Yoshida, J. Agr. Chem. Soc. (Japan) 13:120 (1937).
17. J. Osaki, J. Agr. Chem. Soc. (Japan) 8:1286 (1932).
18. T. Kaneda and S. Tshii, Bull. Japan Soc. Sci. Fisheries 19:171 (1953).
19. J. Green, Ann. N.Y. Acad. Sci. 203:29 (1972).
20. B. Century and M. K. Horwitt, Arch. Biochem. Biophys. 104:416 (1964).
21. R. T. Holman, Progr. Chem. Fats, Lipids 9:1 (1966).
22. R. T. Holman, Progr. Chem. Fats, Lipids 9:611 (1970).
23. L. Soderhjelm, H. F. Wiese, and R. T. Holman, Progr. Chem. Fats, Lipids 9:557 (1970).
24. I. M. Morice, J. Sci. Food Agri. 18:343 (1967).
25. R. Kleiman, F. R. Earle, I. A. Wolff, and Q. Jones, J. Am. Oil Chemists' Soc. 41:459 (1964).
26. E. D. Korn, J. Lipid Res. 5:352 (1964).
27. R. O. Sinnhuber, J. D. Castell, and D. J. Lee, Fed. Proc. (FASEB) 31:1436 (1972).
28. L. A. Witting, Progr. Chem. Fats Lipids 9:517 (1970).
29. O. S. Privett and M. L. Blank, J. Am. Oil Chemists' Soc. 41:292 (1964).
30. H. W. Sprecher, H. J. Dutton, F. D. Gunstone, P. T. Sykes, and R. T. Holman, Lipids 2:122 (1967).
31. H. Schlenk, Progr. Chem. Fats Lipids 9:587 (1970).
32. M. L. Blank and O. S. Privett, J. Lipid Res. 4:470 (1963).
33. H. Schlenk, D. M. Sand, and N. Sen, Biochem. Biophys. Acta 70:361 (1964).
34. C. E. Polan, J. J. McNeil, and S. B. Tove, J. Bacteriol. 88:1056 (1964).
35. R. Viviani, Adv. Lipid Res. 8:267 (1970).
36. P. V. Johnston, O. C. Johnson, and F. A. Kummerow, Proc. Soc. Exp. Biol. Med. 96:760 (1957).
37. P. V. Johnston, C. H. Walton and F. A. Kummerow, Proc. Soc. Exp. Biol. Med. 99:735 (1958).
38. O. S. Privett, L. J. Nutter, and F. S. Lightly, J. Nutr. 89:257 (1966).
39. W. E. M. Lands, M. L. Blank, L. J. Nutter, and O. S. Privett, Lipids 1:224 (1966).
40. L. De Luca, E. P. Little, and G. Wolf, J. Biol. Chem. 244:701 (1969).
41. L. De Luca, N. Maestri, G. Rosso, and G. Wolf, J. Biol. Chem. 248:641 (1973).

42. P. Rodriquez, O. Bello, and K. Gaede, FEBS Lett. 28:133 (1973).
43. Food and Nutrition Board, National Research Council, Recommended Dietary Allowances, 8th Revised Edition, National Academy of Sciences, Pub. No. 2216, 1974.
44. J. A. Olson and M. R. Lakshmanan, in The Fat Soluble Vitamins (H. F. DeLuca and J. W. Suttie, eds.), University of Wisconsin Press, Madison, Wisconsin, 1969, pp. 213-226.
45. R. J. Williams, Vitamins Hormones 1:229 (1943).
46. A. W. Davies and T. Moore, Nature, 147:794 (1941).
47. C. Nieman and H. J. Klein Obbink, Vitamins Hormones 12:69 (1954).
48. H. F. De Luca, Am. J. Clin. Nutr. 28:339 (1975).
49. S. T. Anning, J. Dawson, D. E. Dolby, and J. T. Ingram, Q. J. Med. 17:203 (1948).
50. K. L. Fong, P. B. McCay, J. L. Poyer, B. B. Keele, and H. Misra, J. Biol. Chem. 248:7792 (1973).
51. J. T. Rotruck, A. L. Pope, H. E. Ganther, A. B. Swanson, D. G. Haferman, and W. G. Hoekstra, Science, 179:588 (1973).
52. B. O. Christopherson, Biochim. Biophys. Acta 164:35 (1968).
53. N. Uri, in Autoxidation and Antioxidants (W. O. Lundberg, ed.), Vol. 1, Interscience Publishers, New York, 1961, pp. 55-106.
54. K. S. Chio and A. L. Tappel, Biochemistry, 8:2821 (1969).
55. N. Uri, in Autoxidation and Antioxidants (W. O. Lundberg, ed.), Vol. 1, Interscience Publishers, New York, 1961, pp. 133-169.
56. L. A. Witting, in Modification and Lipid Metabolism (E. G. Perkins and L. A. Witting, eds.), Academic Press, New York, 1975, pp. 1-41.
57. K. Mason, J. Nutri. 23:71 (1942).
58. J. W. Vester and L. R. Williams, Clin. Res. 11:180 (1963).
59. J. G. Bieri and R. P. Evarts, J. Am. Dietetic Assoc. 62:147 (1972).
60. J. G. Bieri and R. P. Evarts, J. Nutr. 104:850 (1974).
61. L. A. Witting, J. Am. Oil Chemists' Soc. 52:64 (1975).
62. H. J. Kayden and L. Bjornson, Ann. N.Y. Acad. Sci. 203:127 (1972).
63. M. S. Losowsky, J. Kelleher, B. E. Walker, T. Davies, and C. L. Smith, Ann. N.Y. Acad. Sci. 203:212 (1972).
64. J. T. Matschiner and W. V. Taggart, Anal. Biochem. 18:88 (1967).
65. J. T. Matschiner and J. M. Amelotti, J. Lipid Res. 9:176 (1968).
66. J. W. Suttie, in The Fat Soluble Vitamins (H. F. DeLuca and J. W. Suttie, eds.), University of Wisconsin Press, Madison, Wisconsin, 1969, pp. 447-462.
67. T. Moore, in The Fat Soluble Vitamins (R. A. Morton, ed.), Pergamon Press, Oxford, 1970, p. 238.
68. K. Rodahl and T. Moore, Biochem. J. 37:166 (1943).
69. P. E. Kolattukudy, Lipids 5:259 (1969).
70. A. P. Tulloch, Lipids 5:247 (1969).
71. C. R. Smith, M. O. Bagby, R. L. Lohmar, C. A. Glass, and I. A. Wolff, J. Org. Chem. 25:218 (1960).

72. F. D. Gunstone, J. Chem. Soc. 1611 (1954).
73. R. C. Badami and F. D. Gunstone, J. Sci. Food Agr. 14:481 (1963).
74. F. R. Earle, A. S. Barclay, and I. A. Wolff, Lipids 1:325 (1966).
75. W. H. Tallent, D. G. Cope, J. W. Hagemann, F. R. Earle, and I. A. Wolff, Lipids 1:335 (1966).
76. C. R. Smith, T. L. Wilson, R. W. Bates, and C. R. Scholfield, J. Org. Chem. 27:3112 (1962).
77. H. Paul and C. M. McCay, Arch. Biochem. 1:247 (1942).
78. W. C. Stewart and R. G. Sinclair, Arch. Biochem. 8:7 (1945).
79. A. Dolev, W. K. Rohwedder, and H. J. Dutton, Lipids 2:28 (1967).
80. E. Vioque and R. T. Holman, Arch. Biochem. Biophys. 99:522 (1962).
81. D. C. Zimmerman and B. A. Vick, Plant. Physiol. 46:445 (1970).
82. T. Galliard, D. A. Wardale, and J. A. Matthew, Biochem. J. 138: 23 (1974).
83. W. H. Tallent, J. Harris, I. A. Wolff, and R. E. Lundin, Tetrahedron Lett. 4329 (1966).
84. L. A. Appelqvist, J. Am. Oil Chemists' Soc. 44:206 (1967).
85. J. A. Robertson, W. H. Morrison, III, and D. Burdick, J. Am. Oil Chemists' Soc. 50:443 (1973).
86. V. R. Bhalerao, M. G. Kokatnur, and F. A. Kummerow, J. Am. Oil Chemists' Soc. 39:28 (1962).
87. R. H. Anderson and T. E. Huntley, J. Am. Oil Chemists' Soc. 41: 686 (1964).
88. N. Matsuo, in Lipids and Their Oxidation (H. W. Schultz, E. A. Day, and R. O. Sinnhuber, eds.), Avi Publishing Co., Inc., Westport, Conn., 1962, pp. 321-359.
89. D. V. Whipple, Proc. Soc. Exp. Biol. Med. 30:319 (1932).
90. W. J. Lennarz, Adv. Lipid Res. 4:175 (1966).
91. R. Kleiman, F. R. Earle, and I. A. Wolff, Lipids 4:317 (1969).
92. P. K. Faure, Nature 178:372 (1956).
93. J. J. Macfarlane, F. S. Shenstone, and J. R. Vickery, Nature 179:830 (1957).
94. R. A. Phelps, F. S. Shenstone, A. R. Kemmerer, and R. J. Evans, Poultry Sci. 44:358 (1965).
95. F. S. Shenstone and J. R. Vickery, Poultry Sci. 38:1055 (1959).
96. J. E. Nixon, T. A. Eisele, J. H. Wales, and R. O. Sinnhuber, Lipids 9:314 (1974).
97. E. Allen, A. R. Johnson, A. C. Fogerty, J. A. Pearson, and F. S. Shenstone, Lipids 2:419 (1967).
98. P. K. Raju and R. Reiser, J. Biol. Chem. 242:379 (1967).
99. C. R. Smith, Jr., Progress Chem. Fats Lipids 11:139 (1970).
100. R. L. Shriner and R. Adams, J. Am. Chem. Soc. 47:2727 (1925).
101. A. M. M. Abdellatif and R. O. Vles, Nutr. Metab. 12:285 (1970).
102. A. M. Abdellatif, Nutr. Rev. 30:2 (1972).
103. A. M. M. Abdellatif and R. O. Vles, Nutr. Metab. 15:219 (1973).

104. G. Rocquelin, B. Martin, and R. Cluzan, Proceedings of International Conference on the Science, Technology and Marketing of Rapeseed and Rapeseed Products. Ste. Adele Quebec, 1970, p. 405.
105. J. K. G. Kramer, Lipids 8:641 (1973).
106. J. L. Beare-Rogers, E. A. Nera, and B. M. Craig, Lipids 7:46 (1972).
107. J. L. Beare-Rogers, E. A. Nera, and B. M. Craig, Lipids 7:548 (1972).
108. L. N. Katz, J. Stamler, and R. Pick, Fed. Proc. 15:885 (1956).
109. E. A. Ahrens, Jr., W. Insull, Jr., R. Blomstrand, J. Hirsch, T. T. Tsaltas, and M. L. Peterson, Lancet 2:943 (1957).
110. R. Reiser, Am. J. Clin. Nutr. 26:524 (1973).
111. Final Report, National Diet-Heart Study, Circulation 37:Suppl. 1 (1968).
112. R. I. Levy, Fed. Proc. 30:829 (1971).
113. A. Keys, J. T. Anderson, and F. Grande, Lancet 2:959 (1957).
114. L. W. Kinsell, Am. J. Clin. Nutr. 12:228 (1963).
115. L. A. Witting and L. Lee, Am. J. Clin. Nutr. 28: 571,577 (1975).
116. J. Bitman, L. P. Dryden, H. K. Goering, T. R. Wrenn, R. A. Yoncoskie, and L. F. Edmondson, J. Am. Oil Chemists' Soc. 50:93 (1973).
117. N. P. Wong, H. E. Walter, J. H. Vestal, D. E. Lacroix, and J. A. Alford, J. Dairy Sci. 56:1271 (1973).
118. R. Ellis, W. I. Kimoto, J. Bitman, and L. F. Edmondson, J. Am. Oil Chemists' Soc. 51:4 (1974).
119. L. F. Edmondson, R. A. Yoncoskie, N. H. Rainey, R. W. Douglas, Jr., and J. Bitman, J. Am. Oil Chemists' Soc. 51:72 (1974).
120. E. L. Hove and P. L. Harris, J. Am. Oil Chemists' Soc. 28:405 (1951).
121. K. J. Moulton, R. E. Beal and E. L. Griffin, J. Am. Oil Chemists' Soc. 50:450 (1973).
122. G. R. List, C. D. Evans, R. E. Beal, L. T. Black, K. J. Moulton, and J. C. Cowan, J. Am. Oil Chemists' Soc. 51:239 (1974).
123. L. H. Going, J. Am. Oil Chemists' Soc. 45:632 (1968).
124. A. Vioque, R. Gutierrez, M. A. Albi, and N. Nosti, J. Am. Oil Chemists' Soc. 42:344 (1965).
125. A. Crossley and A. Thomas, J. Am. Oil Chemists' Soc. 41:95 (1964).
126. C. D. Evans, E. N. Frankel, P. M. Cooney, and H. A. Moser, J. Am. Oil Chemists' Soc. 37:452 (1960).
127. L. A. Baumann, D. G. McConnell, H. A. Moser, and C. D. Evans, J. Am. Oil Chemists' Soc. 44:663 (1967).
128. R. G. Ackman, S. N. Hooper, and D. L. Hooper, J. Am. Oil Chemists' Soc. 51:42 (1974).
129. M. Morita and M. Fujimaki, Agr. Food Chem. 21:860 (1973).

130. M. D. Morris, I. L. Chaikoff, J. M. Felts, S. Abraham, and N. O. Fansah, J. Biol. Chem. 224:1039 (1957).
131. J. D. Wilson, J. Lipid Res. 5:409 (1964).
132. S. M. Grundy, E. H. Ahrens, Jr., and J. Davignon, J. Lipid Res. 10:304 (1969).
133. A. C. Ivy, T. Lin, and E. Karvnen, Am. J. Physiol. 183:79 (1953).
134. E. R. Cousins, J. Am. Oil Chemists' Soc. 40:206 (1963).
135. B. F. Daubert, J. Am. Chem. Soc. 66:290 (1944).
136. B. F. Daubert, C. J. Spiegel, and H. E. Longnecker, J. Am. Chem. Soc. 65:2144 (1943).
137. L. R. Mahoney, Angew. Chem. 81:555 (1969).
138. W. Boguth and H. Niemann, Biochim. Biophys. Acta 248:121 (1971).
139. A. S. Csallany, Int. J. Vitamin Nutr. Res. 41:376 (1971).
140. O. S. Privett and C. Nickell, J. Am. Oil Chemists' Soc. 33:156 (1956).
141. O. S. Privett, J. Am. Oil Chemists' Soc. 36:507 (1959).
142. A. E. Johnston, K. T. Zilch, E. Selke, and H. J. Dutton, J. Am. Oil Chemists' Soc. 38:367 (1961).
143. M. G. Kokatnur, J. G. Bergman, and H. H. Draper, Anal. Biochem. 12:325 (1965).
144. R. F. Pascke and D. H. Wheeler, J. Am. Oil Chemists' Soc. 25:278 (1949).
145. S. S. Chang and F. A. Kummerow, J. Am. Oil Chemists' Soc. 30:251 (1953).
146. S. S. Chang and F. A. Kummerow, J. Am. Oil Chemists' Soc. 30:403 (1953).
147. S. S. Chang and F. A. Kummerow, J. Am. Oil Chemists' Soc. 31:324 (1954).
148. L. A. Witting, S. S. Chang and F. A. Kummerow, J. Am. Oil Chemists' Soc. 34:470 (1957).
149. W. Y. Cobb and E. A. Day, J. Am. Oil Chemists' Soc. 42:420 (1965).
150. W. Y. Cobb and E. A. Day, J. Am. Oil Chemists' Soc. 42:1110 (1965).
151. S. S. Chang, K. M. Brobst, H. Tai, and C. E. Ireland, J. Am. Oil Chemists' Soc. 38:671 (1961).
152. S. S. Chang, R. G. Krishnamurthy, and B. R. Reddy, J. Am. Oil Chemists' Soc. 44:159 (1967).
153. A. M. Gaddis, R. Ellis, and G. T. Currie, J. Am. Oil Chemists' Soc. 38:371 (1961).
154. R. J. Horvat, W. H. McFadden, H. Ng, W. G. Lane, and A. D. Shepherd, J. Am. Oil Chemists' Soc. 43:350 (1966).
155. R. J. Horvat, W. H. McFadden, H. Ng, A. Lee, G. Fuller, and T. H. Applewhite, J. Am. Oil Chemists' Soc. 46:273 (1969).
156. R. J. Horvat, W. H. McFadden, H. Ng, D. R. Black, W. G. Lane, and R. M. Teeter, J. Am. Oil Chemists' Soc. 42:1112 (1965).

157. H. Ng, R. J. Horvat, A. Lee, W. H. McFadden, W. G. Lane, and A. D. Shepherd, J. Am. Oil Chemists' Soc. 45:708 (1968).
158. R. W. Keith and E. A. Day, J. Am. Oil Chemists' Soc. 40:121 (1963).
159. T. Kawada, R. G. Krishnamurthy, B. D. Mookherjee, and S. S. Chang, J. Am. Oil Chemists' Soc. 44:131 (1967).
160. R. G. Krishnamurthy and S. S. Chang, J. Am. Oil Chemists' Soc. 44:136 (1967).
161. P. W. Meijboom and J. B. A. Stroink, J. Am. Oil Chemists' Soc. 49:555 (1972).
162. G. Hoffmann and P. W. Meijboom, J. Am. Oil Chemists' Soc. 45:468 (1968).
163. R. G. Seals and E. G. Hammond, J. Am. Oil Chemists Soc. 47:278 (1970).
164. B. D. Mookherjee and S. S. Chang, J. Am. Oil Chemists' Soc. 40:232 (1963).
165. S. Patton, I. J. Barnes, and L. E. Evans, J. Am. Oil Chemists' Soc. 36:280 (1959).
166. R. G. Scholz and L. R. Ptak, J. Am. Oil Chemists' Soc. 43:596 (1966).
167. T. H. Smouse and S. S. Chang, J. Am. Oil Chemists' Soc. 44:509 (1967).
168. D. A. Lillard and E. A. Day, J. Am. Oil Chemists' Soc. 41:549 (1964).
169. R. F. Matthews, R. A. Scanlan and L. M. Libbey, J. Am. Oil Chemists' Soc. 48:745 (1971).
170. E. Schauenstein, J. Lipid Res. 8:417 (1967).
171. J. E. Kinsella, S. Patton, and P. S. Dimick, J. Am. Oil Chemists' Soc. 44:202 (1967).
172. J. E. Kinsella, S. Patton and P. S. Dimick, J. Am. Oil Chemists' Soc. 44:449 (1967).
173. L. Williamson, J. Appl. Chem. (London) 3:301 (1953).
174. E. N. Frankel, C. D. Evans, and J. C. Cowan, J. Am. Oil Chemists' Soc. 37:418 (1960).
175. J. C. Cowan, J. Am. Oil Chemists' Soc. 39:534 (1962).
176. E. Rossman, Fettchem. Umschau 40:117 (1933).
177. R. F. Paschke and D. H. Wheeler, J. Am. Oil Chemists' Soc. 32:473 (1955).
178. J. A. MacDonald, J. Am. Oil Chemists' Soc. 33:394 (1956).
179. R. A. Eisenhauer, R. E. Beal, and E. L. Griffin, J. Am. Oil Chemists' Soc. 40:129 (1963).
180. L. T. Black and R. A. Eisenhauer, J. Am. Oil Chemists' Soc. 40:272 (1963).
181. J. P. Friedrich, J. C. Palmer, E. W. Bell, and J. C. Cowan, J. Am. Oil Chemists' Soc. 40:584 (1963).

182. R. A. Eisenhauer, R. E. Beal, and E. L. Griffin, J. Am. Oil Chemists' Soc. 41:60 (1964).
183. R. E. Beal, R. A. Eisenhauer, and E. L. Griffin, Jr., J. Am. Oil Chemists' Soc. 41:683 (1964).
184. R. E. Beal, R. A. Eisenhauer, and V. E. Sohns, J. Am. Oil Chemists' Soc. 42:1115 (1965).
185. J. P. Friedrich, J. Am. Oil Chemists' Soc. 44:244 (1967).
186. H. Lange and J. D. von Mikusch, Fette, Seifen, Anstrichm. 69:752 (1967).
187. R. B. Hutchinson and J. C. Alexander, J. Org. Chem. 28:2522 (1963).
188. A. G. McInnes, F. P. Cooper, and J. A. MacDonald, Can. J. Chem. 39:1906 (1961).
189. W. R. Michael, Lipids 1:359 (1966).
190. W. R. Michael, Lipids 1:365 (1966).
191. W. R. Eckert, Fette, Seifen, Anstrichm. 70:329 (1968).
192. A. N. Sagredos, Fette, Seifen, Anstrichm. 69:707 (1967).
193. A. N. Sagredos, W. R. Eckert, W. D. Heinrich, and J. D. von Mikusch, Fette, Seifen, Anstrichm. 71:877 (1969).
194. H. Scharmann, W. R. Eckert, and A. Zeman, Fette, Seifen, Anstrichm. 71:118 (1969).
195. T. Wieske and H. Rinke, Fette, Seifen, Anstrichm. 69:503 (1969).
196. A. Zeman, H. Scharmann, and W. R. Eckert, Fette, Seifen, Anstrichm., 71:283 (1969).
197. A. Zeman and H. Scharmann, Fette, Seifen, Anstrichm. 71:957 (1969).
198. G. Billek and O. Heinz, Fette, Seifen, Anstrichm. 71:189 (1969).
199. T. F. Bradley and W. B. Johnston, Chem. Ind. 32:802 (1940).
200. A. L. Clingman, D. E. A. Rivett, and D. A. Sutton, J. Chem. Soc. 1088 (1954).
201. R. F. Paschke, L. E. Peterson, and D. H. Wheeler, J. Am. Oil Chemists' Soc. 41:723 (1964).
202. D. H. Wheeler and J. White, J. Am. Oil Chemists' Soc. 44:298 (1967).
203. A. K. Sen Gupta, Fette, Seifen, Anstrichm. 70:153 (1968).
204. A. L. Clingman, D. E. A. Rivett, and D. A. Sutton, Chem. Ind. 798 (1953).
205. K. B. Norton, D. E. A. Rivett, and D. A. Sutton, Chem. Ind. 1452 (1961).
206. A. K. Sen Gupta and H. Scharmann, Fette, Seifen, Anstrichm. 70:265 (1968).
207. C. Boelhouwer, L. T. Tien, and H. T. Waterman, Research 6:55S (1953).
208. A. L. Clingman and D. A. Sutton, J. Am. Oil Chemists Soc. 30:53 (1953).

209. R. F. Paschke, L. E. Peterson, S. A. Harrison, and D. H. Wheeler, J. Am. Oil Chemists' Soc. **41**:56 (1964).
210. S. A. Harrison, L. E. Peterson, and D. H. Wheeler, J. Am. Oil Chemists' Soc. **42**:2 (1965).
211. K. Figge, Chem. Phys. Lipids **6**:159 (1971).
212. K. Figge, Chem. Phys. Lipids **6**:178 (1971).
213. J. Petit and G. Bosshard, Bull. Soc. Chim. France 293 (1952).
214. J. Petit and G. Bosshard, Bull. Soc. Chim. France 618 (1952).
215. L. W. Wantland and E. G. Perkins, Lipids **5**:187 (1970).
216. A. Crossley, T. D. Heyes, and B. J. F. Hudson, J. Am. Oil Chemists' Soc., **39**:9 (1962).
217. O. C. Johnson and F. A. Kummerow, J. Am. Oil Chemists' Soc. **34**:407 (1957).
218. D. Firestone, W. Horowitz, L. Friedman, and G. M. Shue, J. Am. Oil Chemists' Soc. **38**:253 (1961).
219. E. G. Perkins and L. A. Van Akkeren, J. Am. Oil Chemists' Soc. **42**:782 (1965).
220. T. Miyakawa, Fette, Seifen, Anstrichm. **66**:1048 (1964).
221. S. Ota, N. Iwata, and M. Morita, Yukagaku **13**:210 (1964).
222. S. Ota, A. Mukai, and I. Yamamoto, Yukagaku **13**:264 (1964).
223. J. G. Endres, V. R. Bhalerao and F. A. Kummerow, J. Am. Oil Chemists' Soc., **39**:159 (1962).
224. L. A. Wishner and M. Keeney, J. Am. Oil Chemists' Soc. **42**:776 (1965).
225. T. Kawada, R. G. Krishnamurthy, B. D. Mookherjee and S. S. Chang, J. Am. Oil Chemists' Soc. **44**:131 (1967).
226. R. G. Krishnamurthy and S. S. Chang, J. Am. Oil Chemists' Soc. **44**:136 (1967).
227. K. Yasuda, B. R. Reddy and S. S. Chang, J. Am. Oil Chemists' Soc. **45**:625 (1968).
228. B. R. Reddy, K. Yasuda, R. G. Krishnamurthy, and S. S. Chang, J. Am. Oil Chemists' Soc. **45**:629 (1968).
229. S. S. Chang and A. Hsieh, J. Am. Oil Chemists' Soc. **51**:526A (1974).
230. N. R. Artman and J. C. Alexander, J. Am. Oil Chemists' Soc. **45**:643 (1968).
231. L. W. Wantland and E. G. Perkins, Lipids **5**:191 (1970).
232. E. G. Perkins and J. R. Anfinsen, J. Am. Oil Chemists' Soc. **48**:556 (1971).
233. N. R. Artman and D. E. Smith, J. Am. Oil Chemists' Soc. **49**:318 (1972).
234. K. Aitzetmueller, Fette, Seifen, Anstrichm. **75**:14 (1973).
235. K. Aitzetmueller, Fette, Seifen, Anstrichm. **75**:256 (1973).
236. W. C. Harris, E. P. Crowell, and B. B. Burnett, J. Am. Oil Chemists' Soc. **50**:537 (1973).
237. H. Inoue, K. Kazuo, and N. Taniguchi, J. Chromatogr. **47**:348 (1970).

238. E. G. Perkins, R. Taubold and A. Hsieh, J. Am. Oil Chemists' Soc. 50:223 (1973).
239. D. Firestone, J. Am. Oil Chemists' Soc. 40:247 (1963).
240. E. G. Perkins and F. A. Kummerow, J. Am. Oil Chemists' Soc. 36:371 (1959).
241. T. Ohfuji and T. Kaneda, Lipids 8:353 (1973).
242. M. M. Paulose and S. S. Chang, J. Am. Oil Chemists' Soc. 50:147 (1973).
243. M. R. Sahasrabudhe and V. R. Bhalerao, J. Am. Oil Chemists' Soc. 40:711 (1963).
244. T. P. Labuza, H. Tsuyuki, and M. Karel, J. Am. Oil Chemists' Soc. 46:409 (1969).
245. K. H. Tjhio, T. P. Labuza, and M. Karel, J. Am. Oil Chemists' Soc. 46:597 (1969).
246. A. F. Mabrouk, J. Am. Oil Chemists' Soc. 41:331 (1964).
247. J. E. Coleman, J. W. Hampson, and D. H. Saunders, J. Am. Oil Chemists' Soc. 41:347 (1964).
248. L. R. Dugan, Jr., and P. W. Landis, J. Am. Oil Chemists' Soc. 33:152 (1956).
249. H. T. Slover and L. R. Dugan, Jr., J. Am. Oil Chemists' Soc. 34:333 (1957).
250. J. R. Chipault and G. R. Mizuno, J. Am. Oil Chemists' Soc. 41:468 (1964).
251. J. P. Kavalam and W. W. Nawar, J. Am. Oil Chemists' Soc. 46:387 (1969).
252. L. J. Janicki and H. Appledorf, J. Food Sci. 39:715 (1974).
253. S. P. Rock and H. Roth, J. Am. Oil Chemists' Soc. 41:228 (1964).
254. T. P. Dornseifer, S. C. Kim, E. S. Keith, and J. J. Powers, J. Am. Oil Chemists' Soc. 42:1073 (1965).
255. E. Yuki, Yukagaku 16:654 (1967).
256. S. P. Rock and H. Roth, J. Am. Oil Chemists' Soc. 43:116 (1966).
257. D. Firestone, S. Nesheim and W. Horowitz, J. Assoc. Offic. Agr. Chem., 44:465 (1961).
258. A. E. Waltking and H. Zmachinski, J. Am. Oil Chemists' Soc. 47:530 (1970).
259. I. P. Freeman, F. B. Padley and W. L. Shappard, J. Am. Oil Chemists' Soc. 50:101 (1973).
260. J. W. Ryder and G. P. Sullivan, J. Am. Oil Chemists' Soc. 39:263 (1962).
261. E. N. Frankel, C. D. Evans, H. A. Moser, D. G. McConnell, and J. C. Cowan, J. Am. Oil Chemists' Soc. 38:130 (1961).
262. L. L. McKinney, J. C. Picker, Jr., F. B. Weakley, A. C. Eldridge, R. E. Campbell, J. C. Cowan, and H. E. Biester, J. Am. Chem. Soc. 81:909 (1959).

263. D. Firestone, W. Horowitz, L. Friedman, and G. M. Shue, J. Am. Oil Chemists' Soc. 38:418 (1961).
264. G. R. Higginbotham, A. Huang, D. Firestone, J. Verrett, J. Ress, and A. D. Campbell, Nature 220:702 (1968).
265. A. D. Rich and A. Greentree, J. Am. Oil Chemists' Soc. 35:284 (1958).
266. G. Van Den Bosch, J. Am. Oil Chemists' Soc. 50:487 (1973).
267. F. B. White, J. Am. Oil Chemists' Soc. 33:495 (1956).
268. S. S. Rini, J. Am. Oil Chemists' Soc. 37:512 (1960).
269. C. T. Zehnder and C. E. McMichael, J. Am. Oil Chemists' Soc. 44:478A (1966).
270. R. J. Fiala, J. Am. Oil Chemists' Soc. 36:375 (1959).
271. S. S. Chang, Y. Masuda, B. D. Mookherjee, and A. Silveira, Jr., J. Am. Oil Chemists' Soc. 40:721 (1963).
272. D. R. Merker and L. C. Brown, J. Am. Oil Chemists' Soc. 33:141 (1956).
273. C. D. Evans, H. A. Moser, G. R. List, H. J. Dutton, and J. C. Cowan, J. Am. Oil Chemists' Soc. 48:711 (1971).
274. G. R. List, R. L. Hoffmann, H. A. Moser, and C. D. Evans, J. Am. Oil Chemists' Soc. 44:485 (1967).
275. T. J. Sullivan, J. Am. Oil Chemists' Soc. 52:3A (1975).
276. R. B. Alfin-Slater, A. F. Wells, L. Aftergood, and H. J. Deuel, Jr., J. Nutr. 63:241 (1957).
277. W. Q. Braun, J. Am. Oil Chemists' Soc. 37:598 (1960).
278. E. W. Eckey, Ind. Eng. Chem. 40:1138 (1948).
279. S. Zalewski and A. M. Gaddis, J. Am. Oil Chemists' Soc. 44:576 (1967).
280. F. Jacobson, R. Nergaard, and E. Mathiesen, E. Tids. Hermetkind 27:225 (1941).
281. A. S. Privett, R. B. Pringle, and W. D. Farlane, Oil Soap 22:287 (1945).
282. E. W. Crampton, F. A. Farmer, and F. M. Berryhill, J. Nutr. 43:431 (1951).
283. E. W. Crampton, R. H. Common, F. A. Farmer, F. M. Berryhill, and L. Wiseblatt, J. Nutr. 43:533 (1951).
284. E. W. Crampton, R. H. Common, F. A. Farmer, F. M. Berryhill, and L. Wiseblatt, J. Nutr. 44:177 (1951).
285. E. W. Crampton, R. H. Common, F. A. Farmer, A. F. Wells, and D. Crawford, J. Nutr. 49:333 (1953).
286. E. W. Crampton, R. H. Common, E. T. Pritchard, and F. A. Farmer, J. Nutr. 60:13 (1956).
287. R. H. Common, E. W. Crampton, F. Farmer, and A. S. W. DeFreitas, J. Nutr. 62:347 (1957).
288. A. F. Wells and R. H. Common, J. Sci. Food Agr. 4:233 (1953).

289. G. A. Nolen, J. C. Alexander, and N. R. Artman, J. Nutr. 93:337 (1967).
290. A. H. Roffo, Bol. Inst. med. exp. estud. cancer 14:589 (1938).
291. A. H. Roffo, Bol. Inst. med. exp. estud. cancer 15:407 (1939).
292. A. H. Roffo, Prensa Med. argent. 26:619 (1939).
293. A. H. Roffo, Bol. Inst. med. exp. estud. cancer 19:503 (1942).
294. A. H. Roffo, Bol. Inst. de med. exp. estud. cancer 20:471 (1943).
295. A. H. Roffo, Bol. Inst. de med. exp. estud. cancer 21:1 (1944).
296. A. H. Roffo, Am. J. Digest. Dis. 13:33 (1946).
297. D. Melnick, J. Am. Oil Chemists' Soc. 34:578 (1957).
298. D. Melnick, J. Am. Oil Chemists' Soc. 34:351 (1957).
299. D. Melnick, F. H. Luckman, and C. M. Gooding, J. Am. Oil Chemists' Soc. 35:271 (1958).
300. C. E. Poling, W. D. Warner, P. E. Mone, and E. E. Rice, J. Nutr. 72:109 (1960).
301. E. E. Rice, C. E. Poling, P. E. Mone, and W. D. Warner, J. Am. Oil Chemists' Soc. 37:607 (1960).
302. C. E. Poling, W. D. Warner, P. E. Mone, and E. E. Rice, J. Am. Oil Chemists' Soc. 39:315 (1962).
303. C. E. Poling, E. Eagle, E. E. Rice, A. M. A. Durand, and M. Fisher, Lipids 5:128 (1970).
304. G. A. Nolen, J. Nutr. 103:1248 (1973).
305. R. H. Wilson, F. DeEds, and A. J. Cox, J. Cancer Res. 1:595 (1941).
306. R. W. Engel and D. H. Copeland, Cancer Res. 12:905 (1952).
307. L. A. Witting, T. Nishida, O. C. Johnson, and F. A. Kummerow, J. Am. Oil Chemists' Soc. 34:421 (1957).
308. C. E. Bunce and K. W. King, J. Nutr. 98:168 (1969).
309. F. A. Kummerow, in Lipids and Their Oxidation (H. W. Schultz, E. A. Day, and R. O. Sinnhuber, eds.), Avi Publishing Company, Inc., Westport, Conn., 1962, pp. 294-320.
310. P. Dubouloz, J. Fondari, and C. Lagarde, Biochim. Biophys. Acta 3:371 (1949).
311. P. Dubouloz and J. Laurent, Compt. rend. soc. biol. 144:1183 (1950).
312. P. Dubouloz, J. Dumas, and J. Laurent, Compt. rend. soc. biol. 145:905 (1951).
313. P. Dubouloz, J. Laurent, and J. Dumas, Bull. soc. chim. biol. 33:1740 (1951).
314. P. Dubouloz and J. Fondari, Bull. soc. chim. biol. 35:819 (1953).
315. P. Dubouloz and J. Laurent, Bull. soc. chim. biol. 35:781 (1953).
316. J. S. Andrews, W. H. Griffith, J. F. Mead, and R. A. Stein, J. Nutr. 70:199 (1960).
317. J. Glavind and N. Tryding, Acta Physiol. Scand. 49:97 (1960).

318. J. Glavind, E. Sondergaard, and H. Dam, Acta Pharmacol. Toxicol., 18:267 (1961).
319. J. G. Bergan and H. H. Draper, Lipids 5:976 (1970).
320. V. R. Bhalerao, M. Inoue, and F. A. Kummerow, J. Dairy Sci. 46:176 (1963).
321. H. Kaunitz and R. E. Johnson, J. Am. Oil Chemists' Soc. 41:50 (1964).
322. J. J. Gottenbos and H. J. Thomasson, Biblio. Nutr. Dieta 1:110 (1965).
323. S. Lassen, E. K. Bacon and H. J. Dunn, Arch. Biochem. Biophys. 23:1 (1949).
324. O. C. Johnson, E. Perkins, M. Sugai, and F. A. Kummerow, J. Am. Oil Chemists' Soc. 34:594 (1957).
325. G. Kajimoto and K. Mukai, Yukagaku 19:66 (1970).
326. O. C. Johnson, T. Sakuragi, and F. A. Kummerow, J. Am. Oil Chemists' Soc. 33:433 (1956).
327. E. G. Perkins and F. A. Kummerow, J. Nutr. 68:101 (1959).
328. L. Friedman, W. Horowitz, G. M. Shue, and D. Firestone, J. Nutr. 73:85 (1961).
329. N. R. Bottino, J. Am. Oil Chemists' Soc. 39:25 (1962).
330. G. Czok, W. Griem, W. Kieckebusch, K. H. Baessler, and K. Lang, Z. Ernahrunswiss. 5:80 (1964).
331. N. B. Raju, M. Narayan Rao, and R. Rajagopolan, J. Am. Oil Chemists' Soc. 42:774 (1965).
332. N. R. Bottino, R. E. Anderson, and R. Reiser, J. Am. Oil Chemists' Soc. 42:1124 (1965).
333. G. M. Shue, C. D. Douglass, D. Firestone, L. Friedman, and J. S. Sage, J. Nutr. 94:171 (1968).
334. E. G. Perkins, S. M. Vaccha, and F. A. Kummerow, J. Nutr. 100:725 (1970).
335. H. Van Tilborg, J. Debruijn, J. Gottenbos, and G. K. Koch, J. Am. Oil Chemists' Soc. 47:430 (1970).
336. T. Ohfuji, S. Iwamoto, and T. Kaneda, Yukagaku 19:887 (1970).
337. T. Ohfuji and T. Kaneda, Yukagaku 19:486 (1970).
338. N. Matsuo, Eiyo To Shokuryo 25:579 (1972).
339. M. K. Govind Rao, C. Hemans, and E. G. Perkins, Lipids 8:342 (1973).
340. C. Hemans, F. A. Kummerow and E. G. Perkins, J. Nutr. 103:1665 (1973).
341. H. Kaunitz, C. A. Slanetz, R. E. Johnson, H. B. Knight, D. H. Saunders, and D. Swern, J. Am. Oil Chemists' Soc. 33:630 (1956).
342. H. Kaunitz, C. A. Slanetz, R. E. Johnson, H. B. Knight, R. E. Koos, and D. Swern, J. Am. Oil Chemists' Soc. 36:611 (1959).
343. J. Miller and D. R. Landes, J. Food Sci. 40:545 (1975).
344. F. A. Kummerow, J. Food Sci. 40:12 (1975).

345. E. Arffmann, J. Natl. Cancer Inst. 25:893 (1960).
346. A. Lane, D. Blickenstaff, and A. C. Ivy, Cancer 3:1044 (1950).
347. M. Sugai, L. A. Witting, H. Tsuchiyama, and F. A. Kummerow, Cancer Res. 22:510 (1962).
348. J. A. Miller, J. W. Cramer, and E. C. Miller, Cancer Res. 20:950 (1960).
349. E. C. Miller, J. A. Miller, and H. Hartmann, Cancer Res. 21:815 (1961).
350. A. E. M. McLean and W. J. Marshall, Biochem. J. 123:28P (1971).
351. W. J. Marshall and A. E. M. McLean, Proc. Nutr. Soc. 30:66A (1971).
352. W. J. Marshall and A. E. M. McLean, Biochem. J. 122:569 (1971).
353. E. Arrhenius, Xenobiotica 1:487 (1971).
354. J. R. Mitchell, D. J. Jollow, J. R. Gillete, and B. B. Brodie, Drug Metabol. Dispos. 1:418 (1973).
355. National Diet Heart Study, Circulation 37:Supplement 1 (1968).
356. American Medical Association, J. Am. Med. Assoc. 222:1647 (1972).
357. J. L. Marx, Science 187:526 (1975).
358. H. Van Den Bosch, A. J. Slotboom, and L. L. M. Van Deenen, Biochim. Biophys. Acta 176:632 (1969).
359. H. Van Den Bosch, L. M. C. Van Golde, A. J. Slotboom, and L. L. M. Van Deenen, Biochim. Biophys. Acta 152:694 (1968).
360. P. D. Jones, P. W. Holloway, R. O. Peluffo, and S. J. Wakil, J. Biol. Chem. 244:744 (1969).
361. S. V. Pande and J. F. Mead, J. Biol. Chem. 243:352 (1968).
362. R. A. Demel, K. R. Bruckdorfer, and L. L. M. Van Deenen, Biochim. Biophys. Acta 255:311 (1972).
363. B. De Kruyff, R. A. Demel, A. J. Slotboom, L. L. M. Van Deenen, and A. F. Rosenthal, Biochim. Biophys. Acta 307:1 (1973).

CHAPTER 5

MINERALS

Steven R. Tannenbaum
Vernon R. Young

Department of Nutrition and Food Science
Massachusetts Institute of Technology
Cambridge, Massachusetts

I. INTRODUCTION

The minerals in foods comprise a large and diverse group of elements and complex ions. Many of these are nutritionally required by man, and some, particularly the trace elements, may be both essential and toxic. The requirements are listed in Chap. 2 and discussion of the biological roles may be found in several recent surveys [1-4].

The mineral content of foods can vary greatly, depending upon such environmental factors as soil composition or nature of the animal feed. Losses occur not so much through destruction by chemical reaction as through physical removal or combination in forms which are not biologically available. A brief summary of some types of processing losses is available [5].

The primary mechanism of loss of mineral substances is through leaching of water-soluble materials and trimming of unwanted plant parts, as discussed in Chap. 3 (Vitamins). Major losses of all minerals occur in the milling of cereals. It is believed by many investigators that accelerated use of highly refined foods may lead to exacerbation of trace mineral deficiencies. Thus supplementation with trace minerals may become a necessity in the future, but as will be seen, it is complicated by the inherent toxicity of the nutrient.

Of equal importance is the interaction of mineral substance with other constituents of food. Polyvalent anions, such as oxalate and phytate, can form salts with divalent metal cations, which are extremely insoluble and pass through the digestive tract unabsorbed. Thus, the measurement of mineral availability is important and will be discussed in a subsequent section of this chapter.

II. CHEMISTRY

A comprehensive discussion of the chemistry of minerals is beyond the scope of this chapter, and the reader is referred to a basic text on inorganic chemistry. Any mineral contains an anionic and a cationic component, but only fluoride, iodide, and phosphate are anions of particular significance from a nutritional point of view. Fluoride is more often a constituent of water than of foods, and intake is highly dependent upon geographic location. Iodine may be present as iodide (I^-) or as iodate (IO_4^-). Phosphate may exist in a variety of forms, including phosphate (PO_4^{3-}), hydrogen phosphate (HPO_4^{2-}), dihydrogen phosphate ($H_2PO_4^-$), or phosphoric acid (H_3PO_4). The respective ionization constants are:

$$K_1 = 7.5 \times 10^{-3}$$
$$K_2 = 6.2 \times 10^{-8}$$
$$K_3 = 1.0 \times 10^{-12}$$

Iodide and iodate are relatively strong oxidizing agents compared to the other important inorganic anions in foods (e.g., phosphate, sulfate, carbonate).

The cations present a far more diverse and complex set of substances, and their general chemistry can be considered through their general chemistry can be considered through their elemental subgroups. Some of the metal ions are important from a nutritonal viewpoint, others are significant as toxic contaminants.

Magnesium, calcium, and barium can be considered as a group. They exist only in the +2 oxidation state. Although the halides of this group are soluble, the other important salts, including hydroxides, carbonates, phosphates, sulfates, oxalates, and phytates, are very insoluble. In foods which have undergone some bacterial decomposition, magnesium can form the highly insoluble complex $NH_4MgPO_4 \cdot 6H_2O$, also known as struvite.

Copper exists in +1 or +2 oxidation states and forms complex ions. The halides and sulfate are soluble, but the carbonate and phosphate are relatively insoluble. Some other metals with multioxidation states include tin and lead (+2 and +4), mercury (+1 and +2), iron (+2 and +3), chromium (+3 and +6), and manganese (+2, +3, +4, +6, +7). Many of these metals form amphoteric ions, and can act as oxidizing or reducing substances. Copper and iron are of particular significance in their ability to catalyze oxidation of ascorbic acid and unsaturated lipids (see Chap. 3 and 4 on vitamins and lipids).

Many of the metal ions also exist as ligands of organic molecules. Examples include iron bound to heme, copper in cytochromes, magnesium in chlorophyll, and cobalt in Vitamin B_{12}. The biologically active form of chromium is called the glucose tolerance factor (GTF), which is a complex organic form of trivalent chromium and is about 50 times more effective than inorganic chromium^{3+} in the glucose tolerance bioassay. It contains about 65% chromium in addition to nicotinic acid, cysteine, glycine, and glutamic acid. The exact structure is unknown [6]. Chromium^{6+} does not have biological activity.

III. OCCURRENCE

Although the mineral content of foods comes primarily from the soil and water used for production of food or animal feed, other factors are also of importance. Thus, with regard to copper, the following environmental factors have been suggested [7]: copper content of soil; geographic location; season; water source; use of fertilizers, insecticides, pesticides, and fungicides; and diet.

The extreme variability of some of these sources is illustrated for the case of drinking water in Table 1. Also of importance is the use of minerals as additives in foods or as processing aids. Again, this is a highly variable factor.

The federal regulations [9] governing the use of minerals are summarized below (21 Code of Federal Regulations, 1975).

1. § 121.101 (d) (5) GRAS for intended use (i.e., nutrients and/or dietary supplements without other limitations or restriction). The list includes only those compounds that could make a significant contribution to the dietary intake of the mineral.

   Ca : carbonate, citrate, glycerophosphate, oxide, phosphate (mono-, di-, tribasic), pyrophosphate, sulfate salts, calcium phytate.

   Cu : copper gluconate, cuprous iodide.

   P : calcium glycerophosphate, calcium phosphate (mono-, di-, tribasic), calcium pyrophosphate, ferric phosphate, ferric pyrophosphate, ferric sodium pyrophosphate, magnesium phosphate (di-, tribasic), manganese glycerophosphate, manganese hypophosphite, potassium glycerophosphate, sodium phosphate (mono-, di-, tribasic), riboflavin-5' phosphate, phosphoric acid.

TABLE 1 Occurrence of Some Trace Minerals in Drinking Water and Foods

Element	Concentration
Arsenic	0 - 1000 μg/liter, 137-330 μg per day in food
Barium	< 1 mg/liter
Beryllium	< 1 μg/liter
Cadmium	< 10 μg/liter
Chromium	< 100 μg/liter
Cobalt	up to 0.5 mg/kg in green leafy vegetables
Copper	present in plant and animal foods, 1-280 μg/liter
Lead	20-600 μg/liter; 100-300 μg per capita per day from food
Manganese	0.5-1.5 mg/liter
Magnesium	6-120 mg/liter
Mercury	< 1 μg/liter, 10 μg/day intake from food
Molybdenum	< 100 μg/liter, 100-1000 μg/kg of diet
Nickel	1-100 μg/liter, 300-600 μg per day in food
Selenium	< 10 μg/liter, 100-300 μg/kg in cereals, meats, seafoods
Silver	Traces
Tin	1-2 μg/liter, 1-30 mg per day from food
Vanadium	2-300 μg/liter
Zinc	3-2000 μg/liter

Source: National Academy of Sciences-National Research Council [8].

Fe : ferric phosphate, ferric pyrophosphate, ferric sodium pyrophosphate, ferrous gluconate, ferrous lactate, ferrous sulfate, iron (reduced).

I : cuprous iodide, potassium iodide.

Mg: magnesium oxide, magnesium phosphate (di-, tribasic), magnesium sulfate.

Zn : zinc sulfate, zinc gluconate, zinc chloride, zinc oxide, zinc stearate.

K : potassium chloride, potassium glycerophosphate.

Mn: manganese chloride, manganese citrate, manganese gluconate, manganese glycerophosphate, manganese hypophosphite, manganese sulfate, manganous oxide.

Si : aluminum calcium silicate, calcium silicate, sodium silicoaluminate, tricalcium silicate.

Sn : stannous chloride.

2. § 121.101(d) (5) GRAS for intended use (i.e., nutrients and/or dietary supplements with certain limitations).
   I : cuprous iodide (0.01% in table salt), potassium iodide (0.01% in table salt).
   Cu: cuprous iodide (0.01% in table salt), copper gluconate (0.005% in foods).
3. § 121.1073 Food additive regulation on potassium iodide: Source of iodine, provided maximum intake of food during a day will not result in intake of more than 225 μg (300 μg in the case of pregnant women, 45 μg in the case of infants <1 year, 105 μg in the case of children <4 years).
4. § 121.1149 Food additive regulation on kelp: Source of iodine, same as No. 3 above.
5. § 121.10 Fluorine-containing compounds (statement of policy): Preparations containing added fluorine compounds are limited to prescriptions (except dentifrices).
6. § 250.203 Status of fluoridated water and foods prepared with fluoridated water: Use of fluoridated water in food processing will not be regarded as actionable unless the process involves a significant concentration of fluorine from the water.
7. § 201.306 Potassium salt preparations intended for oral ingestion by man: Such preparations (capsules or coated tablets), which contain KCl or other K salts that would supply 100 mg or more K per unit, require a prescription. Label warning must include a statement that coated K tablets should be used only when adequate dietary supplementation is not practicable.
8. § 250.104 Status of salt substitutes: Not all are new drugs. FDA renders opinion on product-by-product basis. Typical products contain potassium, calcium, ammonium, magnesium, and choline salts.
9. § 121.1130 Food additive regulation on ferrous fumarate: Source of iron in foods for special dietary use.
10. § 121.1100 Food additive regulation on iron-choline citrate complex: Source of iron in foods for special dietary use.

Iodine can serve as an example of the type of variability discussed above. First, the iodine content of food is dependent upon geographic location. Thus individuals living near the sea are exposed to relatively high levels of dietary iodide through seafood, dairy products, and vegetable matter. In the latter case, the transfer of iodide to soil occurs largely via atmospheric phenomena, and the consumption of local forage high in iodine leads to high levels in milk.

Second, a major source of iodine in foods is iodized salt, usually in the form of potassium iodide or iodate for increased stability. Third, the loss of various forms of iodine from food has not been extensively investigated, but apparently leaching losses can be extensive. For example, up to 80% of the iodine in fish was lost in preparation by boiling, with lesser amounts lost by processes that do not require contact with excess water [10].

Compilations of analytical data on mineral content are not extensive. Agriculture Handbook No. 8 [11] provides information on calcium, phosphorus, sodium, potassium, and iron. An extensive treatment of the elements in a limited number of foods has been compiled in France [12]. Information on trace elements is scarce, but beginning to appear with increasing frequency: copper [7]; zinc [13-15]; calcium, iron, magnesium, phosphorus, potassium, sodium, and zinc in dairy foods [16]; chromium [17]; lead and cadmium [18]; and mercury [19].

IV. LOSSES AND GAINS IN PROCESSING

As indicated in the previous section, losses of minerals in processing can be severe. The nature of these losses varies as a function of the process and the chemistry and distribution of the minerals. The distribution of chromium in wheat products is indicated in Table 2, and there do not appear to be large differences between products. In contrast, there are major differences in the mineral content of egg white and yolk and for various milk products (see Table 3). Thus, there does not appear to be a general rule for predicting patterns of loss or distribution.

Losses by contact with water, particularly in cooking or blanching, can be considerable. The effect of blanching on the loss of major minerals in spinach is shown in Table 4. The wide range of difference is a function of solubility. In some cases minerals may even be added, as is the case for calcium in this example. The loss of nitrate may even be seen as beneficial from the viewpoint of can corrosion and health problems.

TABLE 2 Distribution of Chromium in Wheat and Wheat Products

Product	Relative Biological Value
Wheat grain	3
Wheat germ	4
Bread, whole wheat	3.6
Bread, white	3

Source: Toepfer et al. [17].

TABLE 3 Distribution of Minerals in Egg and in Dairy Products

Food	mg per 100 g								
	Ca	P	Mg	Na	K	Fe	Cu	Mo	Zn
Whole egg	26	103	5.3	79	54	1.1	29	<20	1
Egg white	1	3	3.1	56	43	0.03	1.6	<10	0.003
Egg yolk	12	43	0.7	3	15	0.36	1.7	<20	0.25
Whole milk	252	197	22	120	348	0.07	12	<10	1.0
Skim milk	259	197	22	134	408	0.07	12	<10	1.1
Cottage cheese	74	159	6	444	89	<0.1	<20	<40	0.4

Source: Gormian [20].

A similar difference with reference to copper is shown for cooking of potatoes in Table 5. The unequal distribution of copper in the peel and the inside are also indicated. A somewhat different pattern can be found for loss of minerals in cooked beans (Table 6). In contrast to spinach (Table 5), calcium is lost from beans to about the same extent as the other major minerals, and the trace elements follow a similar pattern.

TABLE 4 Effect of Blanching on Mineral Loss of Spinach

Mineral	g/100 g		Loss (%)
	Unblanched	Blanched	
Potassium	6.9	3.0	56
Sodium	0.5	0.3	43
Calcium	2.2	2.3	0
Magnesium	0.3	0.2	36
Phosphorus	0.6	0.4	36
Nitrate	2.5	0.8	70

Source: Bengtsson [21].

TABLE 5 Copper Content of Processed Potatoes

Type	mg per 100 g Fresh Weight
Raw	0.21 ± 0.10
Boiled	0.10
Baked	0.18
Chips	0.29
Mashed	0.10
French fried	0.27
Instant, uncooked	0.17
Potato peel	0.34

Source: Pennington and Calloway [7].

Trace elements and minerals can be gained in the course of processing through contact with processing water, equipment, and packaging materials. This is illustrated in Table 7 for the canning of vegetables with and without lacquered tinplate. The distribution of metal between solid and liquid portions of the canned contents is also shown.

TABLE 6 Mineral Content of Raw and Cooked Navy Beans

	mg/100 g		
Mineral	Raw	Cooked	Loss (%)
Calcium	135	69	49
Copper	0.80	0.33	59
Iron	5.3	2.6	51
Magnesium	163	57	65
Manganese	1.0	0.4	60
Phosphorus	453	156	65
Potassium	821	298	64
Zinc	2.2	1.1	50

Source: Meiners et al. [22].

TABLE 7 Distribution of Trace Metals in Canned Vegetables

Vegetable	Can[a]		g/kg	
		Lead	Tin	Iron
Green beans	La, L	0.10	5	2.8
	S	0.70	10	4.8
Haricot beans	La, L	0.07	5	9.8
	S	0.15	10	26
Petit pois	La, L	0.04	10	10
	S	0.55	20	12
Celery hearts	La, L	0.13	10	4.0
	S	1.50	20	3.4
Sweet corn	La, L	0.04	10	1.0
	S	0.30	20	6.4
Mushrooms	P, L	0.01	15	5.1
	S	0.04	55	16

Source: Crosby [23].
[a]La = Lacquered, P = Plain, L = Liquid, S = Solid.

V. AVAILABILITY OF MINERALS IN FOODS

Measuring the total amount of an element in a particular food source or diet provides only a limited indication of its nutritional value. Of greater practical significance is the amount of the element in a food that is "available" to the body. Various chemical, dietary, and host or physiological factors serve to determine the overall utilization of an ingested element. Thus, Bing [24] has suggested that the availability of iron or iron salt in a food is not an inherent characteristic of the substance being assayed, but an experimental value indicating absorption or utilization of the iron source under a particular set of test conditions.

Methods employed to determine availability of minerals include chemical balance studies [25], biological assays in experimental animals [26,27], in vitro tests [28,29] and use of radioactive tracers. Procedures involving the use of radioisotopes have been widely used for determining the true digestibilities of dietary minerals for farm livestock, and this topic has been reviewed [25]. In regard to human nutrition, considerable investigation has been devoted to the problem of the availability of iron in food [30, 31], and more recently, of zinc [32,33].

Based on radioisotopic methods, two approaches have been taken in studies with human subjects. Initially, biosynthetically labeled foods were prepared by growing plants in media containing radio-active iron, or by injecting animals with the radio-active tracers (^{55}Fe and ^{59}Fe) prior to slaughter and preparation of the foods. Test meals containing the labeled food item are then ingested, and the absorption of tracer is determined. This is the so-called "intrinsic label" method. More recently food iron (and zinc) absorption has been studied by use of the extrinsic label method, where the radioactive element is added to the food at the time of ingestion. Studies with the extrinsic tag have been validated for the determination of both iron [31] and zinc [32] bioavailability. This latter method has expanded considerably opportunities to measure the quantitative significance of factors influencing biological utilization in foods and diets consumed by human subjects.

In the case of iron, chemical form is important, with simple ferrous salts being more readily available than ferric salts [34]. Furthermore, the size of the elemental iron particle influences bioavailability [35, 36]. The type of food source affects the availability of iron, which is highest in animal foods and lowest in cereals (see Fig. 1) [37, 38]. Other dietary factors are also important. For example, vitamin C enhances iron absorption; phosphate, and to a lesser extent calcium, reduces food iron assimilation; and bran reduces iron absorption, presumably because of its phytate content. Other dietary constituents, including protein, amino acids, and

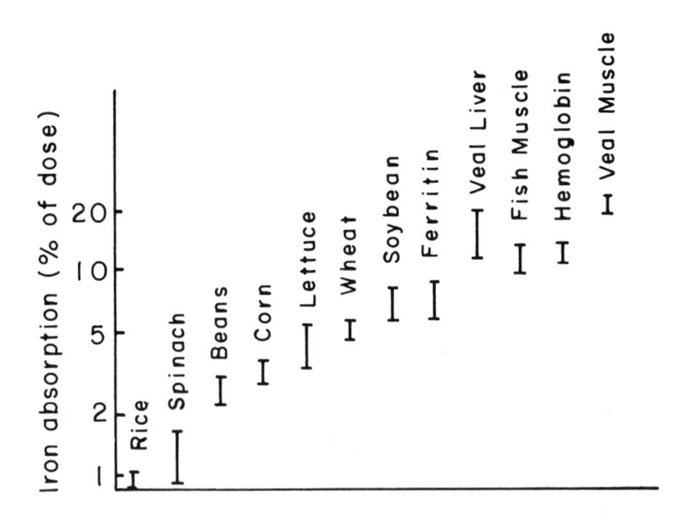

Fig. 1. Iron absorption by adults from a range of foods. Results are expressed as the mean ± the standard error. The dose of iron is 2-4 mg, except for lettuce, 1-17 mg (adapted from Ref. 21).

carbohydrates, have been suggested as having an effect on iron availability [31,39], although the quantitative significance of these factors in human nutrition remains uncertain.

Host or physiological factors that serve to modify the absorption of dietary iron include the iron nutritional status of the subject, with increased iron absorption in iron-depleted subjects or in patients with iron-deficiency anemia. There is some evidence indicating that iron absorption may be greater in women than in men [39] and that iron absorption decreases in children with increasing age [40]. The availability of zinc is similarly affected by various dietary and host factors which have been discussed in a number of reviews [41,42].

The problem of availability of calcium, zinc, and iron has been linked to the presence of phytate (inositol hexaphosphate) in certain foods [43]. Balance studies on zinc in animals have conclusively demonstrated that phytate can impair the absorption of dietary zinc and the reabsorption of endogenously secreted zinc [44]. Reduced retention of iron, copper, and manganese were also found in this study. On the other hand, another study has shown that monoferric phytate has the same availability as ferrous ammonium sulfate [45]. The situation may be further complicated by the role played by phytases in degradation of phytate complexes in the intestine.

The design of rational and effective food fortification programs requires sound information on the availability of minerals in food sources and diets. Furthermore, this information is important for the evaluation of the nutritional properties of food replacers and food analogues. Additional research is required to determine the bioavailability of the various trace elements that are essential in human nutrition and the existence of the various factors affecting mineral availability in the modern diet.

VI. SAFETY OF THE MINERALS

Some minerals have no significance from a nutritional perspective but are of importance because of their toxicity. Mercury and cadmium are discussed in this context in Chap. 9. All minerals can be toxic at some dose, with the range of safety vs. toxicity being highly variable. The relationship between safe and toxic doses of some minerals is shown in Table 8. An explanation of the extreme ranges for a given effect given by the Food and Drug Administration:

> The most important aspect of these tables is the overlap of excess intake ranges that occurs between different degrees of response. This is true for every mineral for which sufficient data exist to make the comparison. The span of values for each response level is frequently very large. For many elements there are examples of small intakes that were very toxic and large intakes that produced no detectable toxic

TABLE 8 Relationships Between Excess Oral Intakes[a] of Some Minerals in Man and Severity of Toxic Responses

Mineral	Response[b]	Toxic effect		
		None	Mild	Severe
Cu	Acute	-	2-16	125-50,000
	Longer	-	-	0.5-4
Co	Longer	-	150	35-600
F	Acute			80-3,000
	Longer	-	3-19	2-160
Fe	Acute	-	-	12-1,500
	Longer	7	-	6-15
I	Longer	8-35,000	1-15,000	1-180,000
Sn	Acute	-	130	23-700
Zn	Acute	13-23	-	8-530
	Longer	2-10	10	-

Source: Food and Drug Administration [9].
[a]Excess oral intake is defined as the actual intake required to produce an effect divided by the required intake or an equivalent reference point of average intake.
[b]Acute: within 24 hr; Longer: from one day through several generations.

effect. Some of the explanations for these dose-reponse relationships are identifiable and are discussed under each individual mineral. In other cases the reason is not apparent and reflects the state of uncertainty about the safety of consuming minerals in excess of requirements.

REFERENCES

1. The Nutrition Foundation, Present Knowledge in Nutrition, Nutrition Foundation, New York, 1976.
2. J. G. Reinhold, Clin. Chem. 21:476 (1975).
3. E. J. Underwood, Trace Elements in the Environment, Academic Press, New York, 1971.
4. E. J. Underwood, in Toxicants Naturally Occurring in Foods, NAS/NRC, Washington, D.C., 1973, pp. 43-87.
5. A. Schroeder, Am. J. Clin. Nutri. 24:562 (1971).

6. W. Mertz, Nutr. Rev. 33:129 (1975).
7. J. T. Pennington and D. H. Calloway, J. Am. Diet. Assoc. 63:143 (1973).
8. Safe Drinking Water Committee, Drinking Water and Health, National Academy of Sciences-National Research Council, Washington, D.C., 1977.
9. Food and Drug Administration, DHEW, Toxicity of the Essential Minerals, Washington, D.C., 1975.
10. M. T. Harrison, S. McFarland, R. McG. Harden, and E. Wayne, Amer. J. Clin. Nutr. 17:73 (1965).
11. B. K. Watt and A. L. Merrill, in Composition of Foods, Agriculture Handbook No. 8, USDA, Washington, D.C., 1963.
12. P. Jaulmes and G. Hamelle, Ann. Nutr. Alim. 25:B133 (1971).
13. J. H. Freeland and R. J. Cousins, J. Am. Dietet. Assoc. 68:526 (1976).
14. K. A. Haeflein and A. I. Rasmussen, J. Am. Dietet. Assoc. 70:610 (1977).
15. E. W. Murphy, B. W. Willis, and B. K. Watt, J. Am. Dietet. Assoc. 66:345 (1975).
16. "Composition and Nutritive Value of Dairy Foods," Dairy Council Digest 47, no. 5 (1976).
17. E. W. Toepfer, W. Mertz, E. E. Roginski, M. M. Polansky, J. Agr. Food Chem. 21:69 (1973).
18. E. A. Childs and J. N. Gaffke, J. Food Sci. 39:853 (1974).
19. M. I. Gomez and P. Markakis, J. Food Sci. 39:673 (1974).
20. A. Gormian, J. Am. Diet. Assoc. 56:397 (1970).
21. B. L. Bengtsson, J. Food Technol. 4:141 (1969).
22. C. R. Meiners, N. L. Derise, H. C. Lau, M. G. Crews, S. J. Ritchey, and E. W. Murphy, J. Agr. Food Chem. 24:1126 (1976).
23. N. T. Crosby, Proc. Inst. Food Sci. Tech, 10:65 (1977).
24. F. C. Bing, J. Am. Diet. Assoc. 60:114 (1972).
25. A. Thompson, Proc. Nutri. Soc. 24:81 (1964).
26. J. C. Fritz, G. W. Pla, B. N. Harrison, and G. A. Clark, J. Assoc. Off. Anal. Chem. 58:902 (1975).
27. B. Momcilovic, B. Belonje, A. Giroux, and B. G. Shah, Nutr. Rept. Internat. 12:197 (1975).
28. H. V. Hart, J. Sci. Food Agr. 22:354 (1971).
29. G. S. Ranhotra, F. N. Hepborn, and W. B. Bradley, Cereal Chem. 48:377 (1971).
30. E. Bjorn-Rasmussen, L. Halberg, and R. B. Walker, Am. J. Clin. Nutri. 25:317 (1972).
31. J. D. Cook, Fed. Proc. 36:2028 (1977).
32. G. W. Evans and P. E. Johnson, Am. J. Clin. Nutri. 30:873 (1977).
33. M. W. Neathery, J. W. Lassiter, W. J. Miller, and R. P. Gentry, Proc. Soc. Exptl. Biol. Med. 149:1 (1975).

34. A. Turnbull, in Iron in Biochemistry and Medicine (A. Jacobs and M. Worwood, eds.), Academic Press, New York, 1974, pp. 369-403.
35. I. Motzok, M. D. Pennell, M. I. Davies, and H. U. Ross, J. Assoc. Off. Anal. Chem. 58:99 (1975).
36. G. W. Pla, B. N. Harrison, and J. C. Fritz, J. Assoc. Off. Anal. Chem. 56:1369 (1973).
37. M. Layrisse and C. Martinez-Torres, in Progress in Haematology, Vol. VII, (E. B. Brown and C. U. Moore, eds.), Grune and Stratton, New York, 1971, pp. 137-160.
38. M. Layrisse and C. Martinez-Torres, Am. J. Clin. Nutr. 25:401 (1972).
39. J. Bowering, A. M. Sanchez, and M. I. Irwin, J. Nutr. 106:985 (1976).
40. J. Schulz and N. J. Smith, Am. J. Dis. Child. 95:109 (1958).
41. J. A. Halsted, J. C. Smith, Jr., and M. I. Irwin, J. Nutr. 104:345 (1974).
42. D. Oberleas, M. E. Muhrer, and B. L. O'Dell, in Zinc Metabolism, (A. S. Prasad, ed.), Charles C Thomas, Springfield, 1966, pp. 225-238.
43. D. Oberleas, in Toxicants Naturally Occurring in Foods, "Phytates," NAS/NRC, Washington, D.C., 1973, pp. 363-371.
44. N. T. Davies and R. Nightingale, Proc. Nutr. Soc. 34:8A (1975).
45. E. R. Morris and R. Ellis, J. Nutr. 106:753 (1976).

CHAPTER 6

PROTEINS AND AMINO ACIDS

J. Claude Cheftel

Laboratoire de Biochimie et Technologie Alimentaires
Université des Sciences et Techniques
Montpellier, France

I. INTRODUCTION

This chapter is mainly concerned with the chemical modifications of the primary structure of food proteins. However, heat and other factors of food processing and storage also affect protein conformation. Such modifications of the secondary, tertiary, and quaternary structures of food proteins may be of great importance, often resulting in the loss of desirable functional and organoleptic properties. From the nutritional standpoint, "denaturation" of food proteins by heat or other treatments is of little concern, but can improve digestibility in some cases. The unfolded denatured protein, although sometimes aggregated and insoluble in water, is frequently more easily digested by proteolytic enzymes than the native protein structure. It is thought that several peptide bonds originally shielded inside the protein become more accessible to proteases. Egg white, collagen, and some vegetable proteins from pulses or oilseeds are not, or only poorly, digested unless they are first denatured by heating. Actually, the first step in digestion also consists of a protein denaturation at the acid pH of the stomach. Acid and pepsin coagulation is certainly important for the proper digestion of milk proteins by infants and young animals; in this special case, previous heat denaturation of milk proteins may be detrimental to their coagulation in the stomach, and therefore to their digestion.

As indicated by Mauron [1], some food proteins may have specific nonnutritive functions to perform and these functions may depend on the native structure of the protein. It is thus thought that some proteins in breast milk such as immunoglobulins, lysozyme, etc., perform certain functions in the digestive tract of the newborn baby, such as transfer of passive immunity, destruction of some microorganisms, activation of others, transport of iron. If the heterologous proteins from cow's milk could perform similar tasks, heat treatment would doubtless inactivate them.

In many cases, the heating of protein-containing foods is highly beneficial for the safety or the nutritional value of the food or of its protein components. This is especially so with many vegetables (e.g., pulses and oilseeds) which contain heat-labile toxic factors, such as soybean trypsin inhibitor, phytohemagglutins, or the enzyme myrosinase. These factors are of a protein nature, and are inactivated by heat denaturation. It must also be remembered that the cooking of vegetables and cereals, by softening fibrous polysaccharides and making starch digestible, increases the digestibility of the protein part in addition to making the food edible and palatable.

From the standpoint of the nutritive value of proteins, only the chemical modifications of the primary structure brought about by processing and storage may have really detrimental effects. Modifications of the side chains of amino acid residues may reduce the content of intact essential amino acids in the protein, and may in many cases also slow down or decrease the release of these amino acids during digestion. Such chemical modifications may, however, have little practical nutritional importance when they affect (a) a food protein which does not contribute significantly to the diet; (b) a small number of amino acid residues; and (c) an essential amino acid which is not the limiting nutritional factor in the diet.

It has been attempted here to review the various chemical modifications of the primary structure of food proteins which may take place during food processing and storage, and impair the nutritive value or safety of food. These modifications have been tentatively classified into three groups according to type of chemical reaction, rather than food processing operation or food class.

The influence of processing and storage on food protein quality has already been reviewed by Bender [2, 3] and, more extensively, by Mauron [1] and Carpenter [4]. In the present chapter special attention has been given to recent studies, and studies which have not been previously reviewed by others.

II. SIMPLE MODIFICATIONS OF AMINO ACID RESIDUES

A. Oxidation

The effects of strong oxidizing agents such as performic or peracetic acids or ozone on the amino acid residues of proteins are well known: sulfur amino acids are oxidized to methionine sulfone and cysteic acid, and tryptophan to N-formylkynurenine; residues of tyrosine, serine, and threonine can also be partly oxidized, but only under more severe conditions [5].

Although protein foods are usually not submitted to strong oxidizing agents, hydrogen peroxide is at present used or suggested for a number of processes, including sterilization of milk (for cheese and for bread manufacture), whey, and containers for milk and other foods; bleaching and detoxifying of various fish or vegetable protein concentrates; and inflation peeling of some seeds. Benzoyl peroxide is used for the bleaching and maturation of wheat flour. In some countries, most uses of peroxides in foods are forbidden by the food additive legislation. However, oxidative compounds such as peroxides, superoxide anion radical and/or quinones may occur naturally in foods as a result of enzymatic or nonenzymatic reactions such as unsaturated lipid oxidation. Physical processes such as γ- or light irradiation in the presence of air, hot air drying, or even lengthy storage, may oxidize various food components, including amino acid residues of proteins.

1. Chemical Effects of Various Oxidizing Agents or Processes

Hydrogen peroxide can oxidize free methionine to methionine sulfoxides (isomers) and/or methionine sulfone, depending on the pH [6].

Mild hydrogen peroxide treatment (up to 0.2 M, at 50°C for 30 min, pH 8), similar to that suggested for milk sterilization, progressively oxidizes most of the methionine residues of casein or milk proteins to methionine sulfoxide [7]. No methionine sulfone appears.

Whey proteins appear to be more sensitive. Methionine sulfone and cysteic acid were formed, and the tryptophan content was lowered as a result of treatment with 0.1 M hydrogen peroxide [8].

It would be useful to investigate whether hydrogen peroxide treatments of different strength and duration also oxidize tryptophan and cysteine/cystine residues in milk.

Free methionine [9,10] or methionine residues [11] can be oxidized into methionine sulfoxide by contact with Mn^{2+}-containing sulfite-oxidizing systems.

In the presence of a sensitizing dye such as riboflavin, photooxidation of residues of tryptophan, histidine and tyrosine and the sulfur-containing amino acids may occur. The histidine reaction is the fastest, especially at a neutral pH. Methionine and tryptophan are said to be the only amino acids readily oxidized below pH 4 [5]. N-Formylkynurenine and/or kynurenine have been identified as reaction products of tryptophan-containing proteins [12].

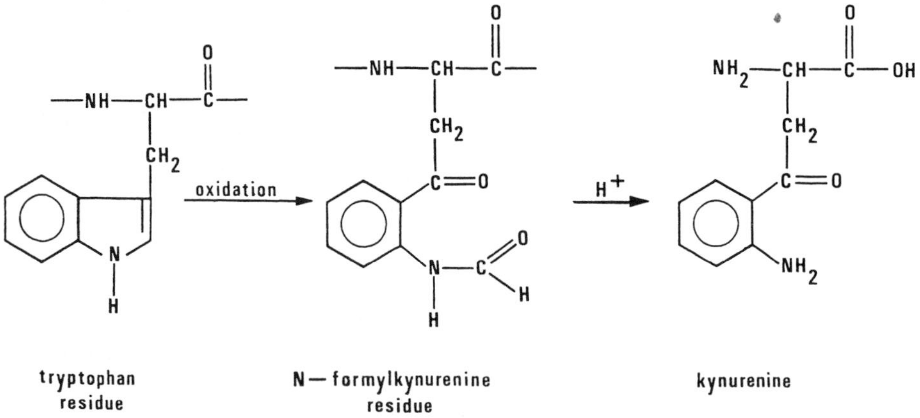

—NH—CH—C(=O)— (tryptophan residue) —oxidation→ —NH—CH—C(=O)— (N—formylkynurenine residue) —H^+→ NH_2—CH—C(=O)—OH (kynurenine)

tryptophan residue   N—formylkynurenine residue   kynurenine

γ-Irradiation of foods in the presence of oxygen gives rise to hydrogen peroxide through radiolysis of water. γ-Irradiation of food proteins is known to induce some radiolysis of sulfur and aromatic amino acids. Volatile sulfur compounds formed may be responsible for off-flavors in irradiated milk, meats, and vegetables. These reactions are reduced when irradiation is performed in the absence of oxygen or in the frozen state. γ-Irradiation of fish and other protein foods at up to 3 Mrads does not appear to lower the protein nutritive value [13,14]. More will be said about the effects of γ-irradiation on model proteins in Sec. IV.B.

Contact with oxidizing lipids (such as methyl linoleate) has been shown to oxidize methionine residues into methionine sulfoxide [11,15,16]. This is probably due to the presence of lipid peroxides. Whether this type of reaction is of importance is debatable, since lipid oxidation probably makes the food rancid and unacceptable before proteins are significantly oxidized (see Sec. IV.B).

Tryptophan destruction occurring during the acid hydrolysis of proteins appears to be of an oxidative nature; it is enhanced by cystine, and to a lesser extent by serine, hydroxyproline, and tyrosine present in the protein hydrolysate. The main oxidation product appears to be β-3-oxindolylalanine [17].

Some oxidative enzymes may modify amino acid residues of proteins. Plant and animal tryptophan pyrrolooxygenases cause the formation of N-formylkynurenine residues, but probably do not play any role in foods. Polyphenoloxidases are known to oxidize some tyrosyl residues of various proteins into residues of dihydroxyphenylalanine quinone. DOPA-quinone may react with ε-amino groups of lysyl residues. It is unlikely that such cross-links occur to a significant extent during the processing of plant protein foods.

Up to 20% of methionine residues have been found [18] to be oxidized to sulfoxide during the production of single cell protein (yeast fermentation). It is not indicated whether this occurs during the aeration of fermentation broth or during the purification of protein.

2. Analytical Methods

Methionine sulfoxides are unstable during acid hydrolysis, giving 25-75% methionine plus variable amounts of methionine methylsulfonium salt, homocysteine, and homocysteic acid [6]. Methionine is also partly unstable during the acid and even enzymatic hydrolysis of some protein preparations, giving rise to methionine sulfoxide. Methionine sulfone, however, is stable during acid hydrolysis and can be determined directly. For this reason, the most accurate method for the determination of the sum of methionine, methionine sulfoxide, and methionine sulfone residues relies on performic acid oxidation to methionine sulfone, acid hydrolysis, and ion-exchange chromatography. Performic acid also oxidizes cystine and cysteine to cysteic acid, which is stable during acid hydrolysis. When carboxymethylation of nonoxidized methionine residues is carried out before performic acid treatment, the sum of methionine sulfoxides and methionine sulfone can be determined after acid hydrolysis [7,19]. Methionine, methionine sulfoxides, and methionine sulfone initially present in a protein can best be determined after alkaline hydrolysis [7,19], although some investigators have observed a partial destruction of methionine sulfoxide under such conditions. With ion-exchange chromatography, cysteic acid, methionine sulfoxides, methionine sulfone, and aspartic acid can be eluted in this order, and determined [7].

Unoxidized methionine residues in a protein can be determined as homoserine lactone, after treatment with cyanogen bromide followed by acid hydrolysis [20].

Unoxidized methionine residues can also be quickly oxidized to sulfoxide by dimethylsulfoxide in the presence of a haloacid. Dimethylsulfide is released and can be determined by gas chromatography. Corrections have to be made for cysteine residues and for ascorbic acid [21].

Unoxidized methionine residues can also be determined by colorimetric methods after partial enzymatic hydrolysis of the protein.

A method for the direct determination of methionine sulfoxide residues in proteins using acetic anhydride has recently been suggested [22].

Some investigators have performed surface analysis by X-ray photoelectron spectroscopy under vacuum of unhydrolyzed samples of skim milk and hydrogen peroxide-treated skim milk [23]. It was possible to determine methionine + cysteine + cystine, methionine sulfoxides, and methionine sulfone + cysteic acid (see Table 1).

Previously, year-old samples of various wheat products were also found by the same investigators to contain notable amounts of oxidized sulfur-containing amino acids (possibly due to contact with oxidized lipids). More correlations of X-ray photoelectron spectroscopy data with data from other analytical methods are needed.

TABLE 1 Sulfur Amino Acid Composition of Oxidized Skim Milk Samples

	Mole % of total sulfur-containing amino acids		
	Met[a] + Cys[b]	Met 0^c	Met 0_2^d + Cysteic Acid
Dry skim milk	78	8.5	13
□ heated	84	8	7
□ treated with 0.015M $H_2 0_2$	61	29	10
□ treated with 0.15M $H_2 0_2$	42	37	21

Source: Walker et al. [23].
[a]Methionine [c]Met sulfoxide
[b]Cysteine [d]Met sulfone

Many of the derivatives of cystine residues which are thought to form upon oxidation at neutral pH are unstable, and cannot currently be determined (see Fig. 1) [24].

3. Nutritional Implications of the Oxidation of Amino Acid Residues

Free cysteic acid, methionine sulfone, and formylkynurenine cannot replace cysteine, methionine, and tryptophan, respectively, in the diet of rats. Free L-cystine mono- and disulfoxides, cysteine sulfenic acid (but not cysteine sulfinic acid) prove capable of partly replacing L-cystine for the growth of rats; free L-methionine sulfoxides (2 stereoisomers) can also partly replace methionine [25, 26]. The nutritional value of methionine sulfoxide appears to depend on the age of the animal [27]. An explanation for this might be the induction, with age, of enzymes able to convert larger amounts of methionine sulfoxide into methionine in vivo. This should be studied further in various animal species. The possible toxicity of excess methionine sulfoxide should also be tested.

Cuq et al. [7] oxidized nearly all of the methionine residues of casein into methionine sulfoxide with hydrogen peroxide. In vitro proteolytic digestion tests with pronase, or with pepsin followed by pancreatin, indicated a sharp decrease in the release of free methionine and little or no release of free methionine sulfoxide [7, 11]. The release of free tryptophan and other amino acids did not change, an observation which may be of significance since it has been shown that oxidation of tryptophan residues in proteins into kynurenine and unknown derivatives renders their in vitro proteolytic release considerably more difficult than the release of the initial tryptophan residues [28]. Similar findings of reduced digestibility in vitro have been reported for proteins where residues of methionine and/or

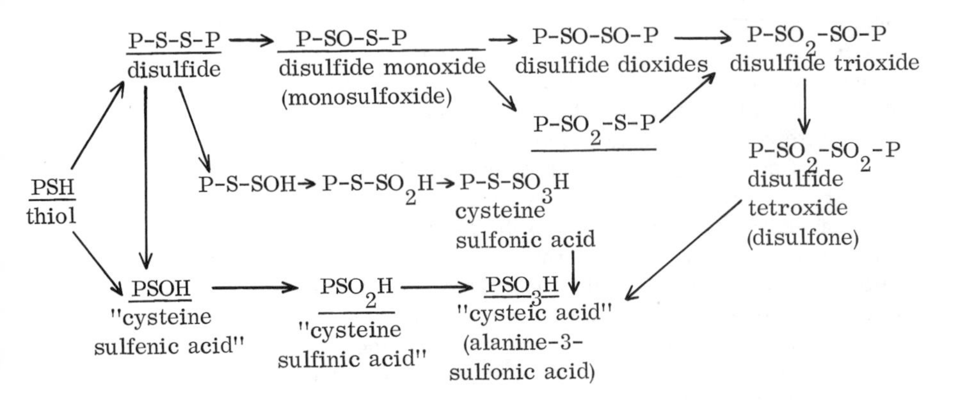

Fig. 1. Oxidation derivatives from cysteine and cystine residues (underlined compounds correspond to isolated derivatives).

cysteine had been oxidized into methionine sulfone and/or cysteic acid [29,30].

In our laboratory, casein with 98% of methionine residues oxidized into methionine sulfoxide, but devoid of methionine sulfone, was fed ad libitum to young rats as the sole source of protein in the diet and at various levels of protein intake (9, 11, and 13% of the diet). Control diets contained normal casein, normal casein supplemented with free methionine, oxidized casein supplemented with free methionine or methionine plus tryptophan, or no protein. Another experiment was performed by pairfeeding. Protein intake, nitrogen digestibility, protein efficiency ratio (PER), net protein utilization (NPU, carcass or nitrogen balance) were determined and their statistical significance calculated. Only a small (10%) decrease in PER and NPU of oxidized casein was detected [11,16]. It thus appears that, in vivo, methionine sulfoxide is released during digestion, absorbed, and to a large extent reduced to methionine and utilized for protein synthesis. Plasmatic and muscular free amino acids were also determined. Free methionine sulfoxide accumulated at a high level in animals fed oxidized casein [11,16]. These findings confirm that the nutritional role of methionine is slightly decreased when the source of methionine consists of methionine sulfoxide residues.

Slump and Schreuder [31] treated casein and fish meal with hydrogen peroxide in the presence of perchloric acid and found that methionine residues had been oxidized to methionine sulfoxide and methionine sulfone. A reduced net protein utilization was determined in rats and attributed solely to the presence of methionine sulfone and cysteic acid.

Anderson et al. [32] found that 0.9 and 2 M hydrogen peroxide treatment of rapeseed flour for 1 hr at room temperature was effective in destroying most of the thyroid-toxic, sulfur-containing glucosinolates. However, it also caused extensive formation of methionine sulfoxide (40-50% of initial methionine residues), methionine sulfone (50-60% of initial methionine), and cysteic acid (ca. 70% of initial half-cystine). The tryptophan content (determined after alkaline hydrolysis) also decreased by up to 20%. Weanling rats were fed a diet containing 10% of casein or rapeseed flour proteins: the PER of casein, of non-thyroid-toxic (water-washed) rapeseed flour, and of a similar flour treated with 0.9 M hydrogen peroxide were 3.5, 4.0, and 1.5, respectively. The decrease in nutritional value of the treated flour could not be completely reversed by methionine supplementation. This was perhaps due to the presence of methionine sulfone, since very small additions of free methionine sulfone to diets were found to depress food intake in weanling rats. Free amino acids were determined in the plasma: the increased concentration of methionine sulfoxide and sulfone was proportional to that found in the H_2O_2-treated proteins; this was not the case for cysteic acid, and it is likely that cysteic acid was further metabolized in vivo into taurine.

Methionine sulfoxide is known to be partly excreted in the urine as N-acetyl methionine sulfoxide, whereas methionine sulfone is excreted either before or after acetylation [33].

In spite of some discrepancies in the reported results concerning the nutritional availability of methionine sulfoxide residues for the rat, it can probably be concluded that mild oxidative treatments do not significantly reduce the nutritional value of food proteins, as long as no methionine sulfone is formed. The absence of formylkynurenine or other oxidation products of tryptophan should also be ascertained. It is felt that it would be useful to study further the nutritional value of methionine sulfoxide residues in animal species other than the rat.

The present data indicate that formation of methionine sulfoxide is unlikely to account for the high (up to 100%) unavailability of methionine observed in various heated protein foods, including fish meals [34-39].

Animal, microbiological, and enzymatic tests have been used for the measurement of this unavailability, which is not revealed by the chemical determination of total methionine [37,40,41]. The cause of this unavailability is still unclear. It is often associated with a low availability of other amino acids and with a general decrease in protein digestibility, both due to the formation of peptide cross-links. It is not known whether methionine residues can be readily alkylated on their methylthio group and participate directly in such cross-links. This will be discussed in Sec. III.C.

B. Desulfurization

Severe heating of moist foods with a low carbohydrate content has been repeatedly found to cause a marked destruction of cystine-cysteine (as seen after acid hydrolysis with or without prior performic acid oxidation), together with a decrease in the availability of many amino acids (see Sec. III.C).

This occurs with meat [42-44]. Some investigators [42] reported a 44% loss in cystine in canned meat after sterilization. Similar losses in cystine were found on heating pork for 24 hr at 110°C [34]. Heating of cod muscle for 27 hr at 116° C also caused cystine destruction [37]. Overheating of soybean protein, as occurred in the old-style mechanical expeller (for oil extraction), also causes cystine destruction to the point where the cystine content of the protein becomes limiting for rat growth [45].

The formation of H_2S and other volatile sulfur-containing compounds (e.g., methyl mercaptan, dimethyl sulfide) has been demonstrated in heated milk and meat, where they contribute to flavor. Similar compounds form when dry bovine plasma albumin is heated [46].

The above mentioned loss of cystine probably occurs through desulfurisation reactions, like that due to alkaline treatments (see Sec. III.D). Such reactions may lead to unstable residues of dehydroalanine capable of reacting with cysteine or lysine residues in the protein, thus creating lysinoalanine and lanthionine protein-protein cross-links (Sec. III.D). Such cross-links have also been recently identified in heated proteins [47] and they contribute, together with the formation of amide protein-protein cross-links, to the other effects of moist heat on proteins, namely decrease in digestibility and in amino acid availability (see Sec. III.C).

Bjarnason and Carpenter [46] suggest that H_2S may also form in heated proteins according to the following reaction:

$$\text{cystine residue} \xrightarrow{H_2O} \text{cysteine residue} + \text{cysteine sulfenic acid residue}$$

$$\downarrow$$

$$H_2S + \begin{array}{c} | \\ NH \\ | \\ H\text{-}C\text{-}\overset{H}{\overset{|}{C}}{=}O \\ | \\ C{=}O \\ | \end{array} \quad \text{aldehyde}$$

C. Isomerization

Alkaline treatments of proteins (see Sec. III.D) may cause the isomerization of various amino acid residues (methionine, lysine, cysteine, alanine, phenylalanine, tyrosine, glutamic and aspartic acids) [48].

Amino acid isomerization is initiated by the dissociation, in an alkaline medium, of the hydrogen on the asymmetric carbon in α-position. The resulting carbanion rearranges by tautomeric change or resonance into form III, which has no more carbon asymmetry:

$$\underset{\text{I}}{R\text{-}\overset{H}{\underset{NH_2}{C}}\text{-}C\begin{smallmatrix}{=}O \\ {-}O^-\end{smallmatrix}} \xrightarrow{OH^-} \underset{\text{II}}{R\text{-}\underset{NH_2}{\bar{C}}\text{-}C\begin{smallmatrix}{=}O \\ {-}O^-\end{smallmatrix}} \longrightarrow \underset{\text{III}}{R\text{-}\underset{NH_2}{C}{=}C\begin{smallmatrix}{-}O^- \\ {-}O^-\end{smallmatrix}}$$

The presence of a negative charge on the carboxylic group opposes α-hydrogen ionization. For this reason, free amino acids are much less easily isomerized than amino acid residues in proteins.

Methionine isomerization has been studied by microbiological analysis in NaOH-digested fish protein concentrate [49]. Complete racemization occurred for strong alkali treatment.

D-lysine residues have been determined in NaOH-treated sunflower protein isolate by enzymatic and microbiological techniques after acid hydrolysis [48]. A marked degree of isomerization was found for treatments in 0.2 M NaOH for 1 hr at 80°C or above, but no isomerization occurred for milder treatments (0.05 and 0.1 M NaOH, 1 hr, 55°C) comparable to those used for vegetable protein purification. A 0.1 M NaOH, 80°C, 16 hr treatment released less than 0.5% of amino acid residues as free amino acids; since lysine isomerization reached 35%, it took place mainly on lysyl residues.

Isomerization of amino acid residues has also been observed on protein treatment with strong acids. However, it occurs less readily in acid than in alkaline solution, and measurable amounts of inversion occur only in concentrated solutions at high temperatures.

Roasting of protein has also been shown to cause amino acid isomerization. Casein, lysozyme, poly(L-alanine), poly(L-glutamic acid) and poly(L-lysine) were heated to dryness at 180-300° C for 20 min under air or nitrogen. Decomposition and isomerization of amino acid residues after acid hydrolysis was investigated by gas chromatography. Aspartic acid, glutamic acid, alanine, and lysine residues were found to be extensively isomerized, and the other amino acid residues, except proline, were also isomerized to a considerable extent at higher temperatures. A gel-filtration study on poly(L-glutamic acid) revealed that on roasting it was broken down into fractions of various molecular weights and that racemization of the glutamic acid residues proceeded more markedly in the smaller fractions. Free amino acids and oligopeptides formed from roasted casein were found to be largely or completely racemized [50, 51].

Isomerization of amino acid residues may partly inhibit the proteolytic digestion of the protein. Studies by Kamath and Berg have shown that free D-lysine and many D-amino acids have almost no nutritional value. Some free D-amino acids, namely methionine, tryptophan, and arginine, can be metabolized by the rat and can partly replace the corresponding L-amino acids, probably because of the presence of D-amino acid oxidases in the organism. However, growth is generally more or less retarded [52, 53]. D-L-methionine is frequently used in animal feeding. D-methionine appears to have little or no nutritional value for man [54].

D. Naturally Occurring Amino Acid Substitutions

Individual amino acids of some unprocessed food proteins, especially of plant origin, are known to be only partly available nutritionally, such as methionine and cystine in beans [55, 56] and in leaf protein concentrates, lysine in wheat, and isoleucine in corn. The reasons for this unavailability are not known, but it is likely that resistance of the proteins to digestion plays a major role (see Sec. III.A).

By contrast, the existence of some naturally occurring amino acid substitutions is well documented, but their nutritional implications have not been studied thoroughly.

Hydroxylysine and hydroxyproline residues are found in collagen. Hydroxylysine has no nutritional value for animals. Phosphoserine and phosphothreonine residues occur in caseins. The carbohydrate units of glycoproteins and mucoproteins are covalently attached to residues of threonine, serine, asparagine, and hydroxylysine. This attachment is resistant to attack by glycolytic and proteolytic enzymes.

Methyllysine, histidine, and arginine have been identified in some proteins. Small amounts of ε-N-monomethyllysine, ε-N-trimethyllysine and 3-methylhistidine were obtained from acid hydrolysates of chick actin and myosin, and an enzyme responsible for the methylation of the amino acid residues was identified [57].

The biological significance of protein methylation remains obscure, although it has been hypothesized that it modifies the activity of some proteins and their resistance to catabolism by proteases [58]. Enzymes which demethylate the methyl protein have not been found. It has been shown that methylation of lysine residues in various proteins (as mono and dimethyl substituted lysine derivatives) does not decrease the in vitro proteolytic digestion of the proteins by trypsin or α-chymotrypsin [59]. However, the esterase activity of pancreatic trypsin on L-lysine methyl ester was greatly depressed when the ε-amino group of lysine was methylated. ε-methylated lysine derivatives are resistant to HCl hydrolysis and can be determined on amino acid analyzers [60,61].

The nutritive value of selectively acylated casein or lysine has been investigated and is reported in Sec. III.C.

III. PROTEIN-PROTEIN INTERACTIONS. FORMATION OF CROSS-LINKS BETWEEN AMINO ACID RESIDUES

A. Naturally Occurring Protein Cross-Links

Many unprocessed protein foods are known to be only partially digestible. The main factors involved are probably: close contact with indigestible polysaccharide cell walls, fibers, and starches; presence of protease inhibitors (generally heat labile); and the protease-resistant nature of protein or peptide fractions (even after heat denaturation) [55,56] (see Sec. III.B). The latter phenomenon may be due to cross-links occurring naturally within or between polypeptide chains.

The best known examples of such cross-links are those of keratin, elastin, and collagen.

1. Keratin

Numerous disulfide bonds in hair and feather keratin make this protein totally insoluble in water and resistant to proteolytic attack. Treatment with reducing and dispersing agents, alkaline, acid or enzymatic hydrolysis, autoclaving, or even ball-milling increase the digestibility to the point that such feather meals can be used as protein feeds.

ε-N(γ-glutamyl)lysine cross-links have been identified in hair and wool keratin and in fibrin. They are formed by transamidases (transglutaminases) which catalyze the replacement of the carboxyamide group of protein-bound glutamine by the ε-amino group of a lysine residue, with the concomitant release of ammonia [62]. ε-N-(γ-glutamyl)lysine was not split after almost complete in vitro proteolysis of fibrin, but was released from adjoining amino acid residues [63]. ε-N-(γ-glutamyl)lysine has nevertheless been found to be a nutritionally available source of lysine (see Sec. III.C).

2. Elastin

Elastin, found in skeletal muscle, is an unusual protein by virtue of its elasticity, insolubility, content of lysine-derived cross-links, and general resistance to proteolysis by many mammalian proteases. Pancreatic elastase, however, readily hydrolyzes and solubilizes elastin.

Elastin contains a number of desmosine and isodesmosine cross-links. These are tetraminotetracarboxylic acids which can cross-link a maximum of four peptide chains in the elastin structure:

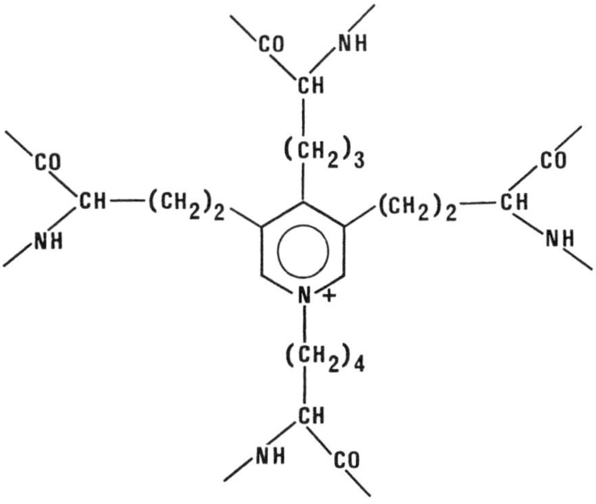

desmosine

A diaminodicarboxylic amino acid, lysinonorleucine, was also isolated from proteolytic digests of elastin.

Biosynthetic studies have shown that desmosines and lysinonorleucine are produced from chemical interactions of side-chain groups of given lysyl residues in the polypeptide chain of elastin [64]. Lysyl or hydroxylysyl residues in elastin or collagen are oxidatively deaminated into the corresponding α-aminoadipic δ-semialdehyde (allysine) by lysyloxidase. Lysine-derived aldehydes then condense spontaneously by aldol and carbonylamine reactions to form the various cross-links [65].

Substituted lysin residues are nutritionally unavailable, as proteolytic enzymes do not break down desmosines nor even release them completely free of attached amino acids.

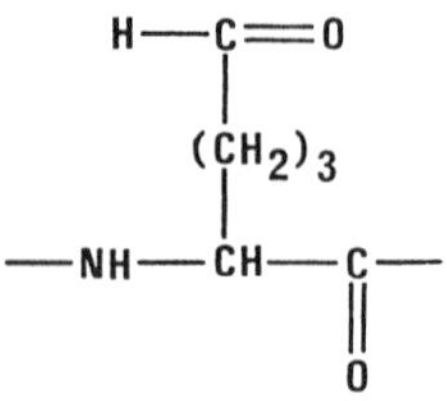

H—C=O
|
$(CH_2)_3$
|
—NH—CH—C—
‖
O

allysyl residue

The various cross-links can be separated, after acid hydrolysis of the protein, by ion-exchange chromatography [66].

3. Collagen

Various types of cross-links occur between the polypeptide chains of the tropocollagen molecules [67]. Apart from hydrogen bonds, which can easily be split by moist heat making meat collagen partly digestible, one can mention (a) ester cross-links between ω-carboxylic groups of aspartic or glutamic acid and hydroxy groups of hydroxyamino acid residues (especially serine); (b) carbonyl-amine cross-links (lysinonorleucine, desmosine, etc., as in keratin); (c) ε-N-(γ-glutamyl) lysine cross-links; (d) phosphodiester cross-links between two hydroxyaminoacid residues; and (e) carbohydrate cross-links, involving galactose or glucose and residues of lysine or hydroxylysine.

Recently a new cross-link was found in cow-skin collagen, which is thought to form from the histidine, allysine, and hydroxyallysine residues of three peptides chains [68]:

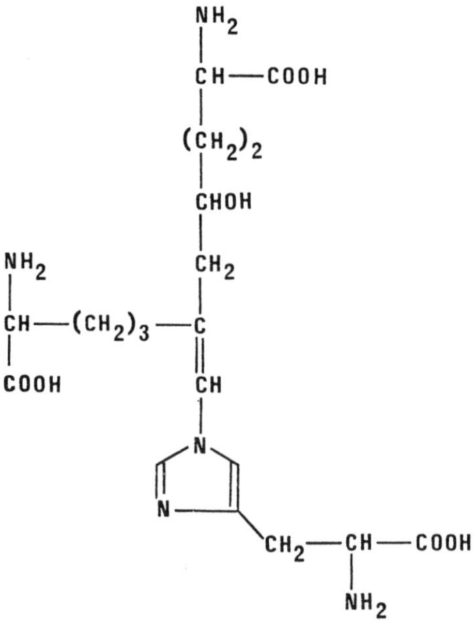

NH_2
|
CH—COOH
|
$(CH_2)_2$
|
CHOH
|
CH_2
|
NH_2—CH—$(CH_2)_3$—C(—COOH)
‖
CH
|
N (imidazole ring: N, N)
—CH_2—CH—COOH
|
NH_2

Other types of cross-links have been shown to occur in other proteins, including those from microorganisms. Imide links between residues of asparagine and glutamine and residues of aspartic or glutamic acid, and also thioester links between sulfhydryl groups and residues of aspartic or glutamic acid, can both be mentioned. Some of these cross-links have also been identified in food proteins treated by heat or other ways.

B. Beneficial Effect of the Denaturation of Vegetable Proteins by Moderate Heat

It has already been reported (Sec. II.D and III.A) that the protein digestibility and the availability of various amino acids are often low in many plant foods [69]. In several of these, especially in legumes, heat brings about an improvement in protein quality, as well as the destruction of growth inhibitors and other toxic factors (see Table 2) [70-72].

The enhancement of the nutritive value of soybean proteins by moderate heat was generally ascribed to the destruction of antitryptic factors and phytohemagglutinins, both of protein nature. From studies on rats and chicks with raw soybean proteins freed from, or enriched in trypsin inhibitors or phytohemagglutinins, the following conclusion has been drawn [70-72]: both trypsin inhibitors and undigestible soybean globulins bind trypsin in the intestine. The resulting hypersecretion of trypsin and other pancreatic enzymes (rich in cystine) adds to the fecal loss of amino acids from the undigested soybean proteins (including trypsin inhibitors), and

TABLE 2 Effect of Heat on the Protein Digestibility and the PER of Legumes Possessing Trypsin-Inhibitor Activity

Legume	Scientific name	Trypsin-inhibitor activity x 10^{-4} units/g	Digestibility[a] (%)		PER[a]	
			raw	heated	raw	heated
Kidney beans	Phaseolus vulgaris	4.25	56	79	loss in weight	0.8
Soybeans	Glycine max	4.15	70	85	1.3	2.4
Pigeon peas	Cajanus cajan	2.77	59	60	0.7	1.6
Cow peas	Vigna sinensis	1.91	79	83	1.4	2.2
Lentils	Lens esculenta	1.78	88	93	0.4	1.2

Source: Liener [70].
[a]Rats.

brings about a severe nutritional deficiency in sulfur-containing amino acids. The poor digestibility of the globulins appears to be the most important factor in growth retardation and fecal loss of nitrogen. Moist heat treatment of raw soybean causes a denaturation of the proteins, inactivating the trypsin inhibitors and making the globulins more digestible (see Table 3). However, it appears that the protein digestibility of a number of soybean and pulse cultivars is differently enhanced by moist heat, since the protein nutritional value of these heated seeds was found not to be proportional to their content in the nutritionally limiting sulfur-containing amino acids [70].

Only the phytohemagglutinins (also called lectins) from some varieties of *Phaseolus* beans have an antinutritional and toxic effect. These phytohemagglutinins are quite resistant to heat. Proper autoclaving or cooking, preceded by soaking in water, is necessary for their complete denaturation and inactivation [72].

Industrial heat treatment of soybean can be done in a number of ways, at various stages of the preparation of full-fat meals or of defatted flours, grits, protein concentrates, or isolates. Experiments have shown that conditions of soybean autoclaving giving the best nutritional value are over 1 hr at 100°C or 30 min at 121°C. This steam processing appears to be sufficient for the proper heat denaturation of proteins, in spite of the low initial moisture content of the bean. Extrusion techniques allow accurate control of moisture content and precise timing of the heat application. However, the protein quality of commercial soy products varies considerably.

Extrusion techniques are used for precooking or toasting of soy flours, texturization of soyflours or concentrates into meat analogs, and cooking or shaping of porous cereals or starch-based snacks. Extrusion provides a good means for the destruction of trypsin inhibitors and other thermolabile

TABLE 3 Effect of Autoclaving[a] and of Methionine Supplementation on the PER of Soybeans

Protein source	PER[b]
Raw soybeans	1.40
Autoclaved soybeans	2.63
Raw soybeans + 0.6% methionine	2.42
Autoclaved soybeans + 0.6% methionine	2.99

Source: Liener [70].
[a]20 Min, 115°C.
[b]Rats.

antinutritional or toxic factors present in soybean and various oilseed flours. The destruction of these factors is generally made easier by high moisture content, high temperature, and long residence time. It then becomes important to check that the heat treatment does not affect the protein nutritional value. Mustakas et al. [73] found the PER of full-fat soy flours to reach a maximum after an extrusion time of 2 min at 135°C (with a moisture content of the flour of 20%). These conditions were found to give an 89% trypsin inhibitor destruction, and a nitrogen solubility index of 21%. The content of available lysine and thiamine were practically unchanged. The high temperature-short time extrusion cooking of high protein-containing mixtures has also been reported as less detrimental than cooking on a drum-dryer. In our laboratory, a protein-enriched snack food prepared by extrusion of a mixture containing 50% corn starch, 20% soy protein concentrate, 20% casein and 10% water was analyzed for its content of available lysine both by the indirect chemical fluorodinitrobenzene (FDNB) method and by bioassay (see Sec. IV.A). Extrusion was performed in a double-screw BC 45 Creusot-Loire extruder. With the highest temperature of 230°C in the last element of the extruder (ca 200° C in the food), chemically unavailable and nutritionally unavailable lysine were found to represent 5 and 10% of total lysine, respectively.

The nutritional value of various soy protein fibers commercially prepared by spinning alkaline protein isolates into an acid bath was evaluated in animals and men by the nitrogen balance method [74-79]. Spun soy protein fibers and a textured food made from them showed similar digestibility and only slightly lower biological values than dehydrated beef, casein, and whole milk proteins, especially when the soy proteins were supplemented with methionine and lysine or fed above the critical level of intake [74]. The spinning process appears to eliminate most of a heat-labile growth inhibitor still present in the initial soy protein isolate. Digestibility values of 92% for the textured food are worth noting, since the spinning process has been said to induce numerous disulfide bonds between tightly packed parallel protein fibrils [80]. Kies and Fox [75] reported PER values in growing rats on the basis of casein (2.5) to be 2.12, 2.82, and 2.37 for TVP, methionine-supplemented TVP and beef, respectively. Spun soy protein and a 5:1 casein-lactalbumin mixture appeared to be equally well utilized in man when supplemented with crystalline amino acids to the same pattern of amino acids [76]. Spun soy protein containing one-third egg albumin also had a high nutritive value for man [78].

Trypsin inhibitor activity was found to persist partially (30% of that in unheated soy flour) in a soy protein isolate, and to a lesser extent in some spun fibers. However, this may be of little nutritional significance, since soybean trypsin inhibitors have but a very limited effect on human trypsin [70-72].

Overheating of defatted soybean flours, as may occur during removal of solvent, partly destroys cystine, lysine, arginine, tryptophan, and serine, making cystine the limiting amino acid. The content of available lysine may also be reduced. The destruction of cystine probably takes place by desulfurization, whereas the destruction of basic amino acids may be due to reactions with sugars contained in the flour.

Heating of peanut seeds was found to increase the PER of the defatted meal and reduce the content of contaminating aflatoxin. However, both wet (40% water) and dry (5% water) treatments at 121°C for 4 hr or at 130° C for 1 hr reduced PER and the content of chemically available lysine [81, 82]. The latter appears to be due to reaction of ε-amino groups of lysine with reducing sugars formed by the hydrolysis of sucrose present in the peanut [83].

The beneficial, then detrimental, effects of increasing the heat treatment of cotton seed protein meals are due to the presence of gossypol and of some oligosaccharides such as raffinose, and the reaction of these compounds with lysine residues (see Sec. IV.C).

The protein quality of sunflower seed meal was found to be increased by a relatively mild heat treatment (1 hr at 100°C). Higher temperatures decreased the nutritional value, probably by lowering the availability of lysine through reaction with chlorogenic acid or carbohydrates [84].

A non-heat treated rapeseed meal contains the enzyme system myrosinase, which hydrolyzes glucosinolates, producing the toxic compounds isothiocyanates, oxazolidinethiones, or nitriles. Heat treatment of the seed or meal is necessary to produce a meal with a high nutritional value, even when the glucosinolate content is low. The optimum temperature and duration of seed treatment for proper inactivation of the myrosinase and of other deleterious factors, and for obtaining a maximum nutritional value (PER), appear to be 100-110° C and 60-15 min at a moisture content of 8% [85]. Higher temperatures or moisture contents, particularly high expeller temperature during oil extraction, bring about a decrease in available lysine. This may be due to Maillard-type reactions with carbohydrates present in the meal. Rapeseed protein concentrate can also be detoxified by water-ethanol extraction of the seeds, cold water extraction, or fermentation.

Heating of wheat flours in a heat exchanger at temperatures of 108-174°C for 2-10 min (with a water content of 13 or 33%) was found to have marked effects only at or above 150° C. The content in lysine, arginine, and cystine fell by 50%, both protein breakdown and aggregation took place, and the in vitro protein digestibility by proteases was reduced [86]. Similar results were obtained with overheated bread crust, puffed wheat, wheat granules, and flakes. No effects were noticed with normal bread crumbs and wheat shreds.

It can be concluded from the studies reviewed above that a moderate heat treatment of many plant protein foods enhances the nutritional value of the proteins. This effect is mainly the result of protein denaturation; that is, toxic protein substances are inactivated, and major protein constituents are made more digestible. However, overheating of these relatively moist vegetable proteins may cause a decrease in the content of the most thermolabile amino acid, cystine, and a partial unavailability of the most reactive amino acid, lysine. Lysine inactivation mainly results from reactions with reducing carbohydrates present in the seed, meal or flour (see Sec. IV.A).

C. ε-N-(γ-Glutamyl)lysine and Other Cross-Links in Heated Animal Food Proteins.

Severe heating in the moist or dry state of pure model proteins or of protein foods with a low carbohydrate content (such as fish and meat) brings about a marked destruction of cystine (see Sec. II.B). The content of other amino acids is usually unchanged, as seen after acid hydrolysis, except for small losses of lysine, and sometimes of methionine [34, 36, 38, 87]. However, the nitrogen digestibility is often severely diminished, together with the availability (40-60% decrease) of a number of amino acids, and the overall nutritional value. Thus when bovine plasma albumin or presscake herring meal (free of lipid peroxides) were heated at 130°C for 27 hr, a sharp decrease in the availability of methionine, tryptophan, arginine, lysine, and even leucine was noticed [88]. The decrease was most marked at 5-14% moisture content. Many comparable results were obtained with cod muscle, meat, and various pure proteins [4, 40, 89].

Why does the severe heat treatment of protein foods with a very low carbohydrate content lower the overall digestibility of the protein and the availability of several amino acids simultaneously? The simplest explanation is that such a treatment gives rise to a number of new protease-resistant cross-links within and between polypeptide chains. Parts of these chains then become biologically unavailable because of the masking of the site of protease attachment. Although this picture is true to a large extent, as reported later, it is appropriate to first consider some of the complex nutritional effects of severe heat treatments.

Miller et al. [36] indicated that the decrease in protein and amino acid digestibility, as measured in rats, cannot account for the much larger decrease in nutritional value (PER, NPU in rats) and in amino acid availability, especially methionine (chick and *Streptococcus zymogenes* tests). The authors therefore assumed that a direct detrimental modification of methionine had occurred and had later been reversed by acid hydrolysis. None of the later experiments has confirmed this hypothesis (see Sec. II.A and later in this section).

Other investigators [40, 90] studied the in vitro release of amino acids by proteases and the microbial and biological availability of essential amino acids in dry-heated cod meal. They demonstrated differences in the rate of release of individual amino acids and postulated that such differences may prevent nutritional supplementation and explain the low availability of methionine and lysine. Further studying the digestion and the metabolism of heat damaged fish proteins in rats, they found abnormally high levels of undigested peptides and also of free amino acids and small peptides in the gut. They suggested that the inefficient digestion came from the resistance of heated proteins to proteases, an inhibition of peptide absorption, and a fecal loss of digestive enzymes simultaneously [89,91]. Finally they found small amounts of peptides containing lysine, aspartic acid, and glutamic acid in the urine, which probably are nutritionally unavailable.

Such peptides may correspond to the ε-N-(γ-glutamyl)lysyl amide cross-links which Bjarnason and Carpenter [46] demonstrated to form from lysyl and glutamine residues with release of ammonia, when bovine plasma albumin was heated for 27 hr at 110-145°C in the presence of 14% water:

$$
\begin{array}{ccccc}
\ldots - NH & - & CH & - & CO - \ldots \\
 & & | & & \\
 & & (CH_2)_2 & & \\
 & & | & & \\
 & & CO & & \\
 & & | & & \\
 & & NH & & \\
 & & | & & \\
 & & (CH_2)_4 & & \\
 & & | & & \\
\ldots - NH & - & CH & - & CO - \ldots
\end{array}
$$

It was first thought that the ε-amino substituted lysine of such cross-links was biologically unavailable, although hydrolysis with hydrochloric acid was able to regenerate lysine. However, Mauron [18] and then Waibel and Carpenter [92] found that ε-N-(γ-glutamyl)-L-lysine can be fully used as a source of lysine by the rat and the chick. This compound is mainly split in the kidney [93].

These findings prompted detailed studies of the nutritional effects of lysine and protein acylation. The results of these studies will be briefly reported here as they partly explain the effects of severe heat processing.

The hydrolytic action of rat and chick kidney acylases on many ε-N-substituted lysine derivatives was investigated. ε-N-formyl and acetyl lysine were split, but the activity appeared to be less with larger acyl groups.

Bjarnason and Carpenter [94] and Mauron [1] studied the availability for the chick and/or the rat of several acylated lysine derivatives; the results are summarized in Table 4. A large amount of ε-N-acetyl-L-lysine and of ε-N-propionyl-L-lysine was recovered from the urine of rats receiving these compounds in the diet.

Lactalbumin (or bovine plasma albumin) in which over three-fourths of the ε-amino groups had been formylated, acetylated, or propionylated (without any cross-link formation) is partly deacylated in vivo, after

TABLE 4 Nutritional Availability of Acylated Lysine Derivatives

Derivative	Utilization as a source of lysine, (%)	Animal
ε-N-formyl-L-lysine	~50	rat
ε-N-acetyl-L-lysine	~50	rat, chick
ε-N-(γ-glutamyl)-L-lysine	~100	rat, chick
ε-N-(α-glutamyl)-L-lysine	~100	rat
ε-N-glycyl-L-lysine	~80	rat
α-N-glycyl-L-lysine	~100	rat
ε-N-(N-acetylglycyl)-L-lysine	0	rat
ε-N-propionyl-L-lysine	0	rat
ε-N-propionyl-L-lysine	~70	chick

ingestion by rats. The nutritional availability of lysine was found to be 77, 67, and 43%, respectively, as compared to the initial lysine content [94]. Deacylation could occur both in the intestine and in the kidney (if acylated peptides were absorbed). A high amount of bound lysine was found in the urine.

It is interesting to note here that acetylation and succinylation of various food proteins, including those of fish and microorganisms, has been suggested as a means of improving solubility and/or functional properties [95]. In addition to the ε-amino groups, SH groups, and OH groups of tyrosine, serine, and threonine may also be acylated.

In a further study, Varnish and Carpenter [96,97] compared four protein samples: fully or nonpropionylated lactalbumin, and chicken muscle autoclaved or not at 116°C for 27 hr. Propionylation decreased the FDNB-reactive lysine content by 100% (the decrease upon autoclaving was only 40%). The amino acid content was unchanged except for a 20% loss in cystine in both cases. Propionylation decreased the nutritional availability (chick growth) of lysine only, by 50%. Autoclaving decreased the availability of most amino acids, including that of chemically inert leucine; a 50% decrease was noted for lysine and tryptophan, 30% for methionine. The digestibilities of crude protein and most individual amino acids were determined by the ileal technique, that is, they were all reduced by 10% by propionylation and 40% by autoclaving. It is of interest to note that the acylation of ε-amino groups, known to prevent proteolytic cleavage by trypsin at the carboxyl end of lysine, does not significantly reduce the protein or the lysine digestibility in vivo. However, there is a large amount of nonavailable lysine in the absorbed peptides. Mauron [1] has similarly indicated that in a sample of overheated milk powder, where the availability of lysine but not of other amino acids (see Sec. IV.A) has been reduced by 75% of its initial value, the digestibility in vitro by pancreatin was reduced by only 12%.

It is therefore clearly apparent that the effects of severe autoclaving cannot be explained simply by an acylation of ε-amino groups of lysyl residues. However, they may be explained by the formation of intra- and intermolecular cross-links of ε-N-(γ-glutamyl)lysine or of ε-N-(δ-aspartyl)lysine since these cross-links do not appear to be digested in the gastro intestinal tract [18]. The presence of many of these cross-links may therefore prevent easy access of proteolytic enzymes by steric hindrance and slow down or even reduce proteolytic attack. Such a mechanism may be partly responsible for the concomitant decrease in availability of several amino acids in severely heated proteins. A time delay in the release of some amino acids, and the absorption of unavailable peptides, could explain why, in some cases, the protein digestibility remains higher than the availability of essential amino acids. As suggested by Carpenter [4], it would be interesting to study the digestibility of proteins where ε-N-(γ-glutamyl)lysine cross-links have been introduced without the use of heat.

Other cross-links may be formed as a result of the heat treatment of protein foods containing little carbohydrate, and may accentuate the steric hindrance effect [46]. Mauron [1] has hypothesized that severe heating could induce the formation of imide, ester, and thioester cross-links:

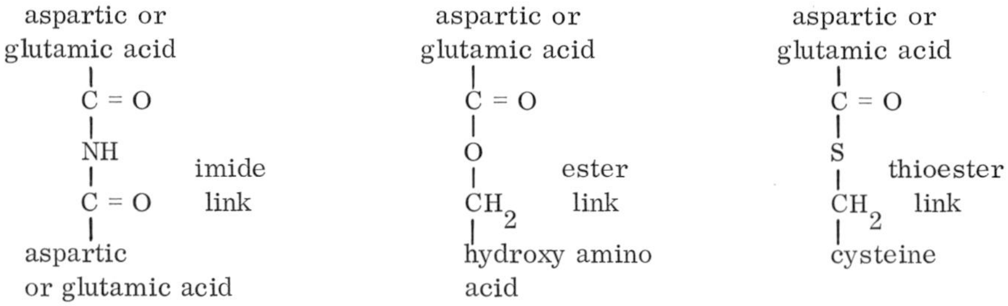

```
  aspartic or              aspartic or            aspartic or
 glutamic acid            glutamic acid          glutamic acid
       |                        |                      |
     C = O                    C = O                  C = O
       |                        |                      |
      NH                        O                      S
       |      imide             |      ester           |    thioester
     C = O     link            CH2      link          CH2      link
       |                        |                      |
  aspartic                   hydroxy amino          cysteine
  or glutamic acid           acid
```

Such links have not yet been found in heated proteins. They would not be resistant to acid hydrolysis.

A recent finding of great interest is the formation of lysinoalanine cross-links known to occur in alkali-treated proteins (see Sec. III.D) during the heat processing of a variety of foods that have never been submitted to a pH above neutrality [47]. These foods include condensed milk, acid casein, cooked chicken thigh, and sirloin steak pan scrapings. Similar results

TABLE 5 Lysinoalanine Content of Heat-Treated Proteins

Protein food or protein	Conditions of heat treatment in the laboratory	Lysinoalanine[a] (μg/g protein)
Strasbourg sausage	no treatment	0
	boiling water, 10 min	50
Evaporated milk	no treatment	700
Canned corned beef	no treatment	<25
Dried skim milk	no treatment	<25
Whipping agents[b]		
Brand A	no treatment	30 000 (20 000)
Brand B	no treatment	5 000
Bovine serum albumin[c]	120° C, 6 hr, pH 6	8 000
Casein[c]	NaOH 0.2M, 75°C, 1 hr	1 000 — 1 500 (1 300)
Sodium caseinate		
Brand A[c]	120° C, 6 hr, pH 6	600
Brand B[c]	120° C, 6 hr, pH 6	200

Source: Sternberg et al. [118].
[a]Determined by thin-layer chromatography. Numbers in parentheses refer to determination by ion exchange chromatography.
[b]From whey proteins spray-dried at high pH.
[c]Pure proteins were heat-treated as a 0.5 or 1% aqueous solution, casein as a 7% solution.

were found by Aymard [98] in our laboratory (see Table 5). It would be useful to investigate if lanthionine and other cross-links found in alkali-treated proteins (see Sec. III.D) occur as a result of heat treatment.

Another question of interest concerns the possibility of a direct alkylation of the methyl-thio group of methionine during the heat processing of proteins [29,39]. The nutritional availability of protein-bound or free methionine sulfonium derivatives is not known. Lipton and Bodwell [99,100] showed that methionine sulfonium derivatives can be desulfurized quantitatively (at 80°C in the presence of dimethyl sulfoxide) into homoserine lactone and a methyl sulfide. This reaction could be used for the determination (after S-alkylation) of methionine residues not initially substituted on the methyl-thio group ("available methionine"?).

$$R-\overset{+}{S}(CH_3)-CH_2-CH_2-CHNH_2-C(=O)-OH \xrightarrow{-H^+} \underset{\text{homoserine lactone}}{CH_2-CH_2-CHNH_2-C(=O)-O\text{ (ring)}} + R-S-CH_3$$

It is now appropriate to question whether or not cross-linking and nutritional damage demonstrated in model systems also take place during domestic or industrial heat processing of protein foods. Meat canning was found to cause very little damage, the decrease in digestibility and nutritive value always being less than 10%, and the reduction in the availability of methionine and lysine even smaller [1]. However, there is a lack of recent data with improved nutritional techniques. Milder heat treatments corresponding to normal domestic cooking did not induce any appreciable loss in nutritive value in meat or fish [101]. It was necessary to keep fish for 35 min in an oven at 230°C to decrease the nutritional value as measured by rat growth.

Nonetheless, fish, meat and bone meals, and — as already reported — oilseed meals, can be seriously damaged when the heat processing is not properly controlled. Well-processed fish meals possess a high lysine and methionine availability, similar to soy meals, whereas meat and bone meals are lower in this respect [102]. Exact knowledge of the availability of these essential amino acids in such meals is necessary for the precise calculation of low-cost rations for poultry and pigs.

Commercial fish meals and flours frequently display a marked variability in protein quality [1,4]. First, the content in lysine and methionine may vary, depending on the proportion of collagen from skin and bones; second, the oil content may be high or low, depending on the type of extraction procedures. Different drying equipment and techniques are also used, such as flame, steam tube, or rotary vacuum dryers, at temperatures ranging from 50 to 500° C. As a result, NPU are found to vary from 90 to 50. Lipid oxidation of the oil, when it occurs, decreases the nutritional value (see Sec. IV.B). Both overheating during drying and long storage of defatted meals are known to reduce the overall digestibility of the protein and the availability of lysine, methionine, and other amino acids. The effect of long storage is not entirely understood. Chemical tests for lysine availability were found to correlate with nutritional value only when severe damage had occurred [4,103].

Finally, it should be mentioned that roasting or frying of protein foods at high temperatures induces numerous breakdowns of peptide bonds and amino acid side chains together with condensation reactions. Various flavors may be produced.

D. Lysinoalanine and Other Cross-Links Caused by Alkaline Treatments

Alkaline treatments of food proteins are being increasingly used for protein solubilization and purification in order to prepare vegetable and other protein concentrates and isolates. Viscous alkaline solutions of soy and other proteins are used for making spun fibers. Partial alkaline hydrolysis has been used as a means of solubilizing fish protein concentrate, feathers, hooves, and hair. Chemical peeling of grain with sodium hydroxide has been advocated. Treatment of contaminated peanut or cottonseed meals with gaseous ammonia has been suggested for aflatoxin destruction. Preservation of fish with ammonia has also been suggested. Alkali cooking of maize is a traditional procedure to increase digestibility. Alkaline processing is used in the manufacture of hominy, pretzels, sausage casings, and some whipping agents. The pH of egg white may reach 9 during refrigerated storage [47,48].

1. Formation of New Amino Acids and Cross-Links

It is well known that severe alkaline processing of proteins provokes a progressive destruction of various amino acids, with cystine, arginine, threonine, and serine being the most sensitive.

Various investigators have shown that some modifications of amino acid residues take place with the simultaneous formation of new amino acids such as lysinoalanine [104,105], lanthionine [106], and ornithinoalanine [107], responsible for intra- or intermolecular cross-links [108].

Lysinoalanine, DL-α-amino-β-(ϵ-N-L-lysyl)propionic acid, has been found in acid hydrolyzates of numerous proteins (including keratin from

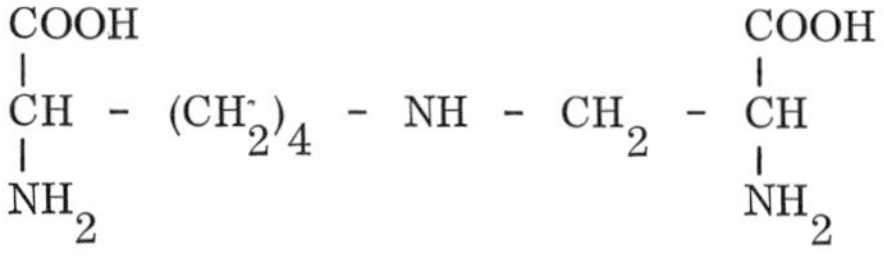

$$\begin{array}{ccccccccc} COOH & & & & & & & & COOH \\ | & & & & & & & & | \\ CH & - & (CH_2)_4 & - & NH & - & CH_2 & - & CH \\ | & & & & & & & & | \\ NH_2 & & & & & & & & NH_2 \end{array}$$

wool and soybean protein isolate) previously submitted to an alkaline treatment [48,109-111]. Lysinoalanine forms through condensation of ϵ-amino groups of lysyl residues with dehydroalanyl residues [104,105].

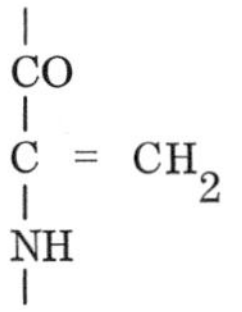

$$\begin{array}{lll} | & & \\ CO & & \\ | & & \\ C & = & CH_2 \quad \text{dehydroalanyl residue} \\ | & & \\ NH & & \\ | & & \end{array}$$

The latter probably come from the alkaline degradation of cystine residues through β-elimination and desulfurization with H_2S formation. They may also come from the decomposition of serine [48] or phosphoserine residues [105,112]. Lysinoalanine has also been found in heated foods that have not been submitted to any alkaline treatment [47] (see Sec. III.C).

Lanthionine has been identified in acid hydrolyzates of wool and silk submitted previously to an alkali treatment [48]. It forms by condensation

$$\begin{array}{c} COOH \\ | \\ CH \\ | \\ NH_2 \end{array} - CH_2 - S - CH_2 - \begin{array}{c} COOH \\ | \\ CH \\ | \\ NH_2 \end{array}$$

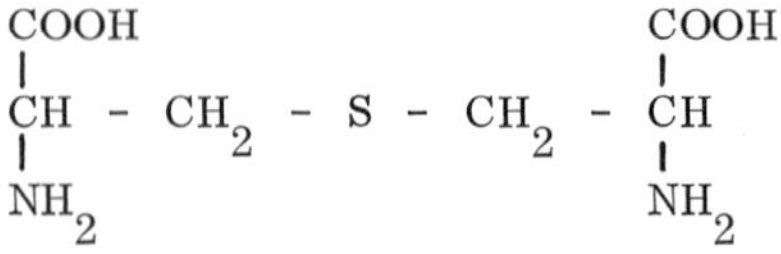

of dehydroalanyl and cysteinyl residues [113].

As suggested by Nicolet and Shin [114,115], a disulfide bond in a cystine residue may split to give rise to both a dehydroalanyl and a cysteinyl residue. Two reaction mechanisms have been suggested:

a) A disulfide ion may be formed, and transformed into a cysteinyl residue, after splitting of H_2S.

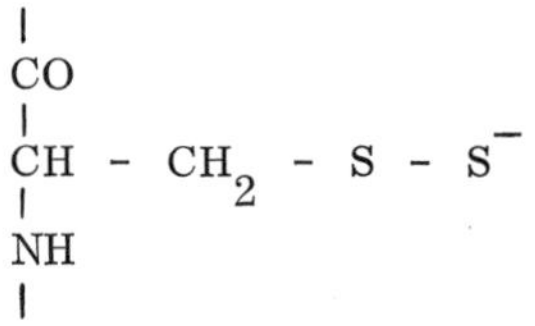

$$\begin{array}{c} | \\ CO \\ | \\ CH \\ | \\ NH \\ | \end{array} - CH_2 - S - S^- \quad \text{disulfide ion}$$

b) The disulfide bond may split into a thiol (cysteine) and a sulfenic acid residue [116]. This may occur in weakly alkaline media. The sulfenic acid residue may decompose into a dehydroalanyl residue.

Ornithinoalanine has been identified in acid hydrolyzates of heat and carbonate-treated wool and silk. It forms by condensation of dehydroalanyl and ornithine residues. The latter may come from alkaline decomposition of arginyl residues.

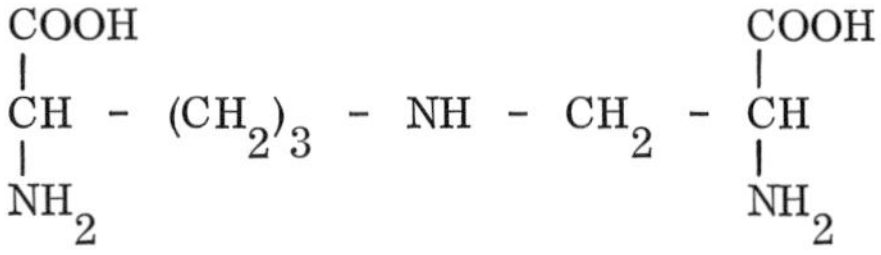

$$\begin{array}{c} COOH \\ | \\ CH \\ | \\ NH_2 \end{array} - (CH_2)_3 - NH - CH_2 - \begin{array}{c} COOH \\ | \\ CH \\ | \\ NH_2 \end{array} \quad \text{ornithinoalanine}$$

Dehydroalanyl residues may also condense with residues from asparagine or glutamine. Such bonds, when submitted to acid hydrolysis, give the dicarboxylic acid and β-aminoalanine.

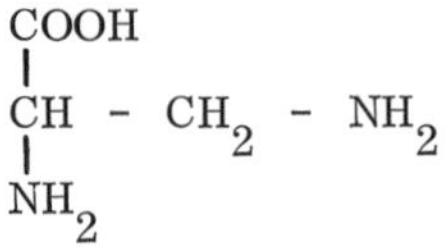

$$\begin{array}{c} COOH \\ | \\ CH \\ | \\ NH_2 \end{array} - CH_2 - NH_2 \quad \beta\text{-aminoalanine}$$

It appears, however, that the latter more frequently derives from the reaction of dehydroalanyl residues with ammonia [109].

Lysinoalanine, lanthionine, ornithine (and possibly ornithinoalanine) and alloisoleucine (from isoleucine) have been identified in a sunflower protein isolate treated with 0.05-1 M NaOH (55-80°C, 1-16 hr) [48,117]. The formation of ornithine and the destruction of arginine are exactly proportional and appear to be the best indicators for the severity of the alkaline treatment. Lysinoalanine was present in the initial sunflower protein isolate (prepared by mild alkali solubilization).

It is likely that the formation of lysinoalanine, lanthionine, ornithinoalanine, and β-aminoalanine in a protein depend not only on the content of lysine, cystine, etc., but also on the relative spatial position of these amino acid residues.

The chromatographic behavior of most of the inhabitual amino acids in a given ion-exchange chromatography system is given in Fig. 2 [18]. Damage to proteins from alkaline processing is relatively easily assessed by such chromatography after acid hydrolysis. Lysinoalanine and lanthionine can also be determined by thin-layer chromatography [118].

Treatment of peanut meal with gaseous ammonia at 2-3 bars for 15-30 min intended to decompose most of the contaminating aflatoxin was found to reduce the cystine/cysteine content by 10-40%; the reduction was more sensitive to the water content of the meal (6-15%) than to pressure or contact time [119].

2. Nutritional Effects

In most experiments, alkali-treated proteins were found to have a decreased nutritional value. The loss or substitution of cystine, lysine, or other amino acids may account for this decrease, depending on the limiting amino acid in the diet. Amino acid isomerization (see Sec. II.C) may also cause a decrease in availability of some amino acids.

It is likely that inter- or intramolecular cross-links due to lysinoalanine, lanthionine, ornithinoalanine, and β-aminoalanine reduce the accessibility of the protein to proteolytic enzymes. Lysine substitution and isomerization are likely to inhibit the activity of trypsin. The in vitro digestibility of NaOH-treated sunflower proteins (release of free amino acids by pronase) was markedly reduced [48]. It is interesting to note that lysinoalanine was partially released. The nitrogen digestibility of severely NaOH-treated casein (0.2 and 0.5 M, 80°C, 1 hr) was determined on rats and found to be 71 and 47%, respectively, as compared to 90% for intact casein [120]. Lysinoalanine (probably in protein-bound form) could be detected in the kidneys and in the liver, but not in the blood of the rats. However, most lysinoalanine was present in the feces. In vivo protein digestibility of mildly NaOH-treated soybean proteins was much less markedly affected [110].

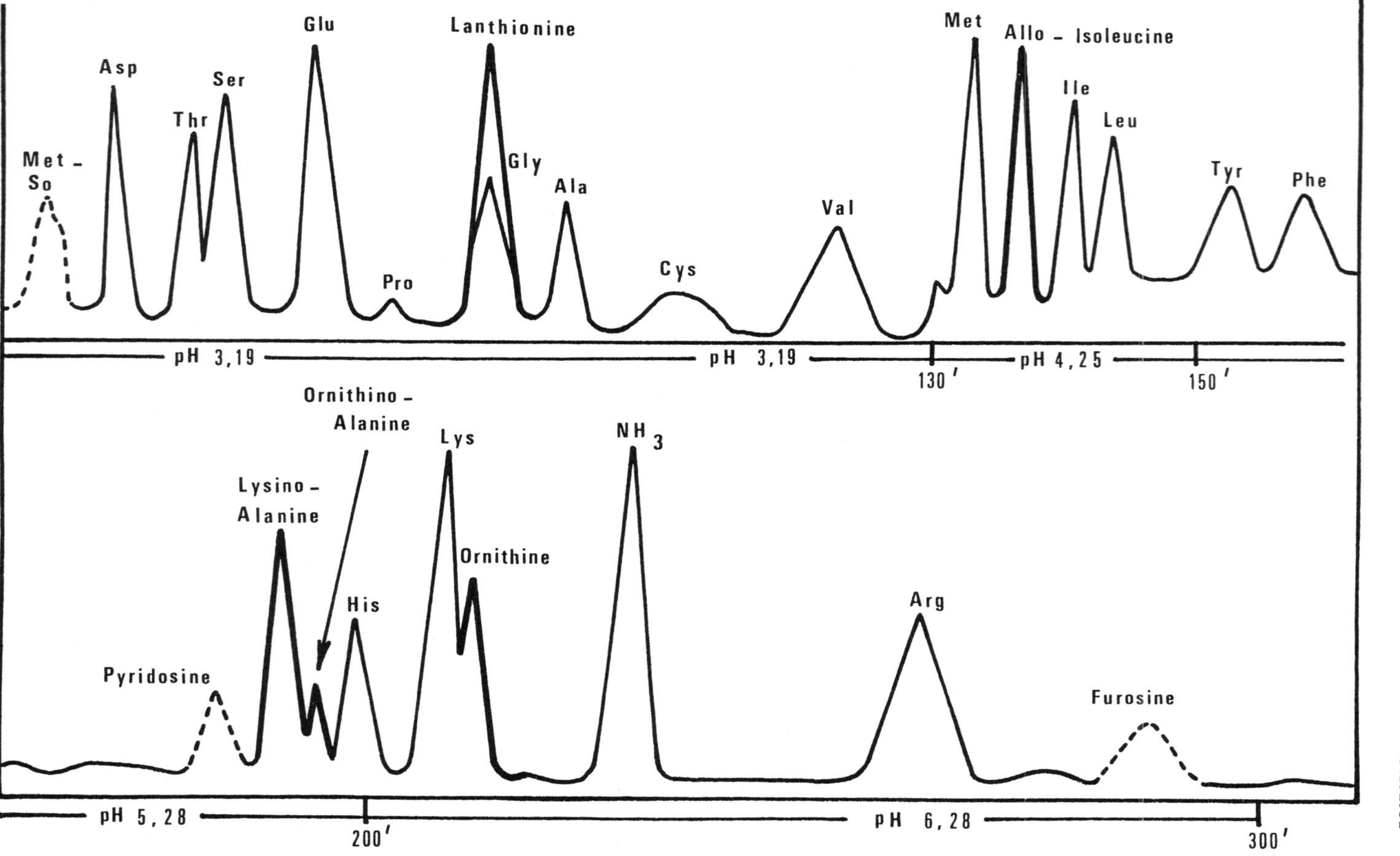

Fig. 2. Separation of unusual amino acids by ion-exchange chromatography (from Ref. 18).

Severely alkali-treated herring meals do not support adequate growth in chicks and induce toxic effects at high levels of feeding [121]. Growth retardation has also been reported in lambs fed NaOH-dispersed soybean protein concentrates [122]. A decrease in the NPU of NaOH-treated soybean proteins was observed in experiments with rats [110].

The possible toxicity of the latter proteins has been studied by feeding rats for 13 weeks with a diet containing 20% of proteins previously treated with NaOH for 4 hr at pH 12.2 and at $40^\circ C$ [110,123]. No histological or clinical particulars were noted, except for increased kidney weight and nephrocalcinosis in females. The latter appeared to be due to high dietary phosphorus and could be inhibited by adding calcium to the diet. However, other investigators feeding weanling rats for 3-4 weeks with a similar diet reported cytoplasmic and nuclear enlargement of renal tubular cells, but no nephrocalcinosis [124]. Similar cytomegalic renal lesions were noted in female weanling rats fed a diet containing up to 0.3% of synthetic lysinoalanine [125,126]. Renal lesions were more marked when more lysinoalanine was given (either by stomach tube or by intraperitoneal injection), showing that absorbed lysinoalanine is the nephrotoxic agent.

The alkali-treated soybean proteins of both groups of investigators contained similar amounts of lysinoalanine residues (0.8 and 0.6% respectively). More recently De Groot et al. [127] also found marked nephrocytomegalia, plus tubular nephrosis and necrosis, when rats were fed acid-hydrolyzed alkali-treated soy proteins or synthetic lysinoalanine. This appeared to be specific to the rat and was not observed with mice, hamsters, quails, dogs, or monkeys.

At this point, it can be tentatively concluded that mild alkaline treatment of protein is acceptable, but that the processing should remain carefully controlled.

IV. BINDING OF NONPROTEIN MOLECULES TO AMINO ACID RESIDUES

A. Reaction with Carbohydrates

1. Maillard Reactions and Nonenzymatic Browning

The main features of the reactions between proteins and carbonyl compounds, particularly reducing sugars, are that they result in the binding of ϵ-amino groups of lysine residues, and that they may occur even during storage at ordinary temperature.

The scheme of Fig. 3 summarizes Maillard reactions and particularly the initial interactions between reducing sugars and ϵ-amino groups of lysine residues, up to the formation of "Amadori products," with little or no browning. Later reactions, which may occur without delay, induce the formation of reactive polycarbonyl compounds, which polymerize and cross-link protein chains, with concomitant intense browning [1,4,67, 128-130].

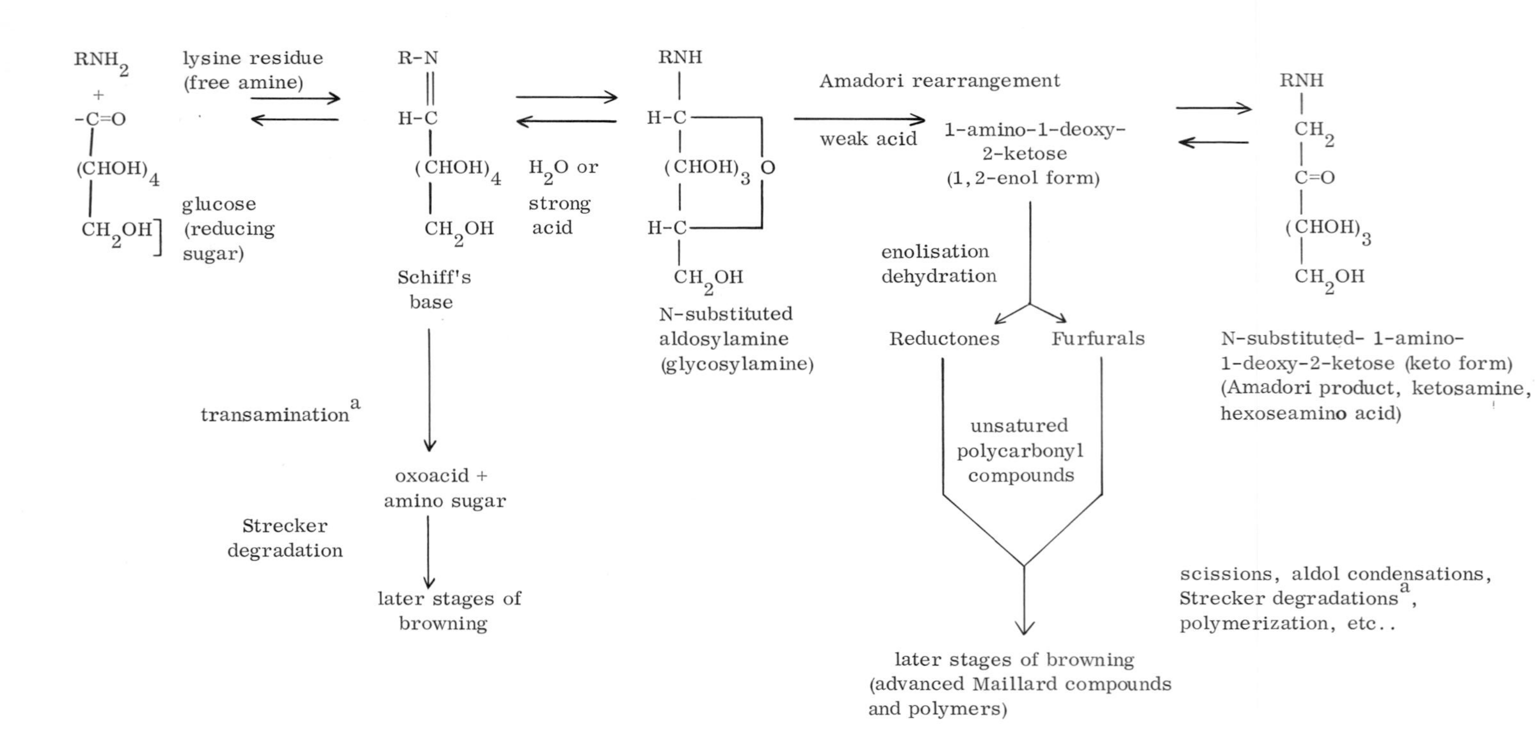

Fig. 3. Simplified scheme of Maillard reactions.

[a]Transamination leading to irreversible destruction of lysine (not regenerated even by HC1 hydrolysis).

Maillard reactions have relatively high energy of activation. Thus, when protein foods containing reducing sugars are moderately heated (as in the drying of milk), Maillard reactions occur in preference to any other type of protein damage. Protein foods containing small amounts of reducing sugar and submitted to a more severe heating (e.g., toasted breakfast cereals, bread crust, some biscuits, oilseed meals, or glucose-protein model systems at high temperature) may undergo both late Maillard-type and protein-protein damage (see Sec. III.C). In such foods, there is not only a loss of available lysine, but also an accompanying decrease in the overall nitrogen digestibility of the protein and therefore in the availability of most amino acids [36].

The water activity of the food, or the relative humidity above the food, influences Maillard reactions. The rate of reaction is maximal at a water activity (a_w) of 0.6-0.8 and decreases at higher concentrations of water (as in liquid milk). It is still appreciable at an a_w as low as 0.2 (corresponding to 5-10% water). This type of deterioration is therefore enhanced during the concentration and the dehydration of various protein foods (milk concentration and drum-drying, dehydration of egg white or whole eggs, drying of oilseed meals). Maillard reactions may also occur during the storage at ambient temperature of some dehydrated or intermediate moisture protein foods. The most efficient browning inhibitor, SO_2, which blocks carbonyl groups, can be added to some foods; for others it is necessary to control carefully temperature, duration, and relative humidity of both processing and storage. In some cases reducing sugars can be removed by fermentation or enzyme action (as in eggs); it may also be beneficial to acidify the food (see below).

Pentoses are usually more active than hexoses in promoting Maillard reactions. The nonreducing disaccharide sucrose will induce browning only after it is inverted to glucose and fructose by heat or by spontaneous hydrolysis in acid medium. However, low pH partly inhibits browning, since the first step of Maillard reactions lies in the condensation of carbonyl groups with nonionized ε-amino groups. Alkaline pH generally enhances Maillard reactions.

From the nutritional standpoint, initial and later Maillard reactions will be considered separately, since they do not induce the same type of damage. It must be understood, however, that both types of reactions may occur almost simultaneously in the same food.

Initial Maillard reactions lead to the substitution of the ε-amino groups of lysine residues and to a marked decrease in the availability of this amino acid. The nutritional value of lysine residues at the stage of glycosylamine is not precisely known, since glycosylamines are quite unstable and undergo Amadori rearrangement. It is supposed that when submitted to HCl hydrolysis, all the lysine from glycosylamines is regenerated into free lysine [1]. Amadori products (mono- or dideoxyketosyllysine) are the main substituted lysine derivatives in food or model proteins undergoing initial Maillard reactions. The ε-N-(1-deoxy-lactulosyl)lysine residues represent 70-75% of Maillard compounds in samples of overheated milk, the remaining 25-30% being mainly in the form of glycosylamine and Schiff's base [130].

Synthetic ε-N-(1-deoxy-D-fructosyl)-L-lysine, corresponding to the reaction of glucose with lysine, is totally unavailable as a source of lysine for the rat. It is extensively absorbed from the intestine and then excreted in the urine [130]. ε-N-(deoxyfructosyl)lysine residues from glucose-protein model systems appear to be partly released during digestion in vitro, absorbed, and then excreted in the urine in the form of unavailable peptides [89,131,132]. This could partly explain the frequent high protein digestibility of Maillard-damaged proteins with a very low content of available lysine. Deoxylactulosyllysine residues from milk proteins reacted with lactose are found in the feces, however, rather than in the urine [130]. ε-N-(Deoxyketosyl)lysyl peptides appear to be partly metabolized by the intestinal flora [130,133]. This metabolism may also explain the apparent high digestibility of proteins in the early stages of Maillard reactions [1,134] (see below). ε-N-(Deoxyfructosyl)-lysine regenerates 45-50% of its constitutive lysine during HCl hydrolysis [135]. The behavior of these initial Maillard derivatives of lysine residues during the main chemical tests used for the determination of lysine availability is reviewed in Sec. IV.A.2.

α-N-Amadori products such as α-N-(1-deoxyfructosyl)-L-methionine or tryptophan or leucine have also been synthesized, and their biological behavior has been studied. They are absorbed, but unavailable to rats [136,137]. This is probably of little nutritional importance, since the formation of these compounds in foods is limited to free or N-terminal amino acids.

In many studies of protein-glucose model systems kept at 37°C for various periods of time at low relative humidities, it was found that the nutritive value of the protein was severely decreased, although its digestibility was much less affected [4]. The addition of lysine to the diet fully restored the nutritional value. Schematically, in such cases of initial Maillard reactions, the protein damage can be exclusively ascribed to the unavailability of lysine residues with attached carbohydrates [138]. Some other amino acid residues, primarily cysteine, may react with reducing sugars, but to a much smaller extent than lysine.

The classic food example of such initial Maillard reactions is drum-dried milk powder. It has been well established by scientists at Nestlé that drum-drying can be markedly detrimental to the nutritive value of milk powder (10-40% drop in available lysine, depending on processing equipment and conditions), whereas properly conducted spray-drying does not decrease the availability of lysine [1] (see Sec. IV.A.2). Whey proteins are specially sensitive. Other examples include biscuits enriched with milk powder, sterilized milk, and evaporated milk [139]. In sterilized but not in pasteurized milk, the availability of lysine decreases by 10-20%; a partial destruction of cystine also occurs. As can be expected, usual nutritional evaluation with rats shows a deficiency of sulfur-containing amino acid when little damaged milk proteins are fed as the only protein source. Lysine unavailability only appears when the damage is more extensive, or when the overall diet is made more limiting in lysine.

In the more advanced stages of nonenzymatic browning, reactive unsaturated polycarbonyl compounds (reductones, etc.) are formed from the reducing sugars. These reactive compounds polymerize and bind simultaneously to α-terminal, ε- and other amino groups of different polypeptide chains, bringing about the formation of colored, high-molecular-weight, highly cross-linked protein-carbohydrate polymers with low solubility, digestibility, and nutritional value. Clark and Tannenbaum [140] submitted such polymers from an insulin-glucose model system (37 days storage at 37°C and 75% relative humidity) to extensive proteolysis. They were able to isolate and determine the partial structure of "limit peptides" with molecular weights up to 8000 (including about 31 sugar residues), thus proving that polymers of carbohydrate residues or derivatives constitute actual cross-links between polypeptide chains. Reactive unsaturated polycarbonyl compounds probably also bind to arginine, histidine, serine, and tryptophan. "Limit peptides" may contain various essential amino acids, including lysine residues with free ε-amino groups, and may be largely undigestible in vivo. This would explain why in proteins in an advanced stage of browning reactions, the overall protein digestibility and nutritional value may be reduced. Several amino acids, and not only lysine, may become unavailable [131,141].

In a study of the nutritional value of protein autoclaved in the presence of various reducing sugars, it was found that egg proteins are much more sensitive to damage than soy proteins or casein [142].

Acid hydrolysis of severely damaged proteins regenerates much less lysine than early stages of Maillard reactions. In addition, during these later stages of nonenzymatic browning splits in the polypeptide chains occur, and free amino acids can be decarboxylated and later deaminated (Strecker degradation).

Late nonenzymatic browning reactions occur extensively in heated or long-stored model systems of proteins with reducing sugars, and also in a number of heated protein-containing foods, especially in baked goods such as various types of breads, biscuits, cookies, cakes, and toasted or puffed breakfast cereals. They are largely responsible for the desirable color and flavor of these foods, and therefore their effect on the nutritional value (quite detrimental in the case of several breakfast cereals) is not considered to be a problem. However, control of baking temperature and sugar content is recommended in order to avoid large decreases in the availability of lysine. Added synthetic lysine is similarly lost. In foods such as milk powders, browning, which may occur during improper storage, renders the product totally unacceptable for consumption. It is likely that browning reactions are also responsible for the moderate decrease in nutritive value and available lysine observed in various cereals and oilseeds stored for several months or years under somewhat inadequate conditions of temperature and relative humidity.

From a toxicological standpoint, Adrian and co-workers [143-146] have shown that some soluble brown melanoidins, prepared by autoclaving mixtures of free amino acids with reducing sugars, are biologically active substances able to stimulate or inhibit rat growth and disturb reproduction when given in the diet. These findings are probably relevant only in the case of foods containing appreciable amounts of free amino acids, such as protein hydrolysates, flavor precursors, amino acid supplemented cereals or some types of single-cell proteins [18]. Compounds from Maillard reactions in heated proteins are only partly digestible and absorbed (see above) and do not influence animal or infant growth, as long as enough available lysine is present or added to the diet [18].

Chichester and co-workers [147,148] have reported a preliminary study of the toxic effects of heated mixtures of amino acids and glucose, of pure browning model compounds such as α-N-(deoxyfructosyl)leucine or tryptophan, of apricot browning pigments, and of browned egg albumin, when added to the diet of rats. Diarrhea, growth retardation, inhibition of protein digestion and of amino acid absorption, modification of liver enzymes, etc., were found to occur. Most of these effects were reversible.

Dillard and Tappel [149] studied the possible occurrence of carbonyl-amine reactions in vivo. It appears that no fluorescent brown products are formed in blood from glucose and plasma proteins. It would be interesting to investigate the possible presence of browning inhibitors in blood.

2. Chemical Methods for the Determination of Lysine Availability

It has been reported in Secs. III.C and IV.A.1 that either through severe heating or through reactions with carbonyl compounds, lysine residues of food proteins can become almost completely unavailable to animals (rat or chick assays). However, the total lysine content, as determined after acid hydrolysis, is never reduced to such a large extent.

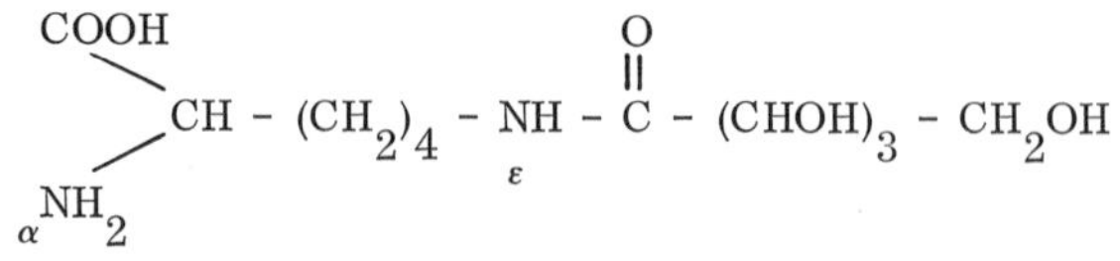

$$\begin{matrix} COOH \\ \quad\diagdown \\ \quad\diagup \\ {}_{\alpha}NH_2 \end{matrix} CH - (CH_2)_4 - \underset{\varepsilon}{NH} - \overset{\displaystyle O}{\overset{\|}{C}} - (CHOH)_3 - CH_2OH$$

lysine deoxyfructose

ε-N-(1-Deoxy-D-fructosyl)-L-lysine, a pure synthetic model compound of initial Maillard reactions (see Sec. IV.A.1) which has no nutritional value as a source of lysine, can regenerate lysine with a yield of approximately 50% on hydrochloric acid hydrolysis [135]. It is likely that many ε-amino substituted lysine residues in heated or otherwise treated proteins similarly regenerate lysine during protein hydrolysis, although they are almost unavailable nutritionally in vivo.

Various investigators have therefore attempted to evaluate the nutritional availability of lysine by the chemical determination of either the ε-N-substituted or, more frequently, the nonsubstituted lysyl residues. Only the principal methods of determination will be briefly reviewed [4,135,138].

a. The Direct Fluorodinitrobenzene Method (Carpenter's Method) [150]. The reaction of free ε-amino groups of lysine residues with FDNB is followed by the colorimetric determination of the ε-N-dinitrophenyllysine (DNP-lysine) released from the protein by acid hydrolysis, and extracted. Various interferences, mainly from carbohydrates, and partial degradation of DNP-lysine during acid hydrolysis can be overcome [151]. Some free ε-amino groups may be sterically inacessible to FDNB, however.

In addition, Mauron and co-workers [1,135], in contrast to Hurrell and Carpenter [138], find that synthetic α-N-formyl-(ε-N-1-deoxy-D-fructosyl)-L-lysine (FFL) and α-N-formyl-(ε-N-1-deoxy-D-lactulosyl)-L-lysine (FLL) partly react with FDNB. These models of initial Maillard reactions between lysine residues and glucose or lactose possess a blocked α-amino group, and probably react with FDNB at the ε-nitrogen and/or at the keto group. When FFL and FLL are treated with FDNB and then submitted to hydrochloric acid hydrolysis, 6-10% of their constitutive lysine is regenerated as free lysine, and 6-30% appears as DNP-lysine (see Fig. 4). In practice, however, this does not hamper the value of the direct method for the evaluation of lysine damage in protein foods which have undergone Maillard reactions [18].

The direct method has been used with success by Mauron and co-workers for the evaluation of lysine availability in scorched roller-dried milk powder. The results were in close agreement with those of in vivo (rat) and in vitro enzymatic (pepsin — "pancreatin" test) evaluations (see Table 6) [152]. Good results were also obtained with this method in a collaborative study for the evaluation of dried fish and meat meals [153].

b. The Indirect FDNB Method. As in the previous method, free ε-amino groups of lysine residues are reacted with FDNB. The difference in free lysine content between the protein acid hydrolyzed before and after reaction with FDNB supposedly corresponds to FDNB-reactive lysine (and therefore hopefully to nutritionally available lysine) [4,138,154,155]. This method (also called "Silcock's" method) gives an accurate result only

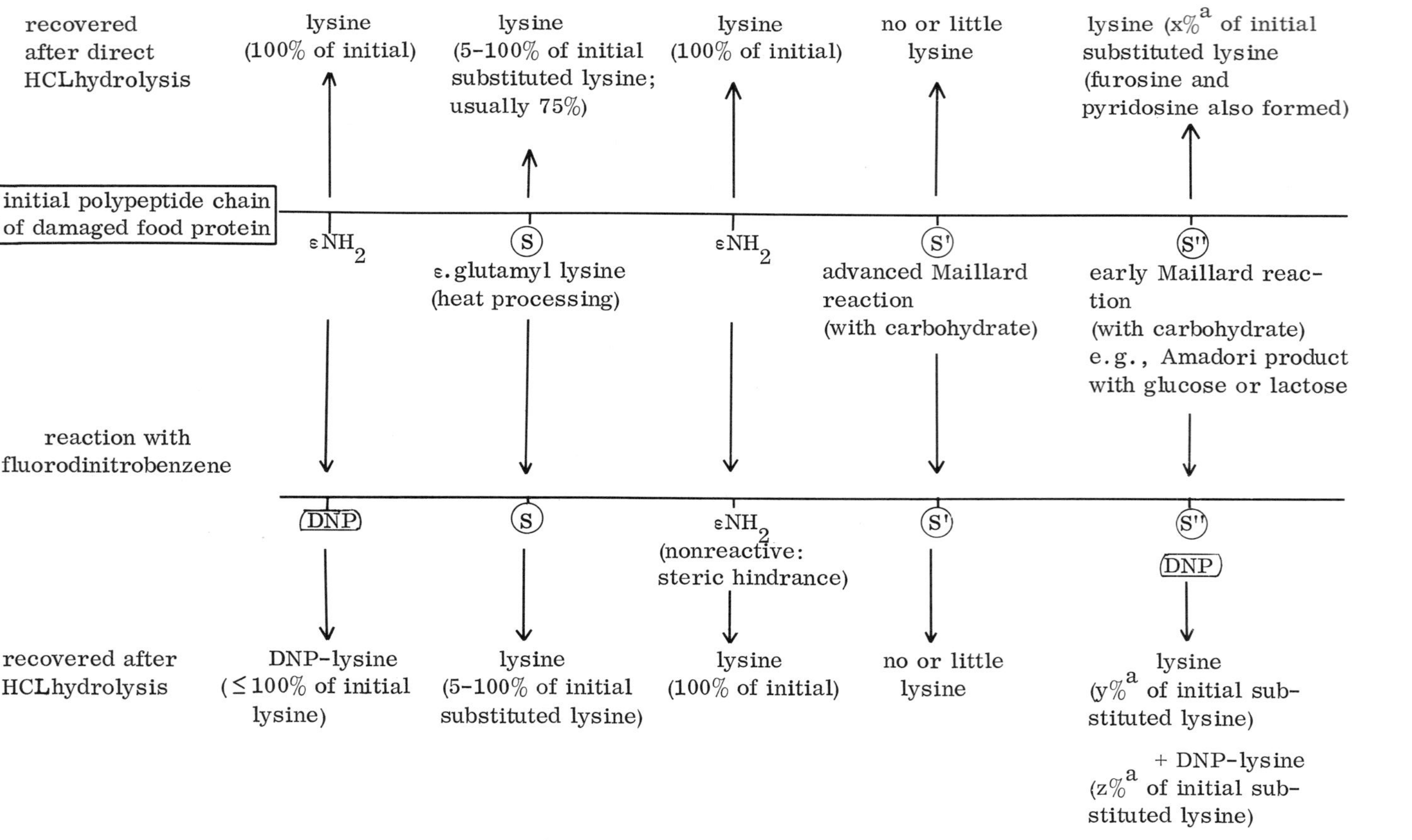

Fig. 4. Behavior of substituted lysine residues during HCl hydrolysis before or after reaction with fluorodinitrobenzene (Regeneration of lysine; formation of DNP-lysine).

[a] FFL $x = 40\text{-}50\%$, $y = 6\text{-}10\%$, $z = 15\text{-}30\%$; FLL $x \cong 50\%$, $y = 6\text{-}10\%$, $z = 6\text{-}17\%$; $y\%$ probably always $< x\%$

TABLE 6 Lysine Content and Availability in Concentrated or Dried Cow's Milk

Method preparation	Total lysine[a] (g/16gN)	FDNB reactive	Available from in vitro proteolysis	Available from rat growth assay
Freeze-dried	8.3	8.4	8.3	8.4
Spray-dried	8.0	8.2	8.3	8.1
Evaporated	7.6	6.4	6.2	6.1
Roller-dried				
(mild)	7.1	4.6	5.4	5.9
(severe)	6.1	1.9	2.3	2.0

Source: Mottu and Mauron [152].
[a]Acid hydrolysis.

if initially substituted lysine residues regenerate the same percentage of lysine upon hydrochloric acid hydrolysis, whether or not the protein has been reacted with FDNB. This is the case with heat-treated animal protein foods of low carbohydrate content, or with either unheated or severely heat-processed vegetable proteins. However, it is not the case for mildly heat-treated reducing sugar-protein mixtures (mildly damaged milk powder, glucose-protein model systems stored at 37°C, etc.) (as demonstrated with FFL and FLL; see Fig. 4) leading to an overestimation of FDNB-reactive and nutritionally available lysine.

c. The Methylisourea Method. The protein is reacted with O-methylisourea. The lysine residues with free ε-amino groups are quantitatively guanidinated into residues of homoarginine, and these are released from the protein by acid hydrolysis, and determined by ion-exchange chromatography (Fig. 2). Homoarginine is generally eluted after arginine and furosine [18,135,138]. This method is lengthy but gives accurate results with all types of lysine substitution, including with derivatives from initial Maillard reactions. It was found to correlate well with the direct FDNB method and with enzymatic and animal tests performed on milk powder and various food proteins [18,138].

d. The Determination of Furosine. Furosine (ε-N-(-2-furosylmethyl)-L-lysine) can be determined by automatic ion-exchange chromatography of proteins initially heated in the presence of carbohydrates. Furosine occurs as one of the breakdown products of FFL and other derivatives from Maillard reactions [130]. Furosine and another basic amino acid, pyridosine, have been used as an index of Maillard-type unavailability of lysine in scorched drum-dried milk powder [130]. When determined with the amino acid analyzer, furosine is eluted between arginine and homoarginine, while pyridosine is eluted shortly before lysine [Fig. 2).

The relative proportions of furosine and pyridosine are constant and related to the amount of lysine which has become nutritionally unavailable in the form of 1-deoxy-2-ketose. Unavailable lysine can be taken as 4.31 x furosine content.

In later stages of Maillard reactions, however, it appears that this relationship does not hold, and that furosine can only be used as a qualitative indicator.

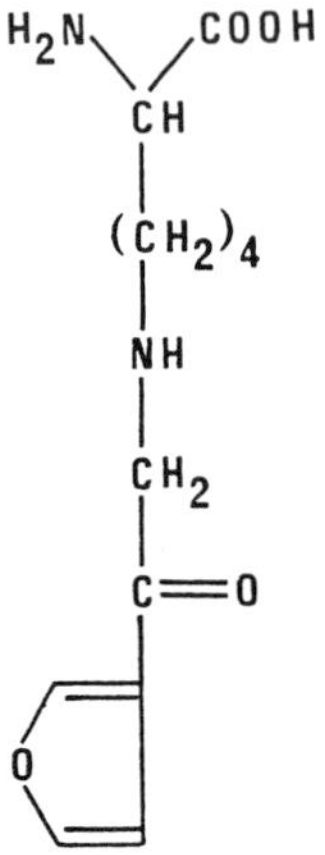

H_2N, COOH, CH, $(CH_2)_4$, NH, CH_2, C=O, O

furosine

H_2N, COOH, CH, $(CH_2)_4$, N, CH_3, HO, O

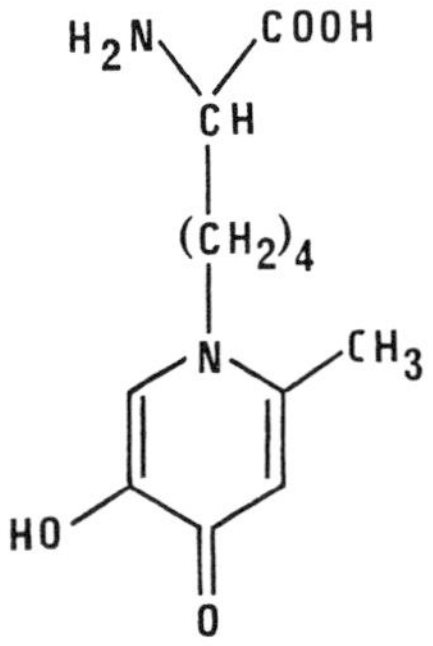

pyridosine

Hurrell and Carpenter [138] conclude from their comparative study that the direct FDNB and the methylisourea methods are the most accurate chemical methods, whatever the type of lysine substitution. However, these two methods may overestimate available lysine in the case of proteins highly cross-linked, either through ε-N-(γ-glutamyl)lysine linkages (severe heating, Sec. III.C) or through late browning reactions (overheating or inadequate storage in the presence of reducing sugars, Sec. IV.A.1). Such proteins may still have a number of FDNB-reactive lysine residues in spite of a poor digestibility and very poor nutritional value [36,131,141].

Other methods suggested [138] for the chemical determination of reactive lysine rely on the use of trinitrobenzene sulfonic acid, sodium borohydride and ethyl vinyl sulfone as reagents for the substituted or unsubstituted lysine residues.

It should also be mentioned that the binding of various types of dyes on proteins has been used for the rapid and inexpensive determination of protein quantity and quality, first for the evaluation of the lysine content of cereals in plant breeding, and more recently for the evaluation of available lysine [156].

Two types of dyes appear as most promising: acid (sulfonic) azo dyes, such as acid orange and amido black, which are adsorbed, at low pH, on the free basic groups of proteins, and reactive dyes, such as Remazol brillant blue (vinyl sulfonic dye), which reacts covalently with free ε-amino and free sulfhydryl groups of proteins, possibly also with hydroxy groups of tyrosine and serine at higher temperatures.

Dye-binding tests must be performed in strictly standardized conditions.

The following indications summarize a study of the correlation between protein dye-binding capacity and content of FDNB-reactive lysine [156]:

1. Proteins severely heated, with little or no sugar present (fish and meat meals; oilseed meals and protein concentrates; model systems including acylated proteins) show a positive correlation between DNP-lysine content and both acid orange and Remazol absorption.
2. Proteins involved in Maillard reactions with carbohydrates (milk powder model systems) show a positive correlation of DNP-lysine with Remazol absorption (less positive for Maillard reactions at early stages) and no correlation with acid orange absorption.

Acid orange binding is not decreased in proteins engaged in Maillard reactions. This can probably be explained by the fact that the ε-nitrogen of deoxyketosyllysine residues is still basic (secondary amine). Remazol blue should be used for such protein foods.

In contrast, ε-nitrogen engaged in ε-N-(γ-glutamyl)lysine amide cross-links is neutral and does not absorb acid orange. Cross-links may also prevent dye penetration between protein chains. In this case of severely heated proteins, acid orange is preferred to Remazol blue because it allows a more rapid determination, i.e., 15 min vs. 7 hr. With acid dyes, the correlation between available lysine and dye-binding is better in animal protein foods, where the ratio lysine/histidine + arginine is high, since the sensitivity of dye-binding to a decrease in the content of free ε-amino groups will also be high. However, the evaluation of fish and meat meals with varying raw materials is difficult, since variations in the content of collagen cause changes in the lysine content and in the lysine/total basic amino acids ratio.

It should be stressed that the use of dyes for the evaluation of lysine availability is still in a preliminary stage of study.

B. Reaction with Lipids

Lipoproteins consisting of noncovalent complexes of proteins and lipids or phospholipids occur widely in living tissues, as cellular and subcellular membranes, plasma lipoproteins, etc. Similar complexes considerably influence the physical and texture properties of various foods such as meat and fish, bread, milk products, and other food emulsions [157]. The lipid constituents can, in most cases, be fully extracted with solvents. The nutritional value of the protein constituents of lipoproteins is not generally affected by the presence of lipids.

There are cases, however, when covalent bonds are known to occur between lipids and proteins. Such interactions take place mainly as a result of the oxidation of unsaturated lipids.

The oxidation of unsaturated fatty acids is a free-radical reaction, the course of which is shown approximately in Fig. 5 [158,159]. The major initial products consist largely of lipid peroxides. As oxidation progresses, breakdown products of peroxides accumulate and eventually the peroxide concentration decreases.

Lipid oxidation and protein-lipid covalent interaction are known to take place in vivo in various aging tissues, where they can damage the integrity of protein polymers including collagen, enzymes, and membranes [160]. They can also take place in some foods and feeds, such as frozen or dehydrated fish, fish meal, oilseeds, etc. Although it is likely that foods having undergone extensive lipid oxidation are organoleptically unacceptable, it also appears that protein may act as a trap for lipid peroxides and their breakdown products. Some degree of protein damage may therefore occur before the food is rendered unacceptable.

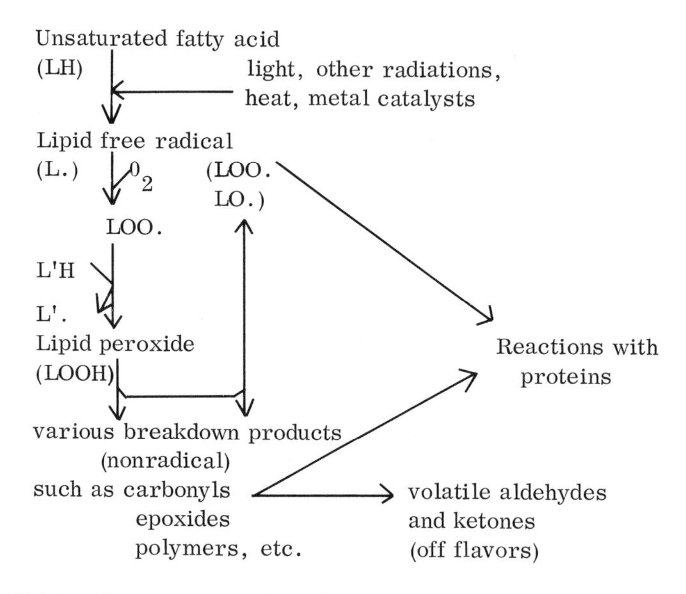

Fig. 5. Schematic representation of reactions linked to lipid oxidation.

The mechanisms of protein damage have been studied in various model systems where a single protein is reacted with an oxidizing lipid (generally an ester of linoleic acid) under given conditions of temperature, relative humidity, aeration, dispersion, length of time, presence of pro- or anti-oxidants, etc. Noncovalently bound lipid constituents are extracted from the mixture with solvents, and the following properties of the protein are determined: content of covalently bound lipid (by determination of nitrogen content); solubility (in water, dilute alkali, hydrogen bond breaking agents); polymerization or scission (by gel chromatography etc.); enzymatic activity (if relevant); color (or fluorescence, indicative of Schiff-base type reactions between amino groups and some carbonyl compounds); amino acid composition; proteolytic digestibility in vitro; nutritional value (PER, NPU, etc.) or toxic effects, after ingestion by animals.

In some studies, electron spin resonance determination of free radicals in the oxidizing (dry) model system has also been performed.

1. Mechanisms of Reaction

Two types of mechanisms appear to be at work in the covalent binding of peroxidizing lipids on proteins and the lipid-induced polymerization of protein: free-radical reactions and carbonylamine reactions.

a. Free-radical Reaction.

Lipid free radicals, such as L·, LO· (alcoxy), or LOO· (peroxy), react with protein (P_H), resulting in:

(1) Addition on to a protein:

$$LOO\cdot + P_H \longrightarrow \cdot LOOP$$

followed by polymerization through cross-linking of protein chains by multifunctional lipid peroxy free radicals:

$$\cdot LOOP + O_2 \longrightarrow .OOLOOP$$

$$\cdot OOLOOP + P_H \rightarrow \rightarrow POOLOOP \text{ (dimer) etc.}$$

This implies that the ratio of lipid incorporated per weight of protein is high [161].

(2) Formation of a protein free radical:

$$LO\cdot + P_H \longrightarrow LOH + P\cdot$$

$$LOO\cdot + P_H \longrightarrow LOOH + P\cdot$$

Protein free radicals have been found by electron spin resonance to be located mainly on the α-carbon atoms, but also on the sulfur atoms of cysteine residues [162]. The concentration of protein free radicals was found to decrease when the water activity of the model system increased [163].

Relatively stable free radicals can also be formed from free amino acids-linoleate peroxides mixtures [162].

The formation of protein free radicals is followed by direct polymerization of protein chains (without lipid incorporation):

$$P\cdot + P \longrightarrow P - P\cdot \quad \text{(dimer)}$$

$$P - P\cdot + P \longrightarrow P - P - P\cdot \quad \text{(trimer), etc.}$$

Roubal and Tappel [164,165] thus studied the reactions (at 37°C) of aqueous solutions of various proteins and enzymes in the presence of peroxidizing fish oil fractions or linoleate and arachidonate. There was a general loss of solubility and of enzymatic activity. Molecular weight determinations indicated the formation of polymers and aggregates, with a simultaneous loss of methionine, histidine, cysteine, and lysine.

The measurement of lipids covalently incorporated into these protein polymers shows that polymerization mechanisms a (1) and a (2) occur simultaneously.

These various effects were found to be similar to those resulting from protein irradiation by γ-rays.

In some conditions, especially in the dry state, protein free radicals may not be able to diffuse and react with each other. Scisson rather than polymerization may then occur in the protein chain. This was observed by Zirlin and Karel [166] in a gelatin-linoleate model system, together with the appearance of new amide groups. The authors hypothesized the following scission:

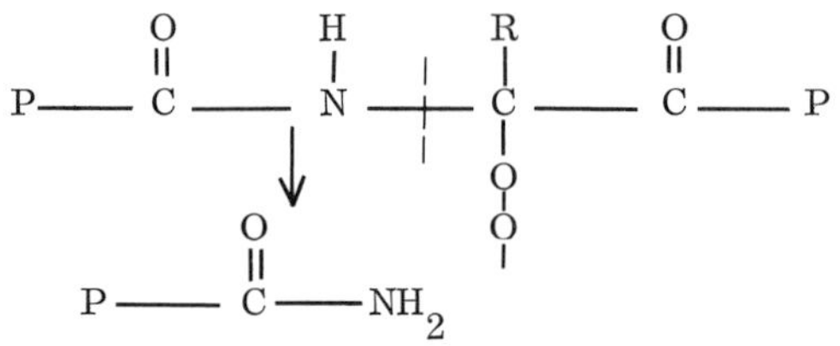

```
          O       H       R       O
          ||      |       |       ||
P ——— C ——— N —┼— C ——— C ——— P
          |               |
          ↓               O
                          |
       O                  O
       ||                 |
P ——— C ——— NH2
```

γ-Irradiation of proteins in the dry state may also provoke chain scissions.

b. Carbonylamine Reactions.

Long or short chain aldehyde derivatives resulting from the breakdown of peroxides of linoleic and other unsaturated fatty acids bind to amino groups of proteins by Schiff-base-type reactions:

$$L\text{-}C{=}O + H_2N - P \longrightarrow L\text{-}C{=}N\text{-}P$$

Bifunctional aldehydes such as malonaldehyde, $CH_2(CHO)_2$, known to form during the peroxidation of many unsaturated lipids, may cause the formation of protein intra- or intermolecular cross-links.

Thus Andrews et al. [167] showed that insulin stored with oxidizing methyl linoleate at $50^{\circ}C$ for 5 days in the dry state, then extracted with ether, became less soluble and more resistant to trypsin, and that the content of FDNB-reactive lysine decreased (acid hydrolysis regenerated lysine). These effects were inhibited in the presence of bisulfites, which react with carbonyl groups. Braddock and Dugan [168] found that salmon myosin incubated in a buffer with oxidizing linoleate became partly insoluble. IR, UV, and visible and fluorescence spectra indicated the presence of C=N groups in extracts from the incubation mixture (similar compounds were found in extracts from frozen stored salmon). There was a marked loss of histidine, lysine, and methionine.

Chio and Tappel [169] reacted ribonuclease with malonaldehyde and observed the appearance of fluorescence, some degree of cross-linking, and losses in lysine, histidine, and tyrosine. The fluorescence was attributed to the formation of a 1-amino-3-iminopropene structure, proving a reaction of malonaldehyde with free amino groups in proteins to give

P-NH-CH=CH-CH=N-P.

Fletcher and Tappel [170] found that peroxidation of fatty acids bound to human serum albumin preparations resulted in the formation of such fluorescence chromophores (also found in tissues from aging animals). Similar chromophores were obtained by incubation of pure bovine serum albumin with oxidizing linoleate or with various carbonyls.

Buttkus [171] has also shown that malonaldehyde reacted with fish myosin at 20, 0, or $-20^{\circ}C$. A loss of lysine, histidine, tyrosine, arginine, and methionine was noted at $20^{\circ}C$; at $-20^{\circ}C$, lysine (but not histidine) was attacked.

Thus aldehydes from oxidized lipids appear to be able to react primarily with amino groups (carbonylamine reactions) but also to react with or attack other groups from various amino acid residues, at least at ordinary temperatures (see also Sec. IV.C).

Such reactions are thought to play a role in the denaturation of protein (decrease in solubility and salt-extractibility; loss of water-binding capacity; hardening of texture) observed in frozen and freeze-dried fish products.

Such carbonylamine reactions also appear to strongly enhance the formation of brown pigments that takes place in tissues or model systems containing protein and oxidizing lipids.

Roubal [172] investigated the relative importance of free radical and carbonylamine reactions in protein damage due to oxidizing lipids. A dry solid model system of peroxidizing fish oils plus muscle or plasma proteins was analyzed for ESR signals, fluorescence, and amino acid losses, after various periods of contact at $37^{\circ}C$ under oxygen. Maximum free radical concentration (and no fluorescence) was observed after 14 hr, while after 72 hr there was maximum fluorescence and very few ESR signals. After 14 hr, there was already a 20-40% loss in methionine, cysteine, tyrosine, alanine, and lysine, while after 72 hr the additional loss was only 15%. The author concludes that damage due to free radicals (in the first period) was more severe than that due to aldehydes from decomposing lipid peroxides.

Matsushita and co-workers [173] were able to confirm the existence of both radical and carbonylamine reactions. They reacted either purified linoleic acid peroxides or a mixture of secondary breakdown products of the peroxides with enzymes, as an in vitro model system for damage which could occur in tissues in vivo. Antioxidants inhibited the action of the purified peroxides but not of their breakdown products. The latter promoted less polymerization than the purified peroxides but gave 425 nm fluorescence characteristic of carbonylamine reaction. Enzyme inactivation, amino acid destruction, and lipid incorporation in the presence of the peroxides or of their breakdown products were investigated in the case of ribonuclease, pepsin, and trypsin, and found to be quite different. These effects certainly depend on the nature of the amino acids present in the outer exposed zones of each enzyme.

The exact nature of amino acid modifications (with or without protein cross-links) by mechanisms a (1) and a (2) is not known, although cysteine, lysine, histidine, and methionine are almost always damaged. Tyrosine, arginine, tryptophan, serine, and proline are often touched.

Methionine and cysteine destruction may be due to oxidation [15] and does not imply any participation to cross-links. Cysteine, however, is known to react with lipid peroxides or with their breakdown products.

Some of the amino acid losses reported here may be due to the presence of lipids at the moment of hydrochloric acid hydrolysis.

The exact reactions of malonaldehyde and other lipid-derived aldehydes with side groups of amino acid residues other than lysine are also not well understood (see Sec. IV.C).

In all cases, it would be interesting to determine the nature of the covalent cross-links responsible for the formation of protein polymers.

It should be mentioned that some investigators have reported the formation of a strong complex (without covalent binding) when oxidizing lipids (maize oil) were incubated with a protein (ovalbumin) [174].

2. Nutritional Effects of Lipid-Protein Reactions

The extensive studies of Yanagita and co-workers [175] have shown that casein (or ovalbumin) reacted with 50-500% oxidizing ethyl linoleate, at 50-60°C for 24 hr in aqueous medium or for 4-14 days in the dry state, had a much reduced nutritive value. Contents in lysine, methionine, histidine, arginine, and other amino acids, content in available lysine, proteolytic digestibility in vitro, digestibility in vivo (rats), PER, biological value, were all repeatedly and markedly reduced. Ingestion of the protein by rats at a 10% level also modified the liver and plasma lipid levels. Reacted ovalbumin contained 7% bound lipids.

In addition to these studies with model systems, the effects of lipid oxidation on the nutritive value of fish meal has been investigated. After storage in air at 25° C for 1 year, herring meal with 15% lipid showed a small 9% decrease in FDNB-reactive lysine (and no decrease when stored under nitrogen or after extraction of the lipids) [176]. Oxidized fat does bind more lysine at higher temperatures, and holding peroxidized herring meals at 100° C for 30 hr caused a 16% decrease in FDNB-reactive lysine (no decrease with nonperoxidized meal) [177]. Commercial fish meals which have been overheated during storage as a result of exothermic lipid oxidation reactions are usually severely damaged, with a low content in available lysine [178,179].

C. Reaction with Aldehydes

In addition to reducing carbohydrates and carbonyl compounds resulting from Maillard-type reactions or lipid oxidation, a number of aldehydes may find themselves in contact with food proteins during processing or storage.

1. Aldehyde Treatment of Feed Proteins

Between 40 and 80% of feed proteins given to ruminants (especially soluble casein or soybean meal) are hydrolyzed and deaminated by the bacterial flora of the rumen. Some ammonia is lost, while the rest is converted into microbial proteins. These are further digested in the lower gut but are often of lower nutritive value than the feed proteins.

Postruminal infusion of feed proteins or amino acids, or intravenous administration of amino acids, especially methionine, lysine, and threonine, has been repeatedly demonstrated to improve the production of wool, meat, and milk. It is therefore of great interest to ensure that high-quality feed proteins reach the intestinal sites of digestion and absorption without preliminary modification in the rumen [180,181].

Mere steaming of groundnut and soybean-feed meals is effective against deamination in the rumen, probably through a reduction of protein solubility. Protection can also be achieved by encapsulation with lipids or by treatment with about 10% plant tannins. Hydrogen bonding between hydroxy groups of the tannins and carboxyl peptide groups decreases the susceptibility of the protein to proteolysis.

Treatment of milk proteins and various vegetable protein meals with formaldehyde, glyoxal, or glutaraldehyde (by spraying) has recently been shown to reduce protein degradation and ammonia formation in the rumen (steer and sheep), and sometimes to increase nitrogen retention and/or wool growth in sheep [180-184].

The optimal processing conditions are not yet fully defined. Treatment of casein or whey proteins with 1% formaldehyde gives good results, while 2-3% formaldehyde is generally used with oilseed meals and grass silage. However, in many cases, no beneficial effects were noted [185-187]. This is particularly the case when initially insoluble protein feeds (such as fish meal) are treated with formaldehyde. Moreover, overtreatment with aldehydes reduces the protein digestibility in the lower gut resulting in protein loss in the feces. The availability of lysine may also be decreased. The best treatment is probably that which gives a maximum intestinal absorption of methionine and the few next nutritionally limiting amino acids [180,181].

A formaldehyde-treated casein adequate for sheep was found not to support growth in rats (PER = 0), unless given as 40% of the whole diet. This low nutritional value was due to a decreased protein digestibility (66% against 93% for the untreated casein), and was not improved by lysine, histidine, and tryptophan supplementation [188]. A lactalbumin concentrate (60% protein w/w) was treated in aqueous solution with up to 1.6% (w/w concentrate) formaldehyde, and then freeze-dried. All the formaldehyde was found to bind to the protein. Protein solubility and digestibility in vitro by trypsin were severely decreased, whereas PER and protein digestibility in vivo (rats) were decreased to a lesser extent. The biological value of the absorbed protein, however, was not impaired.

A four-to-sixfold increase in the concentration of ε-N-methyllysine has been observed in sheep plasma after feeding formaldehyde-treated casein. This derivative was also identified in the treated casein after acid hydrolysis [189]. This shows that formaldehyde treatment induces the formation of some nonreversible links. It has, however, been proved that formaldehyde binding is partly reversed (and that some formaldehyde is released) at an acid pH (such as that of the abomasum, pH 3). This phenomenon appears to depend on the duration of aldehyde treatment.

2. Chemistry of Formaldehyde-Protein Interactions

The chemistry of formaldehyde-treated proteins is not entirely understood. It is known that formaldehyde reacts with ε-amino groups of lysine. This reaction is thought to proceed more to the dihydroxymethyl derivative than to the Schiff base [67]:

$$RNH_2 + \underset{\text{formaldehyde hydrate}}{H\text{-}C(H)(OH)\text{-}OH} \underset{}{\overset{-H_2O}{\rightleftharpoons}} R\text{-}N(H)\text{-}C(H)(H)\text{-}OH \overset{-H_2O}{\rightleftharpoons} \underset{\text{Schiff base}}{R\text{-}N=CH_2}$$

$$R\text{-}N(H)\text{-}CH_2\text{-}OH \overset{+CH_2O}{\rightleftharpoons} \underset{\text{dihydroxymethyl derivative}}{R\text{-}N(CH_2OH)\text{-}CH_2OH}$$

These compounds are degraded during acid hydrolysis.

Residues of cysteine, histidine, arginine, tyrosine, methionine, tryptophan, glutamine, and asparagine may also react with formaldehyde at neutral or alkaline pH [5,67]. Reaction with cysteine takes place at a rapid rate.

It is likely that the decrease in solubility and digestibility caused by treatments of proteins with aldehydes is due to cross-linking between polypeptide chains rather than to simple substitution of ε-aminolysyl or other amino acid residues.

Data on the "tanning effect" of formaldehyde or glutaraldehyde on leather, and on the "firming" of fish meal by formaldehyde, support this hypothesis. It was found that an 0.14% formaldehyde treatment of fish immediately before fish meal processing had a greater biological effect (as measured by chick growth) than the chemical binding effect on lysine (as seen by FDNB reaction or be dye absorption) [103].

In a study [190] of the detoxification of microbial toxin proteins with formaldehyde (to produce immunologically active toxoids), lysine derivatives could be identified, after acid hydrolysis, which corresponded to the covalent cross-linking of one (or two) ε-amino groups of lysyl residues by a methylene bridge to a tyrosyl residue in position 3 (and 5) of the phenol ring (Mannich-type compounds). Such covalent cross-links by a methylene bridge may also occur between ε-amino and imidazole groups [4].

The formation of these and possibly other cross-links in aldehyde-treated proteins should be further studied and correlated with protein digestibility and amino acid availability. The extent of cross-linking and its stability (to pH) probably depends on the prior degree of polymerization of the aldehyde, on the duration and pH of the reaction, and on further treatments such as heating or drying.

3. Other Aldehyde-Protein Interactions

The reactions of glutaraldehyde, pyruvaldehyde, malonaldehyde, etc., with proteins have also been investigated [5,67] (see Sec. IV.B). The chemistry of glutaraldehyde-protein interactions has been the object of much study [191,192], because this aldehyde is widely used for structure-function investigations, protein polymerization, and immobilizing enzymes and antibodies on various supports.

Hexamethylenetetramine (HMT), a bacteriostatic agent used in northern Europe for the preservation of pickled raw fish (pH 4.1-4.5), progressively decomposes into formaldehyde, and 25 mg of HMT per 100 g is considered to be sufficient, as it maintains, for an adequate period of time, a concentration of free formaldehyde $\geq$5mg/100g. Prolonged storage with excessive amounts of HMT lowers the nutritional value of the foods, probably through a reduction in solubility, digestibility and lysine availability. The use of HMT has been advised against, because of its possible toxicity [193-195].

Wood smoke, and more specifically, aldehydic fractions of wood smoke (containing acetaldehyde, formaldehyde etc.) have also been shown to react with ε-amino groups in proteins of cured meats or in model systems; they may reduce the availability of lysine (up to 45% decrease), and the NPU in animals (up to 20% decrease), depending on the concentration, length, temperature, water activity, and pH of treatment [196]. In another experiment, however, salting and smoking were found not to reduce the nutritional value of cod protein [1]. Gossypol, a polyphenol present in cotton seed, may react with the ε-amino groups of the seed proteins through its aldehyde groups [197,198].

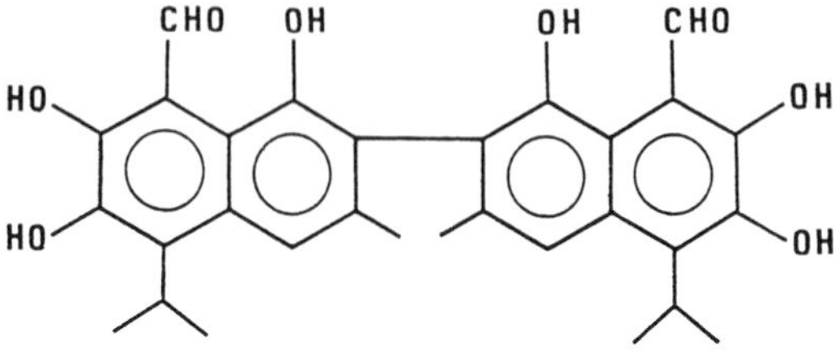

CHO OH OH CHO
HO OH
HO OH

Free glossypol is toxic to animals (inhibition of growth, death in pigs) probably because it binds to endogenous proteins such as enzymes. Gossypol binding may occur during certain processing operations of cotton-seed meals, especially at high temperatures. Bound gossypol is undigestible and much less toxic. Gossypol binding, however, reduces the content of available lysine and often lowers the nutritional value of cotton seed proteins. Relationships between FDNB-reactive lysine, free gossypol, total gossypol, growth, and protein nutritive value may therefore be complex [1]. Cotton seed meal can be detoxified by the extraction of gossypol, or by cooking in the presence of calcium and ferrous ions; such treatments do not reduce the content of available lysine. Varieties of cotton plants free from "gossypol glands" have also been obtained by breeding.

D. Reaction with Quinones

Enzymatic browning involves the oxidation of diphenols to quinones by polyphenol oxidases and molecular oxygen, and the nonenzymatic polymerization of quinones into complex pigment molecules called melanins. Oxidation and pigment formation may take place in the absence of enzymes, at alkaline pH.

Enzymatic browning occurs in certain animal and plant tissues. Only the enzymatic browning of vegetable foods is of technological significance.

It appears that quinones formed by the oxidation of polyphenols present in sunflower seeds or grasses may react with amino acid residues and lower the nutritional value of protein meals or concentrates. It may therefore be useful to prevent polyphenol oxidation and browning during the drying of grasses and the extraction and purification of sunflower seed protein.

Quinones can react with amino groups of free amino acids, and provoke oxidative deamination. Benzoquinone can react with the thioether group of

RSH +

RSH +

excess
quinone

diphenol

amino
acid

noncolored
catecholic
protein

colored
quinonoid
protein

NH_3 +
α keto
acid

methionine [199]. Quinones can also react with amino and sulfhydryl groups of proteins [200-202]. Quinones have been reported to combine with fully acetylated casein [203]. The reaction with sulfhydryl groups may lead to the formation of protein polymers.

Multifunctional melanins can also bind to several proteins simultaneously.

Polyphenol oxidases may oxidize tyrosine residues of proteins. The mere contact with quinones may oxidize methionine and cysteine residues.

The existence of covalently linked phenolic acids in various plant proteins, including sunflower seed protein concentrate, is well documented [204-206]. However, the type of linkage between phenolic acid and amino acids is still not well explained. Van Sumere et al. [205] found that N-ferulylglycyl-L-phenylalanine can be released from some plant proteins by acid hydrolysis but is also degraded as acid hydrolysis proceeds.

Casein allowed to react (in the presence of o-diphenol oxidase) with caffeic acid, isochlorogenic acid, or phenolic compounds from red clover leaves has been found to be inferior in terms of digestibility and biological value in the rat and of available lysine content [207]. Casein model systems incubated with caffeic acid at alkaline pH or in the presence of polyphenol oxydase were found to contain less lysine than control casein after acid hydrolysis. The release of free lysine from such model systems by proteolytic action was also decreased [206].

In a study of various nonlegume leaf protein concentrates (rape, wheat, nasturnium), Allison [208] found significant negative correlations between "nonavailable lysine" (undeaminated by nitrous acid) and nutritional value. "Chloroplastic" protein fractions of both rape and lupin (legume) leaves had a higher content of nonavailable lysine and a lower nutritional value than "cytoplasmic" protein fractions. The latter are known to contain less polyphenol than chloroplastic fractions. Alfalfa (legume) protein preparations had more unavailable lysine when extracted in the absence of reducing agents (sulfites) or in the presence of added polyphenol (chlorogenic acid). It was concluded that the action of polyphenols and quinones on the ε-amino groups and the subsequent polymerization of polyphenols into tannin-protein complexes could render large blocks of amino acids undigestible [208].

However, these conclusions are in contrast with the fact that leaf protein concentrates are rich in lysine and poor in methionine; moreover, a large proportion of methionine or cysteine residues appears to be unavailable in alfalfa and in most other leaf and grass protein concentrates [209, 210]. This deserves further investigation; reaction between quinones and sulfhydryl groups could be partly responsible.

Heat processing (especially above 80°C) of leaf protein concentrates reduces the nitrogen digestibility and the nutritional value; it would also be of interest to determine whether the detrimental effects of overprocessing or overstorage are due to reactions of quinones or of carbohydrates with ε-amino or sulfhydryl groups.

E. Reaction with Nitrites

Nitrites used for the curing of meat and some fish products have three main effects and purposes: (a) nitrosation of the heme pigment of myoglobin, leading to a stabilization of the color; (b) inhibition of the germination and the growth of Clostridium botulinum spores and other microorganisms, possibly through the formation of nitrite-amino acid-metal ion derivatives; and (c) formation of characteristic flavors.

Residual nitrites in meat should not be in excess of 200 ppm, in accordance with the food laws of most countries.

The fate of nitrites in meat or meat-curing model systems composed of myoglobin, nitrite, and ascorbate has been studied [211, 212] with the use of ^{15}N-nitrite (see Fig. 6).

Nitrosomyoglobin formed in cured meat corresponds approximately to 15 ppm of nitrite (10-20% of added nitrite); NO, N_2, and N_2O evolved represent about 5% of nitrite added to meat [213, 214]. The nitrate content increases markedly with the curing period and length of storage, while the residual nitrite content decreases. A complete analytical recovery of ^{15}N from added nitrite was possible in model systems [212] but not in cured meat [211, 215].

1. Formation of Nitrosamines

Nitrites may react with secondary or tertiary amines initially present in foods, or formed through autolysis, bacterial action, or cooking. Some free amino acids such as proline, tryptophan, arginine, histidine, and possibly the corresponding amino acid residues, may constitute such reactive amines. These reactions may take place during the cooking of foods, or even during digestion, at the low pH of the stomach. Some of the resulting nitrosamines or nitrosamides, especially dimethylnitrosamine, are

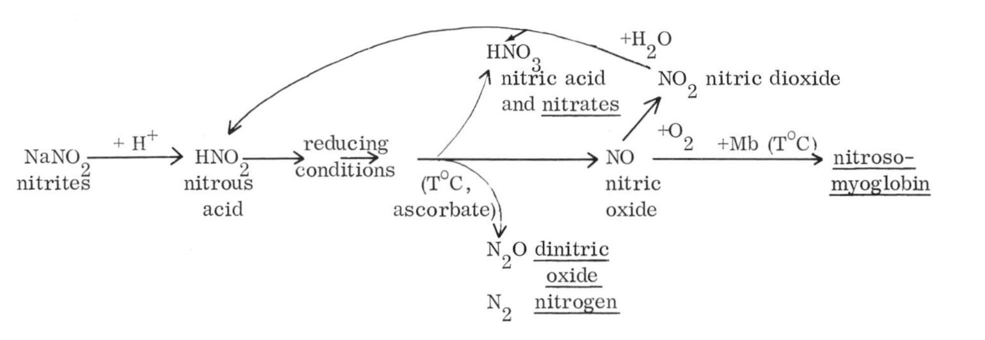

Fig 6. Possible chemical transformation of nitrite in cured meat (stable derivatives are underlined).

known to be potent carcinogens. As an example, nitrosopyrrolidine was found to occur in heated bacon, probably derived from nitrosation of free proline, followed by decarboxylation of nitrosoproline [216].

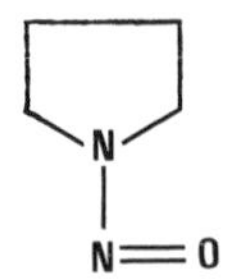

nitrosopyrrolidine

N-Nitrosamines may even form from the reaction of nitrite and some primary amines [217]. N-Nitrosation reactions are at least partially inhibited by compounds which can react with nitrites, such as reducing agents and antioxidants [218].

2. Chemistry of Nitrite-Protein Interactions

Studies of nitrite interactions with amino acid residues in peptides and proteins [5,219] have shown that α- and ε-amino groups are first nitrosated and then split from the molecule. Deamination occurs with nitrogen elimination (Van Slyke reaction). Nitric oxide appears to be the reactive compound in nitrite-protein interaction [220,221].

Nitrosation of tyrosine (on the ortho position with respect to the OH group), histidine, and tryptophan residues can also take place. Nitrosation of tryptophan leads to the formation of a δ-N-nitrosamine. An N-acetyl tryptophan dipeptide has indeed been shown to give a nitrosamine on reaction with nitrites [222]. These reactions are enhanced by heat and low pH and may well take place at the moderately acid pH of most meats.

An in vitro study of nitrite interaction with bovine serum albumin has shown that nitrite reacts with lysine and tyrosine residues. After extensive proteolysis, 6-hydroxynorleucine (formed through ε-amino deamination of lysyl residues), 3,4-DOPA, and 3-nitrotyrosine were identified. The latter two compounds are formed by the oxidation of 3-nitrosotyrosine [223].

Nitrites can react quantitatively with cysteine or with sulfhydryl groups in meat (myosin) at pH 2-3, to give S-nitrosocysteine (which displays antibacterial properties) or nitrosothiols [224]. At the pH of meat and high temperature, only 10-25% of added nitrites form nitrosothiols. Depending on the temperature, these compounds may undergo a redox reaction, resulting in the formation of a disulfide and in the release of nitric oxide, available for further reactions.

Nitrites are normally present in meats at too low a concentration to bring about a significant decrease in the content or availability of lysine, tryptophan, and/or cysteine. It would, however, be of interest to determine if stable covalent bonds can be formed between NO· free radicals and NH groups from peptide links or from arginine, histidine, and tryptophan residues; if this occurred, nitrosamines could be released in the gastrointestinal tract during digestion.

F. Reaction with Sulfites

At high concentration and low pH, undissociated sulfurous acid exerts an antiseptic action. At lower concentrations and a larger range of pH, sulfite ions inhibit enzymatic browning (see Sec. IV.D), nonenzymatic browning (see Sec. IV.A.1), and many oxidation reactions. They are widely used for the stabilization of dehydrated fruits and vegetables, dried fish, fruit juice concentrates, etc. According to many food laws, the concentration of residual free sulfites in food ready for consumption should remain below 200 ppm.

Sulfite ions react with disulfides to form S-substituted thiosulfates (also called S-sulfonates) and thiols [5].

$$\text{P-S-S-P} + SO_3^{-2} \rightleftharpoons \underset{\text{S-sulfocysteine}}{\text{P-S-}SO_3^{-}} + \text{P-S}^{-}$$

Sulfitolysis is enhanced at pH 7 and goes to completion in the presence of compounds which remove the thiol anion. Many cystine residues in proteins, however, especially those of intramolecular bonds, resist the action of sulfites.

S-Sulfonates are unstable in strongly acid or alkaline solutions, and usually decompose to disulfides. Treatment with reducing agents yields cysteine residues. Sulfitolysis does not destroy tryptophan residues.

No detrimental effect of sulfites on the nutritive value of proteins can be inferred from these data.

However, free methionine and tryptophan can be oxidized during the aerobic oxidation of sulfite to sulfate at neutral pH in the presence of oxygen [9,10,225] (see Sec. II.A). The addition of peroxidase and phenol enhances the oxidation. Superoxide anion radicals are involved in the mechanism of the reaction.

High doses of sulfites may exert a deleterious effect on the gastrointestinal tract and may partly inhibit protein digestion. Various foods stored for relatively long periods of time in the presence of sulfites were found to contain a factor toxic to the growth of rats [226]. The nature of this factor is not known.

G. Reaction with Chlorinated Molecules

Trichloroethylene combines with the sulfhydryl groups of proteins under prolonged heating. The S-dichlorovinyl-L-cysteine thus formed, for example, appears to be the toxic factor in trichloroethylene-extracted soybean meal, producing aplastic anemia in calves [227].

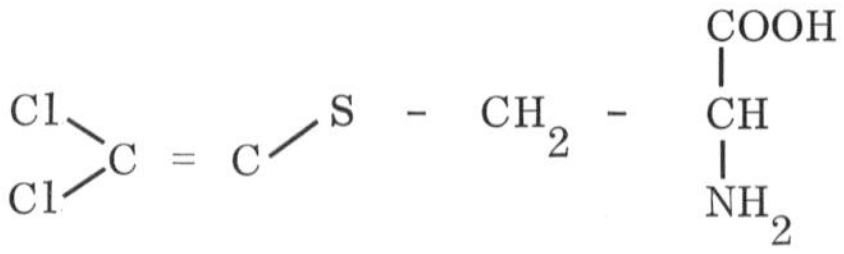

$$\begin{array}{c} Cl \\ Cl \end{array} \!\!>\! C = C \!\!<\!\! ^{S} - CH_2 - \underset{\underset{NH_2}{|}}{\overset{\overset{COOH}{|}}{CH}}$$

S-dichlorovinyl-L-cysteine

Lipid solvents such as methylenechloride (dichloromethane), perchloroethylene (tetrachloroethylene), methylchloroform (1,1,1-trichloroethane), and fluorochlorocarbons do not appear to react with proteins [228].

Especially under slightly alkaline conditions, 1, 2-dichloroethane was found to react with fish proteins, resulting in the partial destruction of histidine and cystine (as seen after acid hydrolysis). The availability of cystine, histidine, and methionine were determined by in vitro digestion with pancreatin and found to be reduced; the proteolytic release of α-amino nitrogen was also decreased [229,230]. These findings suggest that 1, 2-dichloroethane may alkylate sulfhydryl groups of proteins and even cross-link protein chains through such protease-resistant thioether bonds. The 1,2-dichloroethane-extracted fish protein concentrate was fed to young calves as a milk replacer, and it was found to have a lower nutritional value than dried skim milk, even when methionine was added [231].

Agene (nitrogen trichloride), formerly used as a maturing agent for flour, reacts with methionine residues in wheat proteins to form methionine sulfoximine.

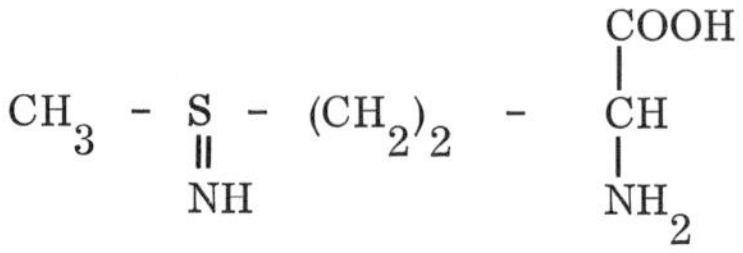

$$CH_3 - \underset{\underset{NH}{\|}}{S} - (CH_2)_2 - \underset{\underset{NH_2}{|}}{\overset{\overset{COOH}{|}}{CH}}$$

methionine sulfoximine

This compound cannot be utilized in vivo and probably interferes with normal protein metabolism [26]. It was also found to provoke toxic effects in dogs and teratogenic effects in mice.

Chlorine from chlorinated waters (used for the transport and blanching of vegetables, and containing about 4 mg free residual chlorine per liter) may react with proteins in foods, particularly with their free amino groups, to give mono- and dichloramines. The reactive form of chlorine is thought to be hypochlorous acid HOCl. Although such reactions are known to inactivate enzymes and probably are responsible for the death of microorganisms, their effect on the nutritional value of food proteins should not be significant.

Peanut protein isolates have been treated with 0.3% sodium hypochlorite at pH 8-9, in order to destroy aflatoxins. The treatment caused a 50% reduction in tryptophan and tyrosine contents, probably through oxidation [232].

CONCLUDING REMARKS

Although this chapter is by no means exhaustive, it is clear that numerous detrimental reactions can take place when food proteins are processed by physical or chemical means. Only a few of these reactions may lead to the formation of toxic derivatives. Many, however, cause a reduction in nutritive value, often through the formation of covalent cross-links between polypeptide chains. This reduction has both biological and economic consequences.

An improved knowledge of these reactions will obviously be of assistance in regulating food processing and storage. A limited degree of damage, however, appears unavoidable, and must be considered together with the beneficial effects of food processing on food safety, flavor, shelf life, and convenience.

ACKNOWLEDGMENTS

This study was supported in part by the Centre National de la Recherche Scientifique, Paris (E.R.A. n° 614 : Modifications biochimiques et nutritionnelles de protéines alimentaires).

REFERENCES

1. J. Mauron, in Protein and Amino Acid Functions (E. J. Bigwood, ed.), Pergamon Press, Oxford, 1972.
2. A. E. Bender, in Evaluation of Novel Protein Products (A. E. Bender, R. Kihlberg, B. Löfqvist, and L. Munck, eds.), Pergamon Press, Oxford, 1970.
3. A. E. Bender, Protein Advisory Group Bull. 2:10 (1972).
4. K. J. Carpenter, Nutr. Abst. 43:423 (1973).
5. G. E. Means and R. E. Feeney, Chemical Modification of Proteins, Holden-Day, Inc., San Francisco, 1971.
6. L. R. Njaa, Acta Chem. Scand. 16:1359 (1962).
7. J. L. Cuq, M. Provansal, F. Guilleux, and C. Cheftel, J. Food Sci. 38:11 (1973).

8. J. K. Munyua, Milchwissenschaft 30:730 (1975).
9. S. F. Yang, Biochemistry 9:5008 (1970).
10. M. Inoue and H. Hayatsu, Chem. Pharm. Bull. 19:1286 (1971).
11. J. L. Cuq, Thesis, University of Sciences and Techniques, Montpellier (in French), 1973.
12. T. Gomyo and M. Fujimaki, Agr. Biol. Chem. 34:302 (1970).
13. T. S. Kennedy and F. J. Ley, J. Sci. Food Agr. 22:146 (1971).
14. F. J. Ley, in SOS/70 Proc. Third Int. Congr. Food Sci. and Technol., Inst. Food Technol., Chicago, 1970.
15. S. R. Tannenbaum, H. Barth, and J. P. Le Roux, J. Agr. Food Chem. 17:1353 (1969).
16. J. L. Cuq, M. Provansal, L. Chartier, P. Besancon, and C. Cheftel, in Proc. IVth Int. Congr. Food Sci. Technol., Madrid, 1974, in press.
17. T. Nakai and T. Ohta, Biochim. Biophys. Acta 420:258 (1976).
18. J. Mauron, Deut. Lebensm. Rundsch. 71:27 (1975).
19. N. P. Neumann, in Methods in Enzymology (S. P. Colowick and N. O. Kaplan, eds.), Vol. XI, Academic Press, New York, 1967.
20. Y. Schechter, Y. Burstein, and A. Patchornik, Biochemistry 14:4497 (1975).
21. S. H. Lipton and C. E. Bodwell, J. Agr. Food Chem. 24:26 (1976).
22. T. L. Lunder, Anal. Biochem. 49:585 (1972).
23. H. G. Walker, Jr., G. O. Kohler, D. D. Zurmicky, and S. C. Witt, in Protein Nutritional Quality of Foods and Feeds (M. Friedman, ed.), Part 2, M. Dekker, Inc., New York, 1975.
24. W. E. Savige and J. A. MacLaren, in Reactions of Organic Sulphur Compounds (N. Kharasch and C. Y. Meyers, eds.), Vol. 2, Pergamon Press, Oxford, 1966.
25. L. R. Njaa, Brit. J. Nutr. 16:571 (1962).
26. D. S. Miller and P. D. Samuel, J. Sci. Food Agr. 21:616 (1970).
27. S. A. Miller, S. R. Tannenbaum, and A. W. Seitz, J. Nutr. 100:909 (1970).
28. L. A. Holt and B. Milligan, Aust. J. Biol. Sci. 26:871 (1973).
29. D. Pieniazek, M. Rakowska, and H. Kunachowicz, Brit. J. Nutr. 34:163 (1975).
30. K. D. Kussendrager, Y. De Jong, J. M. W. Bouma, and M. Gruber, Biochim. Biophys. Acta 279:75 (1972).
31. P. Slump and H. A. W. Schreuder, J. Sci. Food Agr. 24:657 (1973).
32. G. H. Anderson, G. S. K. Li, J. D. Jones, and F. Bender, J. Nutr. 105:317 (1975).
33. R. C. Smith, Biochim. Biophys. Acta 261:304 (1972).
34. G. Donoso, O. A. M. Lewis, D. S. Miller, and P. R. Payne, J. Sci. Food Agr. 13:192 (1962).
35. E. L. Miller and K. J. Carpenter, J. Sci. Food Agr. 15:810 (1964).
36. E. L. Miller, K. J. Carpenter, and C. K. Milner, Brit. J. Nutr. 19:547 (1965).

37. E. L. Miller, K. J. Carpenter, C. B. Morgan, and A. W. Boyne, Brit. J. Nutr. 19:249 (1965).
38. G. M. Ellinger and E. B. Boyne, Br. J. Nutr. 19:587 (1965).
39. D. Pieniazek, M. Rakowska, W. Szkilladziowa, and Z. Grabarek, Br. J. Nutr. 34:175 (1975).
40. J. E. Ford and D. N. Salter, Br. J. Nutr. 20:843 (1966).
41. K. J. Carpenter, I. McDonald, and W. S. Miller, Br. J. Nutr. 27:7 (1972).
42. J. F. Beuk, F. W. Chornock, and E. E. Rice, J. Biol. Chem. 175:291 (1948).
43. M. S. Dunn, M. N. Camien, S. Eiduson, and R. B. Malin, J. Nutr. 39:177 (1949).
44. J. B. Neilands, R. J. Sirny, I. Sohljell, F. M. Strong, and C. A. Elvehjem, J. Nutr. 39:187 (1949).
45. B. J. R. Iriarte and R. H. Barnes, Food Technol. 20:835 (1966).
46. J. Bjarnason and K. J. Carpenter, Br. J. Nutr. 24:313 (1970).
47. M. Sternberg, C. Y. Kim, and F. J. Schwende, Science 190:992 (1975).
48. M. Provansal, J. L. Cuq, and C. Cheftel, J. Agr. Food Chem. 23:938 (1975).
49. S. R. Tannenbaum, M. Ahern, and R. P. Bates, Food Technol. 24:96 (1970).
50. F. Hayase, H. Kato, and M. Fujimaki, Agr. Biol. Chem. 37:191 (1973).
51. F. Hayase, H. Kato, and M. Fujimaki, J. Agr. Food Chem. 23:491 (1975).
52. S. H. Kamath and C. P. Berg, J. Nutr. 82:237 (1964).
53. S. H. Kamath and C. P. Berg, J. Nutr. 82:243 (1964).
54. C. Kies, H. Fox, and S. Aprahamian, J. Nutr. 105:809 (1975).
55. D. Seidl, M. Jaffe, and W. G. Jaffe, J. Agr. Food Chem. 17:1218 (1969).
56. R. J. Evans, D. H. Bauer, K. A. Sisak, and P. A. Ryan, J. Agr. Food Chem. 22:130 (1974).
57. B. Krzysik, J. P. Vergnes, and I. R. McManus, Arch. Biochem. Biophys. 146:34 (1971).
58. W. K. Paik and S. Kim, Science 174:114 (1971).
59. W. K. Paik and S. Kim, Biochemistry 11:2589 (1972).
60. C. G. Zarkadas, Can. J. Biochem. 53:96 (1975).
61. H. W. Lange and K. Hempel, J. Chromatogr. 107:389 (1975) (in German).
62. H. W. J. Harding and G. E. Rogers, Biochemistry 11:2858 (1972).
63. J. J. Pisano, J. S. Finlayson, M. P. Peyton, and Y. Nagai, Proc. Nat. Acad. Sci. 68:770 (1971).
64. J. Thomas, Int. J. Biochem. 2:644 (1971).
65. P. M. Gallop, O. O. Blumenfeld, and S. Seifter, Ann. Rev. Biochem. 41:617 (1972).

66. P. A. Abraham, D. W. Smith, and W. H. Carnes, Biochem. Biophys. Res. Commun. 67:723 (1975).
67. R. E. Feeney, G. Blankenhorn, and H. B. F. Dixon, Adv. Protein Chem. 29:135 (1975).
68. T. Housley, M. L. Tanzer, E. Henson, and P. M. Gallop, Biochem. Biophys. Res. Commun. 67:824 (1975).
69. R. Bressani, L. G. Elias, and R. A. Gomez Brenes, in Protein and Amino Acid Functions (E. J. Bigwood, ed.), Pergamon Press, Oxford (1972).
70. I. E. Liener, in Symposium on Factors in Legume Protein Digestibility, 35th Ann. Meeting Inst. Food Technologists, Chicago, 1975, in press.
71. I. E. Liener, in Symposium on Proteins for Human Consumption, 35th Ann. Meeting of Inst. Food Technologists, Chicago, 1975, in press.
72. I. E. Liener, in Protein Nutritional Quality of Foods and Feeds (M. Friedman, ed.), Part 2, M. Dekker, Inc., New York, 1975.
73. G. C. Mustakas, W. J. Albrecht, G. N. Bookwalter, J. E. McGhee, W. F. Kwolek, and E. L. Griffin, Jr., Food Technol. 20:1290 (1970).
74. R. Bressani, F. Viteri, L. G. Elias, S. De Zaghi, J. Alvarado, and A. D. Odell, J. Nutr. 93:349 (1967).
75. C. Kies and H. M. Fox, J. Food Sci. 36:841 (1971).
76. E. H. Morse, S. B. Merrow, D. E. Keyser, and R. P. Clark, Am. J. Clin. Nutr. 25:912 (1972).
77. M. Korslund, C. Kies, and H. M. Fix, J. Food Sci. 38:637 (1973).
78. R. E. Turk, P. E. Cornwell, M. D. Brooks, and C. E. Butterworth, J. Am. Dietet. Assoc. 63:519 (1973).
79. G. Debry, B. Poulain, and R. E. Bleyer, Ann. Nutr. Alim. 29:159 (1975) (in French).
80. J. J. Kelley and R. Pressey, Cereal Chem. 43:195 (1966).
81. K. Anantharaman and K. J. Carpenter, J. Sci. Food Agr. 20:703 (1969).
82. N. J. Neucere, E. J. Conkerton, and A. N. Booth, J. Agr. Food Chem. 20:256 (1972).
83. K. Anantharaman and K. J. Carpenter, J. Sci. Food Agr. 22:412 (1971).
84. H. E. Amos, D. Burdick, and R. W. Seerley, J. Anim. Sci. 40:90 (1975).
85. E. Josefsson, J. Sci. Food Agr. 26:157 (1975).
86. L. P. Hansen, P. H. Johnston, and R. E. Ferrel, in Protein Nutritional Quality of Foods and Feeds (M. Friedman, ed.), Part 2, M. Dekker, Inc., New York, 1975.
87. E. L. Miller, A. W. Hartley, and D. C. Thomas, Br. J. Nutr. 19:565 (1965).
88. K. J. Carpenter, C. B. Morgan, C. H. Lea, and L. J. Parr, Br. J. Nutr. 16:451 (1962).

89. J. E. Ford and C. Shorrock, Br. J. Nutr. 26:311 (1971).
90. J. E. Ford, in Proteins in Human Nutrition (J. W. G. Porter and B. A. Rolls, eds.), Academic Press, New York, 1973.
91. S. Buraczewski, L. Buraczewska, and J. E. Ford, Acta Biochim. Polonica 14:121 (1967).
92. P. E. Waibel and K. J. Carpenter, Br. J. Nutr. 27:509 (1972).
93. G. Raczynski, M. Snochowski, and S. Buraczewski, Br. J. Nutr. 34:291 (1975).
94. J. Bjarnason and K. J. Carpenter, Br. J. Nutr. 23:859 (1969).
95. J. Spinelli, H. Groninger, B. Koury, and R. Miller, Process Biochem. 10:31 (1975).
96. S. A. Varnish and K. J. Carpenter, Br. J. Nutr. 34:325 (1975).
97. S. A. Varnish and K. J. Carpenter, Br. J. Nutr. 34:339 (1975).
98. C. Aymard, unpublished work, 1976.
99. S. H. Lipton and C. E. Bodwell, in: Protein Nutritional Quality of Foods and Feeds (M. Friedman, ed.), Part 2, M. Dekker, Inc. New York, 1975.
100. S. H. Lipton and C. E. Bodwell, J. Agr. Food Chem. 24:32 (1976).
101. A. P. De Groot, Food Technol. 18:339 (1963).
102. K. J. Carpenter and A. A. Woodham, Br. J. Nutr. 32:647 (1974).
103. K. J. Carpenter and J. Opstvedt, J. Agr. Food Chem. 24:389 (1976).
104. A. Patchornik, and M. Sokolovsky, J. Am. Chem. Soc. 86:1860 (1964).
105. Z. Bohak, J. Biol. Chem. 239:2878 (1964).
106. M. J. Horn, D. B. Jones, and S. J. Ringel, J. Biol. Chem. 138:141 (1941).
107. K. L. Ziegler, I. Melchert, and C. Lurken, Nature 214:404 (1967).
108. R. S. Asquith and J. J. Garcia-Dominguez , J. Soc. Dyers Colour. 84:211 (1968).
109. R. S. Asquith, A. K. Booth, and J. D. Skinner, Biochim. Biophys. Acta 181:164 (1969).
110. A. P. De Groot and P. Slump, J. Nutr. 98:45 (1969).
111. K. Saio and M. Murase, J. Food Sci. Technol. 22:30 (1975).
112. W. Manson and T. Carolan, J. Dairy Sci. 39:189 (1972).
113. J. T. Snow, J. W. Finley, and M. Friedman, Int. J. Peptide Res. 8:57 (1976).
114. B. H. Nicolet and L. A. Shin, J. Am. Chem. Soc. 63:2284 (1941).
115. B. H. Nicolet and L. A. Shin, J. Biol. Chem. 140:685 (1941).
116. J. P. Danehy, in Reactions of Organic Sulphur Compounds (N. Kharasch and C. Y. Meyers, eds.), Pergamon Press, Oxford, 1966.
117. M. Provansal, Thesis, University of Sciences and Techniques, Montpellier, 1974 (in French).
118. M. Sternberg, C. Y. Kim, and R. A. Plunkett, J. Food Sci. 40:1168 (1975).
119. G. Viroben and J. Delort-Laval, personal communication, 1973.
120. C. Cheftel, J. L. Cuq, M. Provansal, and P. Besancon, Rev. Fr. corps gras. 1:7 (1976) (in French).

121. K. J. Carpenter, J. Duckworth, G. M. Ellinger, and D. H. Shrimpton, J. Sci. Food Agr. 3:278 (1952).
122. A. D. L. Gorrill and J. W. G. Nicholson, Can. J. Anim. Sci. 52:665 (1972).
123. L. Van Beek, V. J. Feron, and A. P. De Groot, J. Nutr. 104:1630 (1974).
124. J. C. Woodard and D. D. Short, J. Nutr. 103:569 (1973).
125. J. C. Woodard, Fed. Proc. 32:884 Abs. (1973).
126. J. C. Woodard, D. D. Short, M. R. Alvarez, and J. Reyniers, in Protein Nutritional Quality of Foods and Feeds (M. Friedman, ed.), Part 2, M. Dekker, Inc., New York, 1975.
127. A. P. De Groot, P. Slump, L. Van Beek, and V. J. Feron, Abstr. 35th Ann. Meeting Inst. Food Technol., Chicago, 1975.
128. J. E. Hodge, in Symposium on Foods: The Chemistry and Physiology of Flavors (H. W. Schultz, E. A. Day, and L. M. Libbey, eds.), Avi Publishing Co., Westport, Conn., 1967.
129. T. M. Reynolds, in Symposium on Foods: Carbohydrates and Their Roles (H. W. Schultz, R. F. Cain, and R. W. Wrolstad, eds.), Avi Publishing Co., Westport, Conn., 1969.
130. P. A. Finot, in Proteins in Human Nutrition (J. W. G. Porter and B. A. Rolls, eds.), Academic Press, New York, 1973.
131. J. F. Valle-Riestra and R. H. Barnes, J. Nutr. 100:873 (1970).
132. A. Pronczuk, D. Pawlowska, and J. Bartnik, Nutr. Metabol. 15:171 (1973).
133. H. Erbersdobler, in Proteins in Human Nutrition (J. W. G. Porter and B. A. Rolls, eds.), Academic Press, New York, 1973.
134. M. C. Nesheim and K. J. Carpenter, Br. J. Nutr. 21:399 (1967).
135. P. A. Finot and J. Mauron, Helv. Chim. Acta 55:1153 (1972).
136. V. C. Sgarbieri, J. Amaya, M. Tanaka, and C. O. Chichester, J. Nutr. 103:657 (1973).
137. M. Tanaka, T. C. Lee, and C. O. Chichester, J. Nutr. 105:983 (1975).
138. R. F. Hurrell and K. J. Carpenter, Br. J. Nutr. 32:589 (1974).
139. J. Adrian, Lait 55:24 (1975).
140. A. V. Clark and S. R. Tannenbaum, J. Agr. Food Chem. 22:1089 (1974).
141. A. M. Boctor and A. E. Harper, J. Nutr. 94:289 (1968).
142. J. E. Knipfel, H. G. Botting, and J. M. Mc Laughlan, in Protein Nutritional Quality of Foods and Feeds (M. Friedman, ed.), Part 2, M. Dekker, Inc., New York, 1975.
143. J. Adrian, R. Frangne, L. Petit, B. Godon, and J. Barbier, Ann. Nutr. Aliment. 20:257 (1966) (French).
144. R. Frangne and J. Adrian, Ann. Nutr. Aliment. 21:163 (1967) (in French).
145. J. Adrian and R. Frangne, Ann. Nutr. Aliment. 27:111 (1973) (in French).

146. J. Adrian and H. Susbielle, Ann. Nutr. Aliment. 29:151 (1975) (in French).
147. M. Tanaka, J. Amaya, T. C. Lee, and C. O. Chichester, in: Proc. IVth Int. Congr. Food Sci. Technol., Madrid, 1974, in press.
148. C. M. Lee, C. O. Chichester, and T. C. Lee, in Proc. IVth Int. Congr. Food Sci. Technol., Madrid, 1974, in press.
149. C. J. Dillard and A. L. Tappel, J. Agr. Food Chem. 24:74 (1976).
150. K. J. Carpenter, Biochem. J. 77:604 (1960).
151. V. H. Booth, J. Sci. Food Agr. 22:658 (1971).
152. F. Mottu and J. Mauron, J. Sci. Food Agr. 18:57 (1967).
153. A. W. Boyne, K. J. Carpenter, and A. A. Woodham, J. Sci. Food Agr. 12:832 (1961).
154. A. G. Roach, P. Sanderson, and D. R. Williams, J. Sci. Food Agr. 18:274 (1967).
155. C. K. Milner and D. R. Westgarth, J. Sci. Food Agr. 24:873 (1973).
156. R. F. Hurrell and K. J. Carpenter, Br. J. Nutr. 33:101 (1975).
157. M. Karel, J. Food Sci. 38:756 (1973).
158. H. W. Schultz, E. A. Day, and R. O. Sinnhuber (eds.), Lipids and their Oxidation, Avi Publishing Co., Inc., Westport, Conn., 1962.
159. T. P. Labuza, Crit. Rev. Food Technol. 2:355 (1971).
160. A. L. Tappel, Fed. Proc. 32:1870 (1973).
161. I. D. Desai and A. L. Tappel, J. Lipid Res. 4:204 (1963).
162. M. Karel, K. Schaich, and R. B. Roy, J. Agr. Food Chem. 23:159 (1975).
163. K. M. Schaich and M. Karel, J. Food Sci. 40:456 (1975).
164. W. T. Roubal and A. L. Tappel, Arch. Biochim. Biophys. 113:5 (1966).
165. W. T. Roubal and A. L. Tappel, Arch. Biochim. Biophys. 113:150 (1966).
166. A. Zirlin and M. Karel, J. Food Sci. 34:160 (1969).
167. F. Andrews, J. Bjorksten, F. B. Trenk, A. S. Menick, and R. B. Koch, J. Am. Oil Chemists' Soc. 42:779 (1965).
168. R. J. Braddock and L. R. Dugan, J. Am. Oil Chemists' Soc. 50:343 (1973).
169. K. S. Chio and A. L. Tappel, Biochemistry 8:2827 (1969).
170. B. L. Fletcher and A. L. Tappel, Lipids 6:172 (1971).
171. H. Buttkus, J. Food Sci. 32:432 (1967).
172. W. T. Roubal, Lipids 6:62 (1971).
173. S. Matsushita, J. Agr. Food Chem. 23:150 (1975).
174. K. A. Narayan, M. Sugai, and F. A. Kummerow, J. Am. Oil Chemists' Soc. 41:254 (1964).
175. T. Yanagita and M. Sugano, Agr. Biol. Chem. 39:63 (1975).
176. C. H. Lea, L. J. Parr, and K. J. Carpenter, Br. J. Nutr. 12:297 (1958).
177. C. H. Lea, L. J. Parr, and K. J. Carpenter, Br. J. Nutr. 14:91 (1960).

178. K. J. Carpenter, B. E. March, C. K. Milner, and R. C. Campbell, Br. J. Nutr. 17:309 (1963).
179. J. Bunyan and A. A. Woodham, Br. J. Nutr. 18:537 (1964).
180. G. A. Broderick, in Protein Nutritional Quality of Foods and Feeds (M. Friedman, ed.), Part 2, M. Dekker, Inc., New York, 1975.
181. J. H. Clark, in Protein Nutritional Quality of Foods and Feeds (M. Friedman, ed.), Part 2, M. Dekker, Inc., New York, 1975.
182. A. P. Peter, E. E. Hatfield, F. N. Owens, and U. S. Garrigus, J. Nutr. 101:605 (1971).
183. T. N. Barry and R. N. Andrews, N. Z. J. Agr. Res. 16:545 (1973).
184. D. A. Dinius, C. K. Lyon, and H. G. Walker, J. Anim. Sci. 38:467 (1974).
185. J. F. Nishimuta, D. G. Ely, and J. A. Boling, J. Nutr. 103:49 (1973).
186. J. F. Nishimuta, D. G. Ely, and J. A. Boling, J. Anim. Sci. 39:952 (1974).
187. J. D. Wachira, L. D. Satter, G. P. Brooke, and A. L. Lope, J. Anim. Sci. 39:796 (1974).
188. E. L. Hove and E. Lohrey, J. Nutr. 106:382 (1976).
189. P. J. Reis and D. A. Tunks, Aust. J. Biol. Sci. 26:1127 (1973).
190. B. Bizzini and M. Raynaud, Biochimie 56:297 (1974) (in French).
191. J. W. Payne, Biochem. J. 135:867 (1973).
192. P. Monsan, G. Puzo, and H. Mazarguil, Biochimie 57:1281 (1975).
193. Anon., 6th Report Joint FAO/WHO, Expert Committee on Food Additives, FAO, Rome, 1962.
194. Anon., 10th Report Joint FAO/WHO, Expert Committee on Food Additives, FAO, Rome, 1967.
195. Anon., 15th Report Joint FAO/WHO, Expert Committee on Food Additives, FAO, Rome, 1972.
196. L. B. Chen and P. Issenberg, J. Agr. Food Chem. 20:1113 (1972).
197. R. Bressani, L. G. Elias, and E. Braham, in World Protein Resources (R. F. Gould, ed.), American Chemical Society Publications, Washington, 1966.
198. C. M. Cater and G. M. Lyman, J. Am. Oil Chemists Soc. 46:649 (1969).
199. P. J. Vithayathil and G. S. Murthy, Nature 236:101 (1972).
200. H. S. Mason and E. W. Peterson, Biochim. Biophys. Acta 111:134 (1965).
201. W. S. Pierpoint, Biochem. J. 112:609 (1969).
202. W. S. Pierpoint, Biochem. J. 112:619 (1969).
203. T. Horigome, Chem. Abst. 80:35952 (1974).
204. M. A. Sabir, F. W. Sosulski, and A. J. Finlayson, J. Agr. Food Chem. 22:575 (1974).
205. C. F. Van Sumere, J. Albrecht, A. Dedonder, H. de Pooter, and I. Pe, in The Chemistry and Biochemistry of Plant Proteins (J. B. Harborne and C. F. Van Sumere, eds.), Academic Press, New York, 1975.

206. F. Guilleux, Thesis, University of Sciences and Techniques, Montpellier, 1975 (in French).
207. T. Horigome and M. Kandatsu, Agr. Biol. Chem. 32:1093 (1968).
208. R. M. Allison, in Leaf Protein (N. W. Pirie, ed.), Blackwell Scientific, Oxford, 1971.
209. M. Byers, in Leaf Protein (N. W. Pirie, ed.), Blackwell Scientific, Oxford, 1971.
210. A. A. Woodham, in Leaf Protein (N. W. Pirie, ed.), Blackwell Scientific, Oxford, 1971.
211. R. G. Cassens, J. G. Sebranek, G. Kubberod, and G. Woolford, Food Prod. Dev. 4 (1974).
212. M. Fujimaki, M. Emi, and A. Okitani, Agr. Biol. Chem. 39:371 (1975).
213. G. Woolford, R. J. Casselden, and C. L. Walters, Biochem. J. 130:82 (1972).
214. C. L. Walters and R. J. Casselden, Z. Lebensm. Unters. Forsch. 150:335 (1973).
215. J. G. Sebranek, R. Cassens, W. Hoekstra, W. Winder, E. V. Podebrasdsky, and E. W. Kielsmeier, J. Food Sci. 38:1220 (1973).
216. J. I. Gray and L. R. Dugan, Jr., J. Food Sci. 40:484 (1975).
217. J. J. Warthesen, R. A. Scanlan, D. D. Bills, and L. M. Libbey, Agr. Food Chem. 23:898 (1975).
218. J. I. Gray and L. R. Dugan, Jr., J. Food Sci. 40:981 (1975).
219. A. Kurosky and T. Hofmann, Can. J. Biochem. 50:1282 (1972).
220. A. Mirna and K. Hofmann, Fleischwirtschaft 10:1361 (1969).
221. W. J. Olsman and B. Krol, Proc. 18th Meet. of Meat Res. Workers, Guelph, Canada, 1972.
222. R. Bonnett and R. Holleyhead, J. Chem. Soc. Perkin 1:961 (1974).
223. M. E. Knowles, D. J. McWeeny, L. Couchman, and M. Thorogood, Nature 247:288 (1974).
224. G. Kubberod, R. G. Cassens, and M. L. Greaser, J. Food Sci. 39:1228 (1974).
225. S. F. Yang, Environ. Res. 6:395 (1973).
226. R. J. L. Allen and M. Brook, in Proc. SOS/70 3rd Int. Congr. Food Sci. Technol., Inst. Food Technologists, Chicago, 1971.
227. L. L. McKinney, A. C. Eldridge, and J. C. Cowan, J. Am. Chem. Soc. 81:1423 (1959).
228. J. F. Valle-Riestra, Food Technol. 28:25 (1974).
229. A. B. Morrison and I. C. Munro, Can. J. Biochem. 43:33 (1965).
230. A. B. Morrison and J. M. McLaughlan, in Protein and Amino Acid Functions (E. J. Bigwood, ed.), Pergamon Press, Oxford, 1972.
231. A. B. Makdani, J. T. Huber, and R. L. Michel, J. Dairy Sci. 54:886 (1971).
232. K. R. Natarajan, K. C. Rhee, C. M. Cater, and K. F. Mattil, J. Food Sci. 40:1193 (1975).

CHAPTER 7

TECHNOLOGY OF FORTIFICATION

Benjamin Borenstein*

Roche Chemical Division
Hoffmann-La Roche Inc.
Nutley, New Jersey

I. INTRODUCTION

It is not appropriate in this chapter to argue the value of food fortification to the consumer or industry. The reader is referred to an unusually comprehensive and concise discussion on the thought processes involved in arriving at recommendations for food fortification based on the nutritional status of the U.S. population, dietary habits, and the food supply [1]. This Food and Nutrition Board committee report puts in perspective all aspects of fortification policy including technology, and therefore is strongly recommended to all readers.

The committee proposes that all cereal-grain products be fortified with vitamins A, B_1, B_2, niacin, B_6, and folacin plus the minerals iron, magnesium, calcium, and zinc, if technologically feasible. This last clause seems obvious and perhaps unimportant until one remembers that the successful technology for fortification of rinse-resistant rice with vitamin B_1, niacin, and iron was developed in 1946 [2], but that methodology for adding vitamin B_2 without coloring the rice yellow, and therefore making it

*Current affiliation: CPC International, Inc., Englewood Cliffs, New Jersey.

unacceptable to consumers, has still not been developed today, thirty years later. There are other unsolved technological problems which make fortification a challenging and important area.

Although the FDA has officially defined the terms of fortification, fortified, enriched and restoration [3], in the context of this chapter fortification will be used as a generic term for the addition of vitamins, minerals, and amino acids to foods for the purpose of nutritional improvement. The emphasis will be on vitamins, since the bulk of the literature is in this area. Technological uses of micronutrients, such as vitamin B_1 as a meat flavor, B_2 and iron oxide as coloring agents, α-tocopherol as an antioxidant, and ascorbic acid as an oxygen scavenger, will not be discussed here.

The technical problems in mineral fortification pertain more to organoleptic deficits and bioavailability than to stability of the compounds per se or to methods of incorporation in foods [4]. Large amounts of calcium and magnesium salts are required to obtain nutritionally significant levels of these minerals, and their compounds, e.g., tribasic calcium phosphate, calcium sulfate, and magnesium oxide can produce chalky flavors, opacity, sediment, and color changes in food products. Furthermore, little is known about the bioavailability of these compounds to man. Their poor solubility suggests poor absorption compared to soluble salts such as calcium gluconate, but soluble calcium salts cannot be used in most food applications because of the reactivity and undesirable flavor of calcium ions.

The recommendation by the Food and Nutrition Board that all cereal-grain products be fortified with magnesium, calcium, iron, and zinc has led to increased research on their bioavailability as it relates to fortification technology. The American Institute of Baking has investigated the bioavailability of various zinc and magnesium compounds and their effect on bread-making characteristics. Zinc absorption by rats from zinc acetate, stearate, chloride, exide, and sulfate did not differ much from one another. None of the zinc sources tested (2.2 mg zinc/100 g flour) exerted any adverse effect on loaf volume or on general bread quality [5]. The absorption of magnesium from a variety of compounds did not differ significantly [6] but magnesium oxide, hydroxide, and carbonate (44.1 mg of magnesium per 100 g flour) raised bread pH and adversely affected loaf volume and flavor [7]. This was partially corrected by either pH adjustment with acetic acid or adding the magnesium salt to the dough instead of the sponge in the sponge-dough process.

Both iron and copper compounds catalyze oxidation of fat, vitamin A, and vitamin C. Fortification with these minerals requires thorough evaluation of possible organoleptic changes and effects on product shelf life. The bioavailability of iron and iron fortification is an active research area. Gastrointestinal absorption of iron depends on a number of variables including the oxidation state of the iron (Fe, Fe^{2+}, or Fe^{3+}), whether the iron is bound to organic compounds, the type of binding, the redox potential and pH of the GI tract and the specific anions in the food or foods consumed simultaneously. In effect meal components enter the equilibria pool which determines the ratio of Fe^{2+}/Fe^{3+} present and the rate of iron absorption. Factors which shift the ratio to Fe^{3+} and increase the formation of ferric hydroxide decrease absorption; reducing agents and ions which form soluble iron chelates increase the iron absorption rate. Insoluble compounds such as iron phytates and oxalates are poorly absorbed. Heme iron is well absorbed perhaps as high as 35% in deficient subjects.

To the extent that food processing and storage can cause oxidation of ferrous iron, processing can affect the bioavailability of added iron. Little data is available on the effects of processing on the bioavailability of iron compounds used in fortification.

Fortification with amino acids is not widely practiced. Lysine hydrochloride is commercially available and generally can be added to foods and treated as if it were a moderately stable B vitamin. Methionine presents potentially serious odor and flavor problems in fortification projects.

The technology of fortification involves the selection of the in-process point of addition of micronutrients, the mode of addition, and the market forms of the micronutrients. These decisions are dependent on a number of interrelated factors including stability, bioavailability, organoleptic problems, safety, cost, production practicality, and the reliability of the system.

The literature on fortification technology rarely supplies the details of solutions to problems. Much of the work is proprietary and/or empirical, and vitamin stability results are so dependent on the process specifics and market forms used that extrapolation to other applications should not be made without detailed knowledge of the process and product. Beetner et al. [8], in studying stability of added vitamins B_1 and B_2 during extrusion of corn grits, found that a temperature increase of 40° F (22° C) caused a 21% lower B_1 retention, and that increasing moisture of the grits by 1.5% decreased B_2 retention by 21%. This type of specific processing data is rarely available, but is obviously highly desirable in studying fortification problems. Similarly, even though the kinetics approach to predicting stability of vitamins in foods was used at least as early as 1948 [9], this type of data is rarely published, and the water activity of foods is not measured in most stability studies.

Several recent reviews of fortification and related stability problems are available [10,11]. The mechanics of fortification of rice, wheat, and salt with vitamins and amino acids was discussed by Rubin and Cort [12]. An excellent review on vitamin stability in food processing was prepared by Bender [13], and losses of vitamins and trace minerals in processing and storage were discussed by Schroder [14]. The properties and stability of indigenous and added vitamins and amino acids during food processing has been reviewed [15]. A comprehensive survey of food applications of vitamin A was recently published [16].

II. STABILITY

The more labile vitamins which can present stability problems in food processing and storage are vitamins A, D, B_1, C, pantothenic acid, vitamin B_{12}, and folacin. Although serious degradation of these vitamins can be encountered under adverse conditions, measures can be employed to minimize losses. These include protective coating of individual vitamins, addition of antioxidants, control of temperature, moisture, and pH, and protection from air, light, and incompatible metals during processing and storage. By applying proper technology and using a suitable manufacturing surplus of added vitamins based on a critical evaluation of stability data, it is possible in most food applications to maintain desired label claims for the normal shelf life of food products.

Few data are available on the stability of indigenous or added folacin, pantothenic acid, and biotin in foods compared with the other vitamins.

The stability of added micronutrients generally parallels the stability of indigenous food micronutrients, which is discussed elsewhere in this text. Significant differences, however, can occur in the case of vitamins B_6, A, and E. Vitamin B_6 occurs in food as pyridoxal, pyridoxamine, and pyridoxine. Of these, pyridoxal is the least stable and vitamin B_6 added in the form of pyridoxine hydrochloride would be more stable than native pyridoxal in heat stress processes.

Similarly, thiamine is more stable than thiamine pyrophosphate, which may account for over 50% of the native vitamin B_1 in specific foods. Vitamin E occurs in nature primarily as α-tocopherol, which is highly reactive compared to α-tocopheryl acetate, the most widely used synthetic analog of vitamin E.

Vitamin A occurs in animal source foods primarily as retinyl palmitate and as provitamin A carotenoids in plant products. β-Carotene is the most important indigenous provitamin A carotenoid because of its relatively high vitamin A value and widespread occurrence in foods. In plant products the carotenoids are either complexed with protein, or in solution or colloidal dispersion in lipoid chromoplasts and are extremely stable in almost all food products and processes. A notable exception is the poor stability of carotenoids in dehydrated foods exposed to air. In contrast, both added vitamin A and β-carotene may oxidize very quickly during processing and storage of many products. Therefore, most food applications require coated and/or antioxidant-stabilized market forms of these compounds to obtain adequate stability during shelf storage.

Another pertinent difference between indigenous and added β-carotene is bioavailability. The intestinal absorption of provitamin A carotenoids from foods such as carrots may be less than 30%, and it was recently suggested by the Food and Nutrition Board (FNB) [17] that the average absorption of indigenous food carotenoids should be calculated as one-third of the provitamins ingested, compared with retinol which is assumed to be completely absorbed. Since one international unit of vitamin A activity is equivalent to 0.3 μg retinol and 0.6 μg β-carotene, the overall utilization (intestinal absorption plus convertibility of β-carotene to retinol) of indigenous β-carotene is calculated as one-sixth that of retinol by the FNB. On the other hand, crystalline β-carotene added to margarine is absorbed as well as the USP vitamin A standard and has a provitamin A activity of 1.66 million units per gram [18]. Furthermore, extensive biological studies with rats using the classical USP growth procedure show that water-dispersible gelatin beadlets and emulsions of synthetic β-carotene have full biological vitamin A activity, i.e., 1.66 million units per gram [19]. Therefore, the calculation suggested by the FNB for determining the retinol equivalent of β-carotene in foods is inappropriate for β-carotene added in fortification.

The difficulty of predicting vitamin C stability in food systems is demonstrated in model systems studies by Yu et al. [20], who reported that all amino acids except cysteine reacted with ascorbic acid, resulting in more pronounced color formation than when ascorbic acid was present alone. Cysteine protected ascorbic acid against the formation of the brown pigment. Tryptophan enhanced the browning of ascorbic acid more than the other amino acids. At 72°C, color development caused by ascorbic acid plus tryptophan was further enhanced by the addition of malic acid, citric acid, $FeSO_4$, $CuSO_4$, and Na_2HPO_4, but color formation was reduced by fumaric acid, tartaric acid, succinic acid, $CaCl_2$, $NaHSO_3$, tetrasodium pyrophosphate, sodium tripolyphosphate, sodium tetraphosphate, sodium metaphosphate, and sodium polyphosphate.

The comparative stability of vitamin C in beverages reconstituted and stored under home usage conditions was studied by Beston and Henderson [21] to determine differences in vitamin C stability between orange juice from frozen concentrate, canned orange juice, and formulated beverage powders. Stability curves for 96 hr at 42°F (5.5°C) were very similar for all products. Similarly, the bioavailability of synthetic and natural ascorbic acid provided by orange juice was very close [22]. The slight superiority of synthetic L-ascorbic acid in this study was explained by a slightly higher urinary excretion of the vitamin after consuming orange juice.

The effect on ascorbic acid retention in orange juice of the type of container used, i.e., glass, polyethylene, polystyrene, and cardboard, was studied by Bissett and Berry [23], who confirmed the superiority of glass over oxygen-permeable materials. The significance of packaging materials on vitamin stability is also shown in a kinetics analysis of light-induced riboflavin loss in whole milk [24]. Both these references are atypical in that they supply details on the packaging materials used, wall thickness, etc.

The stability of reduced and total ascorbic acid has been studied in a low moisture dehydrated model food system as a function of water activity, moisture content, oxygen content of the container, and temperature [25]. The kinetics data obtained substantiate prior empirical conclusions that the headspace oxygen dissolves rapidly as the oxygen in the moisture of the product is consumed by reaction with ascorbic acid. Headspace oxygen accelerates ascorbic degradation at a water activity as low as $0.1a_w$.

A comprehensive review on the problems of incorporating vitamin C in baked goods is available [26]. Seib and Hoseney suggest that ascorbate 2-sulfate, a stable analog of ascorbic acid, be used in baking. However, the vitamin value of this compound to primates must first be determined. A high-protein bread product fortified with ten vitamins and six minerals, including vitamin C, has been commercially marketed. In this case, the vitamin C was sprayed on the baked loaves using a proprietary process with satisfactory stability results.

III. FOOD GROUPS

A. Fruit and Vegetable Products

The variety of problems in fortification is perhaps best demonstrated by considering fruit and vegetable products. This large group of foods is diverse chemically, physically (e.g., corn on the cob, diced carrots, sliced string beans, juices, and purees), and in processing technology (e.g., frozen, canned, dehydrated, blanched, refrigerated, concentrated, or fried).

Each product in each form poses its own potential technological difficulties. For example, fortification of canned corn kernels with lysine would require careful preparation to ensure that most of the lysine is not in the canning liquid and thus discarded when used. On the other hand, with cream style corn there is no problem of distribution of nutrients between particles or a liquid likely to be discarded. Fortification of corn products with iron could cause the formation of iron sulfides, which seriously discolor corn products. The addition of insoluble iron salts might circumvent this problem; however, they are not generally favored by nutritionists.

Fortification with iron and most minerals poses many technological problems, e.g., calcium in low-viscosity fruit drinks sediments rapidly and iron catalyzes degradation in fruits containing vitamin C. The forms of iron favored by nutritionists are those most likely to cause problems in fortification, because they catalyze fat autoxidation and vitamin C degradation and lead to discoloration.

Fortification of particulate canned foods with any and all nutrients is difficult because of the distribution of nutrients between the particles and the canning liquid, and the potential waste and deception of the water-soluble vitamins being discarded by the consumer. In theory, it is easier to fortify frozen than canned vegetables. A spray solution or suspension of nutrients can be applied to the frozen food product directly at the packaging line, and the losses during frozen storage would be moderate for most of the vitamins. The losses during home preparation will also be moderate as long as the weight ratio of cooking liquid to solids is low and controlled.

Small-particle-size dehydrated fruits and vegetables (40-200 mesh) are relatively easy to fortify since all the commercial vitamins, amino acids, and minerals are available in dry market forms in this particle-size range. Blended foods containing dehydrated vegetables, e.g., soup mixes with seasonings or salt, are easy to fortify with respect to distribution of the added nutrients. Stability of nutrients in dehydrated foods is greatly dependent on moisture content. Vitamin C can both discolor and degrade at high relative humidity in dehydrated foods. Nitrogen flushing of packaged dehydrated foods improves vitamin C stability [25]. Stable forms of vitamin A for fortification of dehydrated foods are available. Stability of most of the other vitamins in dehydrated foods is satisfactory.

Vitamin C stability in canned foods is dependent on pH and reactants present. Anthocyanins, iron, and copper accelerate vitamin C degradation. For example, the high levels of both anthocyanins and iron in prune juice cause very rapid degradation of added vitamin C. Vitamin B_2 is unstable in glass packs exposed to light. Vitamin B_1 and B_{12} have poor stability at high pH. For this reason, fortification of canned mushrooms with these vitamins would be difficult. Vitamin B_1 can cause off-flavor in some fruit and vegetable products, but the threshold flavor level is dependent on the specific food. The flavor of B_1 is due to trace amounts of sulfhydryl compounds caused by degradation, and off-odors can be produced by analytically immeasurable destruction of B_1. In the case of vegetables, B_1 can also catalyze pyruvic acid degradation to acetoin, which produces off-flavors.

The stability of added vitamins in canned fruits and vegetables generally parallels the stability pattern of the naturally occurring vitamins in the same products with the exceptions previously noted. Processing plus long-term storage of canned and frozen vegetables can be expected to degrade significant amounts (20-50%) of vitamins C, B_1, and B_2 [27]. Although frozen products have a higher initial vitamin C content than canned, storage of 9-12 months tends to equalize the vitamin C levels because of oxidation of ascorbic acid in oxygen-permeable containers even at 0° C. Thus, packaging of frozen foods becomes an important variable in vitamin C stability.

Equally important is the difference in vitamin C stability during storage of vegetables processed in tin vs. glass [27]. The stannous ion is an effective reducing agent and significantly retards ascorbic acid degradation.

Apple sauce has been commercially fortified with vitamin C by metering an ascorbic acid solution directly into the cooker. It is also theoretically possible to dispense ascorbic acid solution, using commercially available equipment, directly into glass or tin containers prior to filling with sauce. The limiting factor in this approach is the rate of diffusion of ascorbic acid throughout the container. It is essential that there be sufficient agitation during the filling, sealing, and cooling operations plus diffusion during the first 2-4 weeks of storage to produce reasonable homogeneity of vitamin C throughout the container so that each serving of sauce meets the label claim.

The decisions required in fortifying juices and drinks with vitamin C are relatively simple. Dry vitamin C can be added on a batch basis, solutions of vitamin C can be continuously metered into the products on the way to the filler, or tablets can be dispersed directly into the final container prior to sealing. The advantage of the last procedure is lower in-process losses of vitamin C, but, the costs of custom vitamin C tablets and automatic dispersing equipment may be higher than the savings resulting from a lower vitamin C overage.

As noted previously, the stability of added vitamin A is not similar to that of the native carotenoid protein complexes which supply substantially all the provitamin A value of fruits and vegetables, and choosing the best market form of vitamin A for a given application is an empirical process. Fortification of low-viscosity fruit beverages with water-dispersible market forms of vitamin A or other fat soluble vitamins is complicated by the tendency of these vitamins to float or ring because of specific gravity differences between the fat and aqueous phases.

B. Particulate Dry Foods

Dry-mix fortification presents minimal technological problems. For example, the reliability of fortifying flour by continuous metering of a dry, free-flowing premix into flour streams was shown by Fortmann et al. [28], who took flour samples at 15-min intervals while processing a carload (1000 cwt.) of bulk flour at a flour mill. The flour was then resampled at the bakery as it was fed from the bakery storage bins into the dough mixers. The vitamin B_1 varied from 2.31-2.56 mg/lb (5.1-5.6 mg/kg) and vitamin B_2 from 1.25 to 1.34 mg/lb (2.75-2.94 mg/kg) in 14 samples at the bakery. Reduced iron analyses were equally satisfactory, ranging from 15.3 to 16.4 mg per lb of flour (33.8 to 36.2 mg per kg).

1. Dehydrated Potato Flakes

It is more difficult to fortify products such as dehydrated potato flakes or breakfast cereal flakes than flour, cake mixes, or other fine-particle dry products. Dehydrated potato flakes can be fortified at the mash stage prior to dehydration. However, two problems exist in this procedure, namely degradation of vitamins A and C and pinking. Degradation of A and C is actually an economic problem since high overages can be used to make up for process losses. Pinking is an unpredictable problem due to vitamin C. Apparently, ascorbate reacts with potato protein causing a pink Schiff-base compound. Pinking occurs erratically and takes time to appear after dehydration. Oxygen is necessary for this reaction since nitrogen flush packaged flakes do not pink until opened and exposed to air.

Flake fortification after dehydration with dry vitamins presents the serious problem of vitamin segregation in the retail package. This problem was solved in a patented process by Pedersen and Sautier [29]. "Vitamin flakes" of approximately the same size and shape as potato flakes are mixed with the potato flakes and are nonsegregating and difficult to detect visually. These vitamin flakes are actually 50-75% fat, m.p. 110-165°F (43.3-73.9°C), containing water-soluble vitamins and minerals. The stability data on vitamins C, B_1, B_2, and niacin shown in this patent are satisfactory. It is essential in this approach that the vitamin flakes be added uniformly to the potato flakes and that they be organoleptically satisfactory.

2. Breakfast Cereals

A different approach to the same type of physical distribution problem was used by Duvall and Stone [30] for the fortification of ready-to-eat cereals. In their process a precooked cereal is coated with a sugar solution, dried, and then coated with dry vitamins while the cereal is hot and tacky. In this patent the preferred procedure is use of fatty coated vitamins to prevent formation of undesirable flavor and to help the vitamins to adhere to the cereal.

According to Brooke [31], heat-labile vitamins such as B1, C, and A are usually sprayed onto toasted breakfast cereals as they leave the oven. Emulsions (or water-dispersible solutions) of vitamin A can be sprayed onto breakfast cereals after toasting, but stability during product storage is greatly dependent upon the composition of the spray solution. A high carbohydrate level in the spray is desirable. Added sugars, in effect, coat the vitamin A and act as an oxygen barrier during storage. Losses during process and cereal storage cannot be specified because of the large number of vitamin A market forms available, and cereal product and process variables. Vitamin A overages above label claim would usually be 30-50% for ready-to-eat cereals.

If a product is to be fortified with both A and D, it is best to add vitamin D at the same stage as vitamin A. The vitamin D market form should be similar in composition and stability to the vitamin A market form so that the vitamin A stability profile can be used to monitor vitamin D. This is necessary because of the extremely poor accuracy of vitamin D assays at food fortification levels, making it almost impossible to study vitamin D stability in foods.

A more in-depth discussion of fortification of cereal products is available [11].

3. Tea

Spraying particulate foods to optimize distribution of added fortificants was recommended by Brooke and Cort [32] for fortification of tea leaves, as distinct from tea dust, with vitamin A. In this study, oil-in-water emulsions of vitamin A palmitate or acetate, diluted in 50% sucrose solution and sprayed on tea leaves, resulted in 90% retention after 6 months storage at 37°C. This is excellent stability for vitamin A, substantiating the fact that coating vitamin A in situ with carbohydrates improves stability.

A further stress for vitamin A in this application is the tea-brewing step. The authors found 100% retention of vitamin A palmitate after cooking tea for 1 hr in boiling water, but only 4% retention using the vitamin A acetate emulsion-spray system. This differential in stability between the two esters is not explained.

4. Potato Chips

Potato chip fortification presents another technological challenge. In the usual potato chip process, whole potatoes are washed, peeled, sliced, washed, fried, salted and/or seasoned, and packaged. There are few suitable in-process points to add vitamins with respect to ensuring adequate physical distribution and minimum degradation. Addition to the frying oil would result in excessive destruction of vitamin A and possibly B_1, and in poor absorption of the water-soluble vitamins by the chips. For development purposes, the salting or seasoning step was chosen as the addition point of vitamins A, C, B_1, B_2, niacin, and E [33]. In this approach a dry premix of all the added micronutrients (using -100 mesh dry market forms to optimize physical distribution) was blended with the proper proportion of salt or seasoning and applied to the fried chips using commercial salters. Controlling the uniformity of salt addition is obviously critical in this procedure, and in some plant experiments the quality control was inadequate. Uniformity of distribution and stability of the B vitamins, C, and E does not appear to be a problem. Vitamin A poses the major problem with respect to uniformity of addition, adhesion to the chips, and stability. Of particular concern is vitamin A stability during light exposure of fortified chips packaged in clear laminates. In one experiment, chips were exposed to fluorescent light with the results shown in Table 1. Although light intensity was not measured, 1-month light exposure in this test is believed to be equivalent to a much longer period of exposure in a supermakret and, hence the stability results are satisfactory. It is important to evaluate stability under typical packaging, distribution, and use conditions in all fortification projects.

5. Textured Soy Protein

In the author's experience, textured soy protein can be fortified either before or after granulation or "texturization" of the soy flour, following the processor's personal preference as to production ease and control. The two major concerns in soy fortification are uniformity of distribution of the added micronutrients and compliance with label claims. The FDA [34] has proposed new micronutrient specifications for these products which include magnesium, iron, zinc, calcium, phosphorus, vitamins A, B_1, B_2, B_6, B_{12}, niacin, pantothenic acid, and folate. The most stable vitamin in this proposed specification is niacin. Riboflavin and B_6 are somewhat less stable; and vitamins A, B_1, B_{12}, and calcium pantothenate present the highest potential losses.

TABLE 1 Vitamin Stability of Fortified Potato Chips Exposed to Fluorescent Lights

Vitamin	Trial A		Trial B	
	Initial	4 wks-RT	Initial	4 wks-RT
A, IU/lb	4000	2750	4485	4160
E, IU/lb	21	23	31	29.8
B_1, mg/lb	2.12	1.67	1.76	1.65
B_2, mg/lb	2.47	2.45	2.17	2.07

A convenient way to monitor fortification and improve physical distribution in soy fortification is to premix all the micronutrients, thus adding 50-100 mg of premixed ingredients to 100 g of soy protein instead of each ingredient separately. This premix can be added directly to the soy flour before granulation by batch mixing, continuous metering, or with the "dough water", if the production process and equipment lends itself to this approach. The major concern when the premix is added prior to texturizing is stability during processing. In our experience, the short processing time required to texturize with heat, pressure, and high moisture does not significantly degrade B_1, B_{12}, and pantothenic acid. Few data are available on the stability of vitamin A and folate in this process since this is a relatively new proposal and it is important that feasibility be determined prior to finalization of nutritional guidelines.

Some processors prefer to add the micronutrients to the finished textured protein via a spray of the premix suspended in vegetable oil. In this approach uniformity of addition is the most serious problem and requires careful engineering and control of the spray system.

The natural micronutrient content of soy protein is high and it is desirable to make a conservative calculation of these levels and adjust the fortification addition levels accordingly. For example, soy protein contains well over the 57 mg magnesium per 100-g level proposed by FDA and there is no reason to add magnesium to comply with this specification.

C. Sausage Products

The premix approach to fortifying sausage products at the stage of adding seasoning and curing salt has been successful with vitamins B_1, B_2, niacin, B_6 and B_{12}. As expected from experience using ascorbates as a cured pigment synthesis accelerator in curing meat products, stability of vitamin C is highly variable depending on the amount of entrained air, cooking variables, and packaging. Vitamin stability during home preparation of fortified weiners is shown in Table 2. In this commercial scale test [35] in-process stability of the six added vitamins was satisfactory and refrigerated wieners, approximately 35 g each, were then heated by three commonly used home procedures with the excellent results shown.

The use of sodium erythorbate, as well as sodium ascorbate, in cured meat products as a cure accelerator can cause analytical confusion in determining the vitamin C content of these products, since both compounds are reducing agents and are indistinguishable in the titrations commonly used for vitamin C. Analysts reporting vitamin C values in cured meats should be certain that the products do not contain erythorbate.

TABLE 2 Vitamin Stability of Fortified Wieners During Home Cooking

Vitamin	Content per 35-g wiener			
	Uncooked[a]	Boiled	Broiled	Fried
A, IU	1500	1500	1500	1500
C, mg	19.8	19.8	17.3	19.3
B_1, mg	0.59	0.55	0.50	0.53
B_2, mg	0.72	0.76	0.71	0.71
B_6, mg	0.86	0.74	0.79	0.82
B_{12}, μg	3.7	3.5	3.4	3.5

[a]As is.

IV. LEGAL ASPECTS

There is an interrelationship between the technology of fortification and federal regulations which is pertinent to this discussion. The FDA has established sampling procedures, reference molecular weights for the vitamins, and classes of nutrients for compliance, and has discussed in the Federal Register both analytical error and the need for overages in fortification.

For compliance purposes two classes of nutrients are defined [36]:

Class I: Added nutrients in fortified or fabricated foods.
Class II: Naturally occurring nutrients.

Under Class I, a food is in compliance if the nutrient content of the composite sample assayed is at least equal to the value for that nutrient declared on the label. Under Class II, the nutrient content must at least equal 80% of the label claim.

The same section of the regulations defines food sampling techniques:

> The sample for nutrient analysis shall consist of a composite of 12 subsamples (consumer units), taken one from each of 12 different randomly chosen shipping cases, to be representative of a lot. . .

In addition to the above, the Division of Mathematics of FDA has issued a pamphlet to explain the detailed procedures to be employed by the FDA to evaluate compliance with nutrition labeling regulations [37].

In both classes of nutrients, no regulatory action by FDA will be based on a determination of a nutrient value which falls below this level by a factor less than the variability generally recognized for the analytical method used in that food at the level involved. This is very important since the accuracy of vitamin assays at low levels in foods can be very poor.

Equally important, reasonable excesses of a vitamin, mineral, or protein over labeled amounts are acceptable within good manufacturing practices. These expressions of FDA philosophy are extremely important in designing stability studies, evaluating stability, and deciding required overages in fortification.

The FDA [38] also established reference molecular weights for nine vitamins (see Table 3), which helps to eliminate confusion in determining label claim compliance. The equivalent weights of some of the commonly used commercial vitamin compounds corresponding to the reference standards are shown below.

1.00 mg thiamine hydrochloride	=	1.00 mg reference B_1
0.97 mg thiamine mononitrate	=	1.00 mg reference B_1
1.21 mg pyridoxine hydrochloride	=	1.00 mg reference B_6
1.12 mg sodium ascorbate	=	1.00 mg reference C
1.09 mg D-calcium pantothenate	=	1.00 mg pantothenic acid

Label claim compliance is required by FDA up to the point of retail sale, not after consumer preparation of the food. Apparently, FDA feels that proving compliance with the variability of home preparation procedures would be very difficult. On the other hand, the food processor has a moral obligation to compensate for obvious vitamin destruction during home preparation, in the author's judgment. For example, if preparation at home of vitamin C fortified gelatin dessert powder causes 20% degradation, appropriate overages should be added.

The FDA established U.S. Recommended Daily Allowances for essentially all nutrients to standardize labeling of foods [36]. This eliminates the use of the Minimum Daily Requirements established by FDA in 1941

TABLE 3 FDA Vitamin Reference Forms

Vitamin	Name	Molecular weight
Vitamin C	L-ascorbic acid	176.12
Folic acid	Pteroyl mono-L-glutamic acid	441.41
Niacin	Nicotinic acid	123.11
Riboflavin	Riboflavin	376.37
Thiamine	Thiamine chloride hydrochloride	337.28
Vitamin B_6	Pyridoxine	169.18
Vitamin B_{12}	Cyanocobalamin	1,355.40
Biotin	D-biotin	244.31
Pantothenic acid	D-pantothenic acid	219.23

for the small group of vitamins and minerals whose nutritional requirements had been determined at that time.

Many other regulations related to fortification have been issued or proposed in 1973 and 1974, but they do not specifically relate to its technological aspects. The reader is urged to keep abreast of new FDA Nutritional Guidelines for Classes of Foods and General Principles of Food Fortification.

REFERENCES

1. Food and Nutrition Board, NAS/NRC, Proposed Fortification Policy for Cereal-Grain Products, Washington, 1974.
2. M. F. Furter, W. M. Lauter, E. De Ritter, and S. H. Rubin, Ind. Eng. Chem. 38:486 (1946).
3. FDA, Fed. Reg. 39 (116):20904 (1974).
4. B. Borenstein, Food Technol. 27(6):32 (1973).
5. G. S. Ranhotra, R. J. Loewe, and L. V. Puyat, Cereal Chem. 54:496 (1977).
6. G. S. Ranhotra, R. J. Loewe, and L. V. Puyat, Cereal Chem. 53:770 (1976).
7. G. S. Ranhotra, R. J. Loewe, T. A. Lehmann, and F. N. Hepburn, J. Food Science. 41:952 (1976).
8. G. Beetner, T. Tsao, A. Frey, and J. Harper, J. Food Sci. 39:207 (1974).

9. S. Brenner, V. O. Wodicka, and S. G. Dunlop, Food Technol. 2:207 (1948).
10. B. Borenstein, Crit. Rev. Food Technol. 2:171 (1971).
11. B. Borenstein, in Wheat: Production and Utilization (G. E. Inglett, ed.), Avi Publishing, Westport, Conn., 1974, pp. 366-383.
12. S. H. Rubin and W. M. Cort, in Protein-Enriched Cereal Foods for World Needs (M. Milner, ed.), American Association of Cereal Chemists, St. Paul, Minn., 1969, pp. 220-233.
13. A. E. Bender, J. Food Technol. 1:261 (1966).
14. H. A. Schroeder, Am. J. Clin. Nutr. 24:562 (1971).
15. B. Borenstein, in Handbook of Food Additives (T. E. Furia, ed.), Chemical Rubber Co., Cleveland, Ohio, 1968, pp. 107-136.
16. J. C. Bauernfeind and W. M. Cort, Crit. Rev. Food Technol. 5:337 (1974).
17. Food and Nutrition Board, NAS/NRC, Recommended Dietary Allowances, 8th ed., 1974, Washington, D.C.
18. W. Marusich, E. De Ritter, and J. C. Bauernfeind, J. Am. Oil Chem. Soc. 34:217 (1957).
19. J. C. Bauernfeind, M. Osadca, and R. H. Bunnell, Food Technol. 16:101 (1962).
20. M. H. Yu, M. T. Wu, D. J. Wang, and D. K. Salunkhe, Can. Inst. Food Sci. Technol. J. 7:279 (1974).
21. G. H. Beston and G. A. Henderson, Can. Inst. Food Sci. Technol. J. 7:183 (1974).
22. O. Pelletier and M. O. Keith, J. Amer. Diet. Assoc. 64:271 (1974).
23. O. W. Bissett and R. E. Berry, J. Food Sci. 40:178 (1975).
24. R. P. Singh, D. R. Heldman, and J. R. Kirk, J. Food Sci. 40:164 (1975).
25. D. B. Dennison and J. R. Kirk, J. Food Sci. 43:609 (1978).
26. P. A. Seib and R. C. Hoseney, Baker's Dig. 48(5):46 (1974).
27. N. B. Guerrant and M. B. O'Hara, Food Technol. 7:473 (1953).
28. K. L. Fortmann, R. R. Joiner, and F. D. Vidal, Baker's Dig. 48(3):42 (1974).
29. D. C. Pedersen and P. M. Sautier, U. S. Patent 3,833,739, Sept. 3, 1974.
30. L. F. Duvall and C. D. Stone, U.S. Patent 3,782,963, Jan. 1, 1974.
31. C. L. Brooke, J. Agr. Food Chem. 16:163 (1968).
32. C. L. Brooke and W. M. Cort, Food Technol. 26(6):50 (1972).
33. H. T. Gordon and B. Borenstein, unpublished work, Hoffmann-La Roche, 1970.
34. FDA, Fed. Reg. 39(116):20894 (June 14, 1974).
35. H. T. Gordon, unpublished work, Hoffmann-LaRoche, 1974.
36. FDA, Fed. Reg. 38(49):6960 (March 14, 1973).
37. FDA, Division of Mathematics Compliance Procedures for Nutrition Labeling, Washington, 1973.
38. FDA, Fed. Reg. 38(148):20717 (Aug. 2, 1973).

CHAPTER 8

PREDICTION OF NUTRIENT LOSSES AND OPTIMIZATION OF PROCESSING CONDITIONS

Marcus Karel

Department of Nutrition and Food Science
Massachusetts Institute of Technology
Cambridge, Massachusetts

I. INTRODUCTION

The nutrient content of foods at the time of consumption depends on the composition of the raw materials, on the history of the food, and, in particular, on conditions of processing, storage, and preparation prior to consumption. Interest in the nutrient content of food has recently increased because of the growing public awareness of nutritional aspects of public health and, at least in part, because of the FDA nutritional labeling program. The food industry and government agencies are spending significant amounts of time and money to update information on the nutrient content of processed foods [1].

If we consider the great variety of processing modifications and the practically infinite number of variations in initial composition, methods of processing, conditions of storage and distribution, and handling histories of individual foods, we can see that estimating the nutritional value of a given food at the time of consumption is an enormously complicated task. Recently, efforts have been initiated to provide a rational basis for such estimations using accelerated storage tests and models of specific food processing operations. This chapter is an attempt to provide a status report on work in a rapidly developing field and to indicate the probable directions in which this work will progress. We shall consider the influence of various environmental factors on nutrient retention and how a mathematical description of this influence, obtained in accelerated tests, may aid in the design and control of food processing operations.

II. KINETICS OF NUTRIENT LOSSES AND DEPENDENCE ON ENVIRONMENTAL FACTORS AND ON FOOD COMPOSITION

A. Overall Considerations

In order to optimize nutrient retention in processes in which environmental factors and/or composition are controllable variables, it is necessary to know the relation between the nutrient destruction rate and these factors. This takes the form shown in Eq. (1).

$$-\frac{dC}{dt} = f(E_1, \ldots, E_n, F_1, \ldots, F_n) \qquad (1)$$

where

C = concentration of the nutrient
E = environmental factor, for instance, temperature
F = composition factor, for instance, pH

It should be noted that the concentration (C) of the nutrient constitutes one of the composition factors (F).

The effect of some of these factors is relatively easy to express in mathematical terms in a form that is similar for very diverse nutrients and food systems. Others are very specific and must be dealt with separately for each particular situation.

B. Temperature

The most common and generally valid assumption is that the temperature-dependence of nutrient destruction will follow the Arrhenius relation:

$$k = k_o e^{-E/RT} \tag{2}$$

where

k = rate constant in the nutrient destruction reaction
k_o = constant independent of temperature
E = activation energy
T = absolute temperature

The Arrhenius relation is usually applicable; its use constitutes the soundest approach to modeling temperature dependence. The activation energy (E) does, however, vary with concentration and other composition factors. Furthermore, when the reaction mechanism changes with temperature, the activation energy may vary.

Functions other than the Arrhenius relation have been occasionally suggested as suitable for correlation of food stability data. Kwolek and Bookwalter [2] considered zero-order food deterioration reactions represented by Eq. (3).

$$Y = Y_o + f(T)t \tag{3}$$

where

Y = index of quality (for instance, nutrient content), and
t = time.

They considered, among others, the following forms of temperature dependence:

linear $$f(T) = a + bT \tag{4}$$

exponential $$f(T) = aT^b \tag{5}$$

hyperbolic $$f(T) = \frac{a}{b-T} \tag{6}$$

They found that the hyperbolic model and the Arrhenius equation gave the best correlation and applied over the widest range of temperatures when tested with data selected from three published stability studies, which included both organoleptic and chemical food quality indexes.

Over a limited temperature range, however, a linear approximation may be used, as Olley and Ratkowsky [3] demonstrated for various refrigerated foods. These studies led them to the conclusion that, in the temperature range of 0 to 6°C, a linear approximation was adequate for predicting the temperature effect on a variety of deterioration reactions in refrigerated fish, beef, and poultry. Furthermore, they claimed that a single proportionality constant is adequate for the many diverse reactions reported in the literature for these items.

In most cases, the use of the Arrhenius equation appears to be the most appropriate method even when other approximations offer equal accuracy over a limited range of temperatures. The inconvenience of using this nonlinear relation rather than linear approximations over a limited range is readily overcome through use of electronic calculators, graphical methods, and other simple techniques. It should be noted, however, that all equations, including the Arrhenius relation, have limited applicability and the range of validity and the influence of other factors on activation energy must be taken into consideration [4,5].

Special consideration must be given to the methods adopted by most food technologists in describing the kinetics of inactivation of microorganisms during pasteurization and sterilization processes, since, as we shall discuss later, a major problem in heat-processing food is to achieve "commercial sterility" with a minimum level of nutrient destruction. It seems necessary, therefore, to compare effects of temperature on nutrient destruction and on inactivation of microorganisms using the same terminology. Unfortunately, most examples in the literature use the Arrhenius equation for prediction of nutrient destruction, but the so-called "thermal death time" relations for inactivation of microorganisms. Fortunately, the two methods are readily interconvertible.

Death of microorganisms because of exposure to a given temperature in the lethal range (above growth temperatures) usually follows first-order kinetics:

$$-\frac{dn}{dt} = kN \tag{7}$$

$$\ln \frac{N}{N_o} = kt \tag{8}$$

where

N = number of viable microorganisms at time t
N_o = number of viable microorganisms at beginning of heating (t=o)
t = time
k = a constant

The value of k depends strongly on many factors, including temperature. Since the survival is logarithmic, it is possible to specify a time corresponding to a fractional reduction in the microbial population. In particular, it is common to calculate a so-called D value (decimal reduction value) corresponding to a reduction of N by a factor of 10. It can be readily demonstrated that D is related to k as shown in Eq. (9) [6].

$$D = \frac{\ln 10}{k} = \frac{2.303}{k} \tag{9}$$

Commercial sterility is usually based on reducing the probability of microbial survival to a given and (very low) level. Reduction by a factor 10^{12} (12D value) is often the basis for such sterility. In any case, it is possible to define a so-called "thermal death time" (TDT) which is defined as the time needed to achieve a given reduction in the probability of survival of microorganisms.

$$\mathrm{TDT} = \frac{-\ln (N_F/N_o) \times D}{2.303} \tag{10}$$

where N_F/N_o is the desired ratio of final to initial concentration of microorganisms.

Inactivation of microorganisms usually follows the Arrhenius equation, and it is possible to use this equation in describing the temperature dependence of k (and obviously of D and TDT, since they are related to k). However, the usual practice in food technology is to use Eq. (11):

$$\log \left(\frac{\mathrm{TDT}_1}{\mathrm{TDT}_2} \right) = \frac{-1}{z} (T_1 - T_2) = \frac{1}{z} (T_2 - T_1) \tag{11}$$

where subscripts 1 and 2 refer to temperatures T_1 and T_2, and z is a constant. It is the usual practice to express T_1 and T_2 in °F, in which case z is the temperature increase in °F required to reduce TDT by a factor of 10. It is, furthermore, a common convention to use a reference temperature T_r (usually 250°F), in which case Eq. (11) becomes

$$\log \left(\frac{\mathrm{TDT}}{\mathrm{TDT}_r} \right) = \frac{1}{z} (T_r - T) \tag{12}$$

When the reference temperature is 250°F, and TDT corresponds to commercial sterility, it is usually called F_o and is expressed in minutes. The effectiveness of heating at other temperatures is then often expressed as the equivalent time (in minutes) at 250°F and is called the F value.

The relationship between z and E shown in Fig. 1 allows the use of a common basis for sterilization and nutrient destruction effects of heating.

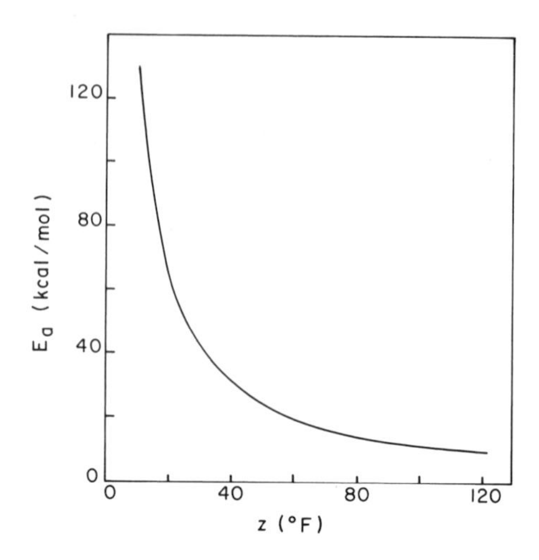

Fig. 1. Relationship between activation energy and z value (from Karel, Fennema, and Lund [6], p. 45).

C. Water Activity and Water Content

Water is a major constituent of foods and plays an important role in controlling rates of deteriorative reactions, including those resulting in nutrient losses. It has been widely recognized that water content, per se, is not always the best parameter to use in expressing the effects of water on reaction rates, and water activity (a) is often used.

$$a = (p_f/p_o)_T \tag{13}$$

where

a = water activity
pf = partial pressure of water in food
p_o = vapor pressure of water at temperature (T)

Onset of growth of microorganisms is often readily correlated with a critical level of water activity, but most chemical reactions are influenced in a more complex manner. This is not altogether surprising because water can act in one of the following roles:

1. Solvent for reactants and/or products
2. Reactant (in hydrolytic reactions)
3. Reaction product (in condensation reactions)
4. Modifier of activities of catalysts and inhibitors

A number of relations have been suggested to describe the dependence of reaction rates on moisture content (or on water activity). Several authors who studied browning of vegetables reported an exponential relation between the rate of browning and moisture content [7,8]. Mizrahi et al. [9] found that browning rates for cabbage could be correlated by Eq. (14).

$$\text{Rate of browning} \propto [1 + \sin(-\pi/2 + m\pi/k^n] \tag{14}$$

where m is moisture content, and k and n are constants.

Oxidation of potato chips was investigated by Quast and Karel [10], who reported that oxidation was related to the function ($a^{-1/2}$), where a = water activity. Sometimes a linear relation between (m) and reaction rate or between (a) and reaction rate provides a satisfactory approximation of the behavior of nutrient retention in dry foods. Beetner et al. [11,12] found that retention of thiamine and riboflavin in extrusion is affected by several variables (including moisture content), each of which could be approximated by a linear relationship that was valid within the narrow range of experimental values studied. Labuza [13] reviewed studies on nutrient losses during dehydration and storage and suggested that the data of several investigators who studied retention of ascorbic acid show a simple correlation of the type shown in Fig. 2, which implies a linear relation between the logarithm of the reaction rate constant and water activity. Wanninger [14] proposed a model in which the reaction rate constant is directly proportional to water concentration. More recent studies [15] on several different systems, however, indicated that ascorbic acid loss can depend on water in a more complex manner than the simple relations mentioned above.

D. Other Environmental and Composition Factors

Partial pressure of oxygen, intensity of light, pH, and concentrations of specific components (for instance, metals) all influence the kinetics of nutrient loss in food. Unfortunately, correlations giving generally valid relations between these variables and rates of nutrient loss are simply not available.

In the case of concentration of the nutrient itself, the order of reaction in food is often similar to that observed in solution. Significant deviations, however, can occur. Ascorbic acid loss has been reported to follow either zero- or first-order kinetics, depending on the food and on conditions of the study. Furthermore, in a model system described by Karel and Mizrahi [15], the apparent rate of ascorbic acid loss also depended on the initial concentration of the vitamin, although the rate followed first-order kinetics at any given initial concentration. In another study, it was observed that loss of amino acids during heating of milk powder followed first-order kinetics for tryptophan but fourth-order kinetics for lysine [16].

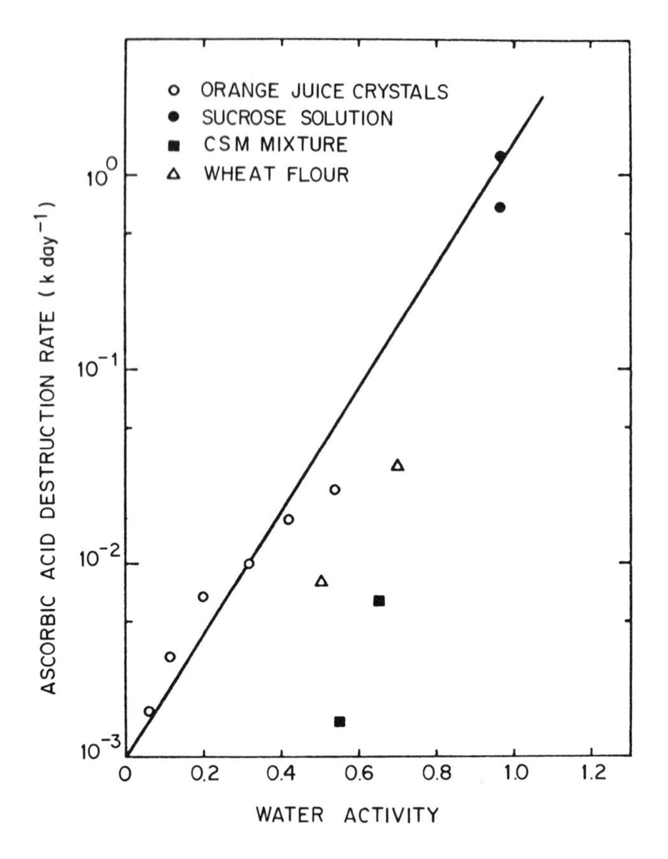

Fig. 2. Ascorbic acid destruction rate as a function of water activity (from Labuza [13], with permission from CRC Press, Inc.).

When the effect of oxygen is considered, it is often simply a question of total amount available for reaction with food components. If this amount is limited to a level that causes no significant effect in the food, and there is no potential for additional oxygen coming into contact with the food, then the rate of reaction is irrelevant. In other cases, the total amount of oxygen potentially able to react with nutrients is, in fact, significant, and consideration must be given to the effect of concentration (or partial pressure) of oxygen on the rate. A general form often found useful in correlating reaction rate with oxygen concentration is given in Eq. (15).

$$R = \frac{[O_2]}{k_1 + k_2 [O_2]} \tag{15}$$

where

R = reaction rate
$[O_2]$ = oxygen concentration
k_1, k_2 = constants

Note that when $k_2[O_2]$ is much smaller than k_1, the rate is linear with oxygen concentration, and when $k_2[O_2]$ is much larger than k_1, the rate becomes independent of concentration. Both of these extremes can occur in the same product, depending on temperature, surface-to-volume ratio, oxygen pressure, and extent of reaction [17-19].

The effects of light on chemical reactions involving nutrients are equally complex. Riboflavin is especially photosensitive and, when exposed to sunlight, loses its vitamin value and activates or sensitizes other components to photodegradation. Ascorbic acid is also quite sensitive to light and interacts with other components during light exposure.

The catalytic effects of light are most pronounced with light of the highest quantum energy, i.e., light in the lower wavelengths of the visible spectrum and in the ultraviolet spectrum. Specific reactions involved in deterioration of foods, however, may have specific wavelength optima; in particular, the presence of sensitizers may shift the effective spectrum substantially. Sensitizers present in food include riboflavin, β-carotene, vitamin A, and peroxidized fatty acids.

In addition to wavelength, the intensity of light and duration of exposure are important. In this respect, we again must distinguish among the sensitivities of different products.

The most penetrating light quanta are those with wavelengths that are least effective per quantum of energy. Thus, ultraviolet rays penetrate less deeply into food than the red-light components. In some cases, the relatively shallow penetration of light into food products is an adequate protection in itself. Consider, for instance, the potential for loss of riboflavin in bread. In breads with well-developed crusts, 95% of the short-wavelength light is absorbed in the outer 2-mm layer of bread, and the amount of vitamin exposed to light is not significant. On the other hand, in liquid products, absorption of light in the outer layer does not guarantee adequate protection because either the light-susceptible nutrient can diffuse into the illuminated zone, or reactive intermediates formed in the illuminated zone can diffuse to react with nutrients shielded from light [6]. Quantitative studies on the effects of light on reactions involving nutrients in foods are very limited and do not lend themselves to the formulation of models for predicting nutrient retention as a function of history of exposure to light [20-25].

III. ACCELERATED TESTS

A. General Considerations

Recent studies have shown that it is feasible to predict changes during processing and storage of foods by combining equations expressing the dependence of nutrient deterioration on environmental factors with equations describing how the environmental factors themselves change with time during processing and storage. An example of application of this approach to "moisture-sensitive" foods is provided by the studies of Aguilera et al. [26] in which nonenzymatic browning in vegetables was predicted on the basis of previously established relations between moisture content and reaction rate, relations between temperature and reaction rate (Arrhenius equation), experimentally established relations between activation energy and moisture content, and mass transport data for moisture content changes in dehydration.

Some recent reviews and research papers treating other examples include Labuza [13] on effects of dehydration and storage; Kramer [27], Charm et al. [28], Herrmann [29] on storage effects, Lund [30] and Teixera et al. [31] on effects of heat processing. Effects of light and oxygen were simulated for liquid foods by Singh et al. [23], and the problems of storage of oxygen-sensitive foods were treated by Quast et al. [32], Quast and Karel [10,33], and Herlitz et al. [18]. Nutrients retained in extrusion were studied by Beetner et al. [12]. A review of nutrient retention in packaged food was provided by Karel and Heidelbaugh [34].

One of the major problems to be resolved is the development of effective accelerated tests. The preceding discussion shows that the dependence of the kinetics of nutrient loss on environmental factors varies from product to product and process to process and must, therefore, be established for each specific case to be evaluated. Unless accelerated tests are used for this purpose, the time involved becomes prohibitively long.

The problem of developing accelerated tests for nutrient stability during processing of foods amounts, therefore, to finding more rapid and convenient ways to obtain the kinetic constants for deterioration reactions, since approaches for rapid evaluation of the mass transfer constants and for temperature distribution are generally available [35,19].

Kinetic data are typically correlated in the form of Eq. (16):

$$\frac{dC}{dt} = f(T, m, O_2, C, t, \ldots,) \tag{16}$$

where the independent variables in parentheses may include, in addition to temperature (T): moisture content (m), oxygen pressure (O_2), nutrient concentration (C), time (t), and other pertinent variables. The extent of reaction is computed by integration of Eq. (16). The currently common method of evaluating the parameters is based on a balanced matrix-type of experiment. Only one variable is changed at a time and usually within a wide range. This method has the advantage of permitting two-dimensional plotting of the data for each variable separately. In that way, the procedure of formulating the mathematical model is simplified. This method is time consuming—hence, the need for acceleration.

Accelerated tests in complicated food systems present a problem because previous kinetic data are scarce or unavailable. Therefore, accelerated tests must be developed in the face of many unknowns. In this respect, foods present more difficult problems than other chemical systems.

Theoretically, there are two general approaches to accelerated tests: the "known model" and the "unknown model" approach. The "known model" approach is based on the previously established form of mathematical expression shown in Eq. (16). The "unknown model" approach, on the other hand, applies to cases in which the pattern is unknown for one or more variables.

The form of the function relating nutrient deterioration to the independent variables shown in Eq. (16) is often known for some of these variables. The variables for which models are often available are temperature and concentration. The Arrhenius equation is usually applicable for temperature dependence, and concentration dependence is expressed most often as "an order of reaction." When the order of reaction is known and the Arrhenius equation applies with a constant activation energy, it is possible to utilize either isothermal or nonisothermal accelerated stability tests using elevated temperatures.

B. Accelerated Tests at Elevated Temperatures

Isothermal accelerated stability tests have been used extensively for several decades by industry as well as by government agencies. Typically, foods are subjected to storage at 37 and 51°C, and various correlations (usually based on the Arrhenius relation or simplifications of that relation, such as the Q_{10} concept) are used to extrapolate the results to the expected storage temperatures. When more accuracy is desired, several elevated storage temperatures are used and the Q_{10} value, or activation energy, is determined experimentally, based on the assumption that it is independent of temperature.

Another approach based on prior knowledge of reaction order and on the assumption of constant activation energy employs nonisothermal stability tests. These tests utilize a programmed temperature change during the reaction. The linear nonisothermal stability tests, for instance, involve a programmed linear rise in temperature as the reaction proceeds. The theory of linear nonisothermal kinetics is based on the relationship between three equations, Eqs. (17-19):

$$-\frac{dC}{dt} = kC^n \tag{17}$$

where

n = order of reaction
k = temperature-dependent reaction rate constant

$$k = k_o e^{-E/RT} \tag{18}$$

where

E = energy of activation
T = temperature in K
k_o = constant
R = molar gas constant

$$T = bt + T_i \tag{19}$$

where

b = heating rate constant K/min
t = time in minutes
T_i = initial temperature

For the case of first-order reactions ($n = 1$), the combined equations can be integrated to give Eq. (20):

$$\ln\left(\frac{C}{C_i}\right) = \frac{-k_i}{b} e^{(E/RT_i)} \int_{T_i}^{T} e^{(-E/RT)} \, dT \tag{20}$$

where

C_i = initial nutrient concentration
T_i = initial temperature
k_i = reaction rate constant at temperature T_i
C = nutrient concentration when temperature equals T

Since $(e^{-1/T})\,dT$ does not have an exact solution, numerical solutions using a digital computer are used to evaluate the integral and to calculate E and k [36]. An analog computer has also been used for this purpose [37]. Zoglio et al. [38] used the graphic technique demonstrated in Fig. 3. A series of model curves is constructed of log $(C/C_i \times 100)$ vs. T, each assuming a given activation energy. The experimental points are plotted in the same graph and compared with the model curves in order to estimate E.

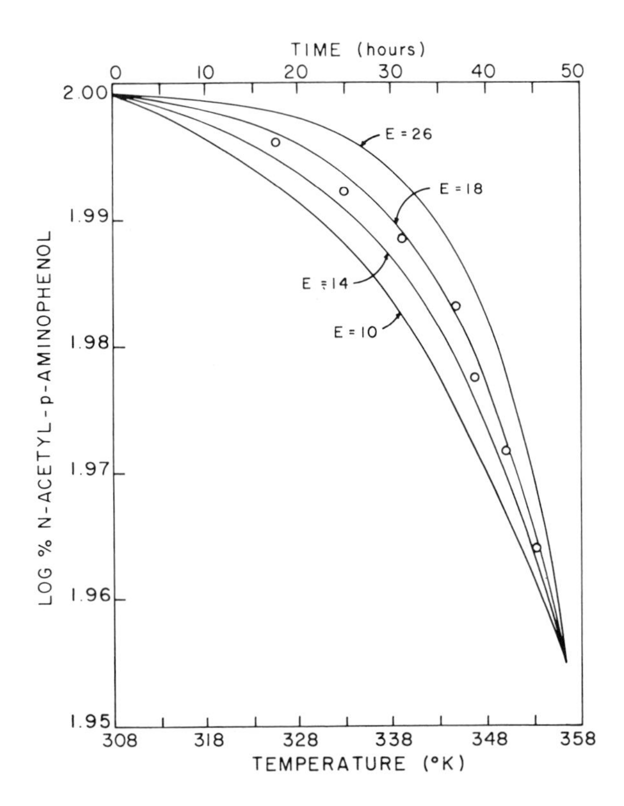

Fig. 3. Nonisothermal hydrolysis of N-acetyl-p-aminophenol in aqueous solution (literature value, E = 16.7 kcal/mol). Solid lines correspond to different assumed values of E (in kcal/mol). Circles represent experimental points. (From Zoglio et al. [38], with permission of copyright holder.)

The method has been applied in evaluations of stability of drugs, dried influenza virus, and solutions of individual nutrients. Another approach to nonisothermal methods is to use nonlinear temperature programming. In fact, one of the earliest nonisothermal studies [39] used a logarithmic temperature rise in investigations of inversion of sucrose and stability of vitamins in solution. Maulding and Zoglio [40] developed a method in which temperature is changed in any manner consistent with analytical findings to give maximum flexibility to the investigator.

The proponents of nonisothermal methods claim that they have certain advantages, such as use of a single experimental unit, the analysis of one set of samples, and potentially shorter overall time for completion of the experiment. The method does, however, present major disadvantages, namely need for a separate experiment to determine the order of reaction, mathematical complexity, applicability of the Arrhenius equation, and, perhaps most importantly, a need for extreme accuracy and stability of the temperature program. Finally, since the method requires rapid equilibration and mixing, it appears applicable only to liquid samples with relatively low viscosity.

C. Accelerated Tests at High Oxygen Pressures

Reactions involving oxidation can sometimes be accelerated by conducting stability tests at high oxygen pressures. The potential for acceleration is not very great because oxidation reactions typically become independent of oxygen concentration above some critical concentration level, which varies with temperature and other conditions. Nevertheless, traditional accelerated tests for oxidation-susceptible oils use exposure to high oxygen pressures to assess the relative stability of such oils. There are, however, no acceptable conversion factors that allow oil stability at low oxygen pressure to be predicted on the basis of tests conducted at elevated pressures, and the value of these accelerated tests is limited [41].

D. Accelerated Tests Using High Moisture Contents

1. "Known Model" Approach

As mentioned previously, accelerated tests are facilitated when the general form of the reaction's dependence on a given environmental factor is known ("known model" approach). This is true in the case of accelerated tests using elevated moisture content to speed up the reaction. Mizrahi et al. [42], using dehydrated cabbage as a model food, have shown that once the general form of the dependence of nonenzymatic browning on moisture content is known and the relation of activation energy to moisture content established, an accelerated test can be used to predict stability with substantial accuracy. By conducting tests at high temperatures can high moisture contents, they were able to predict stability from tests requiring only ten days, instead of the 200 days that would be needed to obtain stability data under actual storage conditions.

However, "known models" for the effects of moisture content on nutrient destruction are very limited, and, if accelerated tests using moisture content exaggeration are to be applied, the "unknown model" approach must be used.

2. "Unknown Model" Approach

There are two theoretical ways to overcome the lack of a model, that is, employing accelerated tests designed for model formulation or using a "no model" method. The first method suggests the use of an appropriate accelerated test to develop the model, whereas the second gives a possible ways of obtaining the desired information for specific applications without formulation of a model.

Recently, Mizrahi and Karel [43, 44] have developed a method useful in predicting the storage stability of foods packaged in containers permeable to water vapor. The method does not require knowledge of the dependence of the kinetic model on moisture content. The method can be applied to dehydrated foods when the index of deterioration is dependent only on moisture content, which changes continuously during storage, and on temperature. We shall discuss the method in more detail later.

IV. POTENTIAL FOR OPTIMIZATION OF NUTRIENT RETENTION IN SPECIFIC FOOD PROCESSING OPERATIONS

A. Heat Sterilization and Other Thermal Processes

One of the tenets of food processing is that high-temperature, short-time processes result in minimal damage to product quality (provided that enzyme activity is effectively eliminated). The superiority of high-temperature, short-time processes is that the energy of activation needed to kill microorganisms is much greater than that needed to destroy nutrients. In any process predicated on achieving a given level of destruction of microorganisms, the higher the temperature, the lower the loss of nutrients. This can be shown readily by comparing spore inactivation with a hypothetical first-order vitamin inactivation reaction [5].

For spore inactivation we have:

$$\ln\left(\frac{N_F}{N_o}\right) = -A_1 e^{(-E_s/RT)} t_c \tag{21}$$

and for the vitamin inactivation:

$$\ln\left(\frac{V}{V_o}\right) = -A_2 e^{(E_v/RT)} t_c \tag{22}$$

where

N = number of spores
V = vitamin concentration
A_1, A_2 = constants
E_s = activation energy for spore destruction
E_v = activation energy for vitamin inactivation
t_c = process time
subscript F refers to the maximum permissible final concentration of spores
subscript zero refers to initial concentrations

Therefore, at any temperature:

$$\ln\left(\frac{V}{V_o}\right) = \ln\left(\frac{N_F}{N_o}\right)\left(\frac{A_2}{A_1}\right)\left[\exp\left(\frac{E_s}{RT} - \frac{E_v}{RT}\right)\right] \tag{23}$$

As the temperature increases, the time to achieve the desired N_F/N_o ratio decreases and, since $E_s > E_v$, the retention of nutrients is improved. For instance, in the case of thiamine and a 10D commercial sterility process (reduction of N_o by a factor of 10^{10}) for Bacillus stearothermophilus, Tannenbaum [5] calculated that, if the process were conducted at 100°C, 99.99% of the thiamine would be lost, but, if it were conducted at 140°C, only 2% would be lost. This improvement in retention is the result of an assumed activation energy of 22 kcal/mole for thiamine loss and of 69 kcal/mole for spore destruction. Lund (in Ref. 6) compiled activation energies for several nutrients in a number of heat-processed foods. The reported values for thiamine range from 20 to 29.4 kcal/mole. Data for other nutrients are more limited, but the reported values for other vitamins were in the same general range (below 30 kcal/mole). Destruction of spores, on the other hand, requires activation energies ranging from 53 to 153 kcal/mole.

Similar conclusions were reached by Nehring [45], who analyzed the interaction of chemical and microbiological effects of heat processes on the basis of published D and E values. Nehring considered E values of 12 to 25 kcal/mole as typical for chemical reactions and values of 75 to 105 kcal as typical for microbial inactivation. Thermal processing is made feasible by the fortunate fact that at high temperatures D values for nutrients are 100 to 1000 times greater than those for microorganisms. Thus the D value for thiamine at 250°F (121°C) is 150 min, but, for Clostridium botulinum, it is only approximately 0.15 min.

It must be noted, however, that the above model has important limitations, including the following:

1. Nutrient retention in processes based on destruction of toxins or enzymes, rather than microorganisms, may not be optimal at high temperatures because the activation energy for toxin or enzyme inactivation may actually be smaller than that for nutrient destruction. For instance, inactivation of peroxidase in whole peas has an activation energy of 16 kcal/mole and that of the Staphylococcal enterotoxin B in milk 25.9 kcal/mole. These energies are of the same magnitude as those for nutrient destruction.

2. Activation energies for nutrient destruction in concentrated solutions and in solid foods are often higher than those observed in dilute solution. The same is often but not always true of energies of activation for "death" of microorganisms.

3. Activation energies and reaction rate constants for nutrient destruction are very sensitive to reaction conditions, including pH, concentration, etc.

4. The analysis assumes that there are no temperature gradients within the food and that come-up times and cooling times are of little significance. Although these assumptions approximate reality in liquid foods, they do not apply to conduction-heating foods or foods in which individual particles or pieces are heated by conduction. In conduction-heating foods, each point in the food container has a different heating and cooling history. Methods for calculating both effectiveness of sterilization and nutrient retention in a given process require, therefore, the integration of local effects with respect to time and location within the food container. This is a complicated and often tedious process that requires the use of high-speed computers.

A beginning has been made in attacking the problem of prediction of nutrient retention and process optimization in thermal procedures involving conduction-heating foods [31,46,47]. The most significant work has been reported by researchers at the University of Massachusetts [31,47-50].

Teixera et al. [31] developed a program that allows the simulation of retention of nutrients in conduction-heating foods. The simulation requires that the following data are available:

1. Can dimensions
2. Thermal diffusivity of the food (or equivalent data describing heat penetration, such as "f_h" value used in heat process calculation)
3. D and Z values for the microorganism used to determine the process
4. The required level of "sterility" (N_F/N_o) and initial spore count
5. D and Z values for the loss of nutrient under study (it should be obvious to the reader that E and k are readily convertible to D and Z)
6. Conditions or process (retort temperature and initial food temperature)

Teixera et al. [31] performed the simulation for one of the very few nutrients for which adequate data are available, namely thiamine in green bean puree, and showed that optimum retention was achieved by processing at 248°F (120°C). Above and below this temperature, thiamine retention was reduced. This optimum process was close to conventional techniques, rather than being at high temperature. This behavior is a result of the integration of the nonuniform temperature distribution existing during the process. The authors also simulated the retention of nutrients with different z values and showed that the optimum retort temperatures were shifted to higher temperatures as the z values increased (or as the E values decreased). A change in D value or in k value for a first-order reaction, as described in Eq. (17), does not change the position of the temperature optimum, provided the can size and the heat transfer properties remain the same.

In subsequent investigations, Teixera et al. [49] conducted a validation study in which their mathematical models were tested by experimental analysis of thiamine retention in heat-processed pea puree. They obtained good agreement between predictions and experimental results. Most recently, Teixera et al. [50] reported on approaches to optimization of heat processes. They again used thiamine as the model nutrient; their microorganism determining the process requirement was *Bacillus stearothermophilus*, and the assumed thermal diffusivity of the hypothetical food was taken to be 0.0143 in^2/min. The magnitudes of other kinetic constants assumed in the study were:

D_{246} value for thiamine = 178.6 min
z value for thiamine = 18°F
D_{250} value for *B. stearothermophilus* = 4 min
z value for *B. stearothermophilus* = 46°F

Teixera et al. [50] used the above values in computer simulations designed to evaluate nutrient retention as functions of two different approaches to process optimization:

1. Constant can size (no. 2 can) but different retort temperature programs (designed to maximize thiamine retention)
2. Constant retort temperature at 250°F (121°C) but different can geometries maintaining the same constant volume (equivalent to a no. 2 can)

They found that programming the surface temperature is only marginally effective in improving thiamine retention. Increasing the surface-to-volume ratio of the can, on the other hand, is highly effective in improving thiamine retention. Furthermore, for these more favorable geometries, programming the retort temperature does further improve retention. This is, of course, to be expected, since when the surface-to-volume ratio becomes very large, the conditions become similar to those existing in convection heating in which, as we have seen, the optimum nutrient retention is achieved by high-temperature short-time processing.

The major impediment to further progress in the optimization of heat processes with respect to nutrient retention is lack of data on kinetics of loss of various nutrients under conditions approximating those existing in thermally processed foods. The mathematical and computational methodology is well ahead of the availability of data that are needed for the application of these methods.

B. Dehydration Processes

The problem of estimating nutrient losses during dehydration is conceptually not much more difficult than the similar problem of estimating them in thermal processing. In thermal processing for foods heating by conduction, the estimation is made by knowing the temperature at every point in the food at any time. The extent of reaction for suitably short-time intervals at every location is calculated, and integrating these extents of reaction with respect to distance gives the total extent in the food. In dehydration, in addition to temperature, the moisture content varies with time and location (hence the concentration of nutrients as well), and the extent of the reaction causing nutrient loss must therefore be calculated taking into account effects of temperature and moisture at each point.

The problem of optimization is actually more complicated than the above brief concept. The objective of dehydration is the reduction of water content to some desirable low level, and this can be achieved by an infinite number of combinations of three variables: surface temperature, partial pressure of water in surrounding atmosphere, and time. The temperature range that can be used is much wider than in thermal processing, and the process can be conducted at subfreezing temperatures (freeze-dehydration).

Two limiting cases may be recognized, which make the analysis less cumbersome than is implied by the general statement.

1. In air dehydration, it is usually possible to assume that the temperature gradient within the food is negligible. Thus, it is possible to calculate the moisture-temperature profiles by assuming that moisture content varies with time and location but that temperature varies with time and not with location in the food.

2. In freeze-dehydration within the "ice-free" layer (deterioration in the frozen layer during drying may often be neglected), it may be assumed that the temperature varies with time and location but that the moisture gradients are small. In fact, in freeze-drying it is often permissible to simplify the situation even further and to assume that as long as any ice is present, the temperature difference across the dry layer remains unchanged.

Work on actual simulation of deterioration during dehydration has so far been very limited. Kluge and Heiss [51] measured rates of nonenzymatic browning in a model food system (glucose, glycine, and cellulose) and used these rates to develop relations for the calculation of browning occurring in drying. Labuza [13] suggested using computers to predict the extent of nutrient losses in dehydration. Aguilera et al. [26] evaluated the extent of nonenzymatic browning in dehydration of potato slabs using data on kinetics of nonenzymatic browning in potatoes published by Hendel et al. [52] and some simplifying assumptions concerning moisture and temperature distribution in the potato slab. The rate of nonenzymatic browning depends, of course, on both moisture content and temperature. Hendel et al. [52] observed that at any moisture and temperature the browning rate was constant, and the degree of browning increased linearly with time.

The data of Hendel et al. [52] were used to evaluate the dependence of activation energy on moisture content. The activation energy was assumed to decrease linearly with moisture content from a value of approximately

43 kcal/mole at zero moisture to 26 kcal/mole at 16% moisture and to remain constant at this value at higher moisture contents.

The moisture-temperature distribution of any drying condition was derived using idealized assumptions including the following:

1. Slab geometry.
2. No shrinkage.
3. Constant condition of air temperature, humidity, and velocity.
4. Drying in the falling-rate period is considered with moisture content = 3.5 g/g at the beginning of that period.
5. Moisture distribution in potato slabs given by Eq. (24) and the average moisture given by Eq. (25).
6. A diffusion coefficient independent of concentration but depending on temperature as shown in Eq. (26).
7. Equilibrium moisture contents (m_e) for each air temperature condition were based on isotherms for the potato-water system that were published by Görling [53].
8. Temperature was assumed not to vary with position but with time.

The moisture distribution throughout the potato slab is given by:

$$m(x) = m_o + \frac{4}{\pi}(m_1 - m_o) \sum_{n=0}^{\infty} \frac{(-1)^n}{(2n+1)} \cos \frac{(2n+1)}{2L} \pi \times \exp\left[-\frac{\pi^2}{4} \frac{(2n+1)^2}{L^2} Dt\right] \tag{24}$$

The mean moisture content is given by:

$$\frac{m - m_o}{m_1 - m_o} = \frac{8}{\pi^2} \sum_{n=0}^{\infty} \frac{1}{(2n+1)^2} \exp\left(-\frac{(2n+1)^2}{L^2} \frac{\pi^2}{4} Dt\right) \tag{25}$$

where

$m(x)$ = moisture content at location x in the slab,
L = half thickness of the slab.

The dependence of diffusivity on temperature is given by

$$D = D_o e^{-E_D/RT} \tag{26}$$

where

D = diffusivity (cm^2/min)
D_o = constant (assumed equal to 66 cm^2/min)
E_D = activation energy for diffusion (assumed equal to 7500 cal/mol)

The temperature depends on time:

$$\ln \frac{T_a - T_w}{T_a - T} = \mu t \tag{27}$$

where μ = slope of the line ln (m/m_1) vs. time.

A computer program was used to calculate the browning rate by first calculating the moisture and temperature distribution in the potato slab and then computing the corresponding browning rate. The results of the simulation were plotted [26] and showed clearly that there is a difference between the total browning calculated by summing up browning at each location and browning computed by using an average value of moisture for each drying time.

An advantage of the approach discussed here is that once a computer program is written and constants are calculated, it is possible to simulate rapidly various conditions of drying and to determine optimum conditions. The major limitations on the approach arise from the paucity of kinetic data on nutrient deterioration reactions and on physicochemical properties of foods. Some work in progress at M.I.T. and elsewhere, however, promise to provide more data and also more simplified approaches to nutrient loss prediction during drying.

V. OPTIMIZATION OF STORAGE AND PACKAGING CONDITIONS

A. General Conditions

The work on prediction of storage life as a function of efficiency of packaging measures and of environmental storage conditions was initiated on a large scale after World War II. Initial work was stymied by the complexity of the factors involved in this problem. The general conclusion was often voiced that the complexity of these factors precluded laboratory prediction of shelf life. Preliminary evaluation of packaging problems on the basis of laboratory tests, however, can greatly simplify the overall task of eliminating obvious failures and pinpointing likely successes. Such an analysis requires that answers be obtained to the following questions:

1. What are the optimal conditions for storage of the particular food product in terms of the major significant environmental factors?
2. What are the expected external environmental conditions to which the package is likely to be exposed?
3. What barrier properties will be required in order to maintain the internal environment at the desired optimal conditions?

Recently, Karel [35] described an approach to analyzing changes in foods during processing and storage on the basis of combining kinetic data on food deterioration (for instance, nutrient loss) with the data on mass transfer and, in case it applies, also heat transfer.

In the case of analysis of storage and packaging problems, this approach may be summarized as follows:

1. Properties of food that determine quality depend on the initial condition of the food and on reactions that change these properties with time. These reactions, in turn, depend on the internal environment of the package. It is assumed that the deteriorative mechanisms limiting shelf life, and their dependence on environmental parameters (oxygen pressure, water activity, temperature), can be described by a mathematical (though not necessarily analytical) function.

2. Maximum acceptable deterioration level can be determined by correlating objective tests of deterioration with organoleptic or toxicological parameters.

3. The internal environment depends on the condition of the food, on package properties, and on external environment. It is assumed that changes in environmental parameters can be related to food and package properties.

4. Barrier properties of the package can, in turn, be related to internal and external environments.

5. The various equations can be combined and solved with or without the aid of a digital computer. The solutions predict storage life or required package properties for a given storage.

B. Optimization of Storage and/or Packaging of Moisture-Sensitive Foods

As discussed previously, many of the reactions that cause quality deterioration depend on moisture content. We can distinguish between two types of dependence: reactions for which a definite critical moisture content may be postulated below which the reaction rate is considered negligibly small, and reactions that proceed at all water contents at significant rates that depend on moisture content.

The analysis of the problem involving the first type of dependence is relatively simple [54]. If a linear isotherm relating moisture content to partial pressure of water is assumed, and the storage conditions are considered to be constant with respect to temperature and relative humidity, then the safe storage time may be related simply to food and package properties as follows [6]:

$$t_c = \frac{xs\delta}{bAP^\circ} \ln\left(\frac{m_e - m_i}{m_e - m_c}\right) \qquad (28)$$

where

t_c = permissible storage time
x = thickness of packaging material
b = permeability of the packaging material
δ = proportionality constant between moisture content and water activity
s = weight of solids in the package
A = area of package
P^o = vapor pressure of water at storage temperature
m_e = moisture content in equilibrium with outside humidity
m_i = initial moisture
m_c = maximum permissible moisture content

If the needed storage time is known, the required packaging protection may be calculated as follows:

$$\frac{b}{x} = \frac{s\delta}{AP^\circ} \ln\left(\frac{m_e - m_i}{m_e - m_c}\right) \qquad (29)$$

Karel, Fennema, and Lund [6] give numerical examples for this type of analysis. Mizrahi and Karel [9] have also demonstrated the feasibility of simulating storage behavior of moisture-sensitive foods that deteriorate because of reactions depending on moisture content throughout a range of moisture contents, provided that the general form of the kinetic behavior is established.

Mizrahi et al. [9] concentrated on the behavior of one system, freeze-dried cabbage. The mathematical model was based on a combination of kinetic data for the browning reaction, the sorption properties of the cabbage, and the permeability characteristics of the packages. Thus, the following functions had to be developed:

1. A function relating extent of browning to time of storage and to moisture content
2. A function relating moisture content within the food to partial pressure of water (sorption isotherm)
3. Function(s) relating change of moisture content in the samples to properties of the package, the food, and the environment

It was determined that simple functions relating the above parameters could be used to give excellent predictions of browning in packaged cabbage stored at 37°C. A subsequent study demonstrated that tests at elevated temperatures and increased moisture contents could be used to obtain the data necessary for the prediction of storage life of dehydrated cabbage, reducing the time required to evaluate package requirements from one year to two weeks [42].

Work currently underway at M.I.T. on ascorbic acid retention in dehydrated tomato juice uses a very similar approach, and preliminary results indicate its usefulness [43]. Similar computer simulations of nutrient retention in dry foods were also reported by Purwadaria et al. [55] and suggested in a review by Labuza [13].

The tests of Mizrahi et al. [9,42] were conducted by determination of the kinetic model and subsequent simulation of storage behavior. Recently, a method was developed to accelerate stability tests for moisture-sensitive products that does not require prior knowledge of the kinetic model of effects of moisture on rate of deterioration [43,44].

The accelerated tests are based on monitoring quality changes in the product, which undergoes rapid deterioration because of a high, albeit controlled, rate of moisture gain. The no-model accelerated tests were formulated and tested on a system with the following properties: (a) the moisture content of the product is increased continuously with time because of storage conditions in which external water activity, a_e, is always larger than activity in the package, a_i; (b) the deterioration rate in the dehydrated product is known to depend only on the momentary moisture content; and (c) the order of reaction, n, is known. Under these circumstances the kinetic equation reads:

$$\frac{dc}{dt} = c^n f'(m) \tag{30}$$

where

c = concentration, and
t = time.

For the sake of simplicity, an index of deterioration (D) was defined as:

$$D - D_o = \int_{c_o}^{c} - \frac{dc}{c^n} \tag{31}$$

where D_o is a reference value at c_o (which, in many cases, could be arbitrarily assigned as zero).

For the two common cases of zero and first order of reaction, D assumes the values of $D - D_o = c_o - c$, and $D - D_o - \ln c_o/c$, respectively.

$$D - D_o = \int_0^t f'(m)\,dt \tag{32}$$

When moisture is introduced into the food system at a constant rate b, i.e.,

$$m = m_o + bt \tag{33}$$

then dt can be replaced by dm/b in Eq. (6) to yield:

$$D - D_o = \frac{1}{b} \int_{m}^{m} f'(m)\,dm = \frac{1}{b}\,[f(m) - f(m_o)] \tag{34}$$

where $f(m) = \int f'(m)\,dm$.

For cases in which b, m_o, and m are kept constant, the change in the index of deterioration is inversely proportional to b. Therefore, when moisture is introduced into two samples at two different but constant rates b_1 and b_2 and when both samples have the same m_o and m, the index of deterioration ratio will be:

$$\frac{(D - D_o)_2}{(D - D_o)_1} = \frac{b_1}{b_2} \tag{35}$$

Therefore, an experiment in which accelerated deterioration is produced by rapid moisture gain can be used to predict the extent of deterioration in samples in which moisture is gained very slowly.

The simplest and most convenient way to control accurately the rate of moisture introduction is to use a controlled-permeability package. In such cases, the rate of moisture gain is no longer constant, although the moisture gain curve can be broken into constant rate periods, at least for relatively small increments of moisture gain. Mizrahi and Karel [43] demonstrated that the above procedure is applicable to prediction of nonenzymatic browning and of loss of ascorbic acid in storage of dehydrated vegetables.

In subsequent studies, Mizrahi and Karel [44] expanded the isothermal "no model" method of accelerated stability testing that was developed for isothermal storage of moisture-sensitive dehydrated products packaged in water-vapor-permeable containers to include storage at different temperatures. The expanded method can be applied to dehydrated products when moisture content changes continuously during storage and when the deterioration rate is only dependent on moisture content and temperature. Knowledge of how the deterioration rate depends on moisture is not required, but it is assumed that at each moisture level the temperature dependence is given by the Arrhenius equation. The method is based on accelerating the deterioration process by subjecting the product to high rates of moisture uptake at elevated temperatures.

The mathematical procedure of predicting the extent of deterioration for any combination of package-storage conditions is based on a transformation of data to an arbitrarily determined reference moisture gain curve and an extrapolation to other temperatures by the Arrhenius equation.

The method successfully predicted the loss of ascorbic acid in stored tomato powder and the extent of browning in dehydrated cabbage.

The predicted value for ascorbic acid retention at low temperatures obtained through extrapolation of data from high-temperature tests was compared with actual values obtained when tomato powder packaged in 3-mil-thick polyethylene was stored at 23°C over water ($a_e \simeq 1$). The results, shown in Fig. 4, indicate very good prediction, even when the extrapolation was based on the parameters evaluated from data obtained

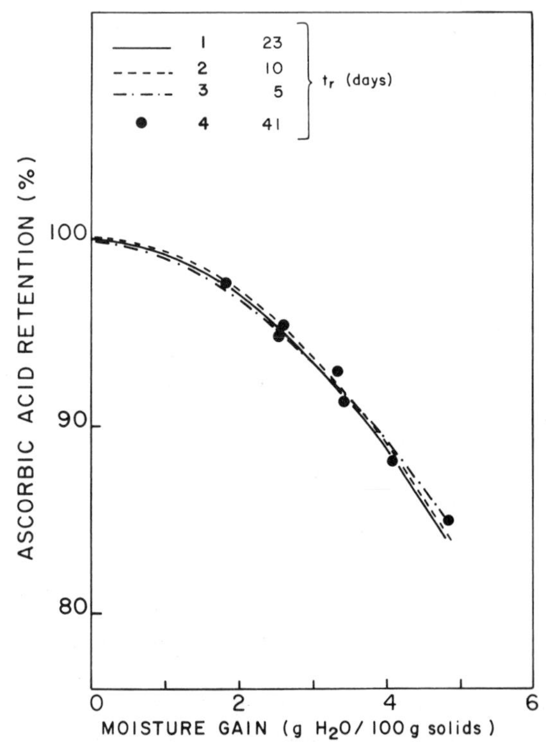

Fig. 4. Comparison of experimental results with the predicted retention of ascorbic acid obtained by extrapolation from different temperatures; t_r is the time required to obtain data for the specific prediction. Curve 1 was calculated from data at 51, 44, 37, and 28°C; curve 2 from data at 51, 44, and 37°C; curve 3 from data at 51 and 44°C; and curve 4 represents the experimental data at 23°C (a_e = 1.0). (From Mizrahi and Karel [43].)

at 51 and 44°C. In this case, the highest observed acceleration factor was about 8. At low rates of moisture gain and at low temperatures, this factor would be greater but less accurate.

C. Oxygen-Sensitive Foods

Simulation of oxygen-initiated oxidation in food has been the subject of several recent studies, but none of them have reached the stage where they can be conveniently applied to optimization of storage conditions and packaging of foods. A major hindrance to progress is the lack of convenient accelerated tests. Many free-radical reactions involving oxygen have relatively low activation energies, and acceleration by increasing temperature is not very effective. As discussed previously, acceleration of stability tests by increasing oxygen pressure is rarely effective.

The following recent studies, however, deserve particular mention, since they may lead to the eventual development of methods to optimize storage of oxidation-susceptible foods.

1. Herlitze et al. [18], Quast and Karel [10,33,56], and Simon et al. [57] pioneered the approach of combining kinetic equations that relate the oxygen dependence of oxidation to mass transport equations describing oxygen transport into the package and within the food in order to simulate the progress of oxidation during storage of dry foods. These studies were based on the initial development of kinetic models of rate dependence on oxygen and other environmental factors. The most complicated storage problem studied so far is that of dehydrated foods which deteriorate upon either absorption of water or oxidation. Quast and Karel [10,33] chose potato chips as a suitable model of this type of food and established that they showed the following behavior when stored at 100°F (37.8°C).

a. There was a maximum level of oxygen absorption (approximately 1200 μliter of oxygen per gram) above which the product began to oxidize and become rancid extremely rapidly.
b. There was a maximum water content corresponding to an equilibrium relative humidity of 32%, which could be considered as an acceptability limit.
c. The water content could be related to the equilibrium relative humidity.

The rate of oxidation was a function of the following variables: partial pressure of oxygen, equilibrium relative humidity, and the amount of oxygen already absorbed. The rate depended also on the intensity of environmental light, but all experiments were conducted in the dark or in the presence of diffuse light only. Quast et al. [32] tested a variety of kinetic models and were able to develop an equation relating rate of oxidation to the independent variables. This equation was based on empirical data as well as on theories of lipid oxidation. They then analyzed potato chips packaged in semipermeable containers. As a result of simultaneous infiltration of oxygen through the packaging material and its consumption by the food, the oxygen pressure changed with time. At the same time, water was absorbed by the food. By solving simultaneously a set of equations, the authors were able to simulate successfully storage behavior of potato chips. They concluded that this simulation is particularly effective for rapidly evaluating what effect a change in packaging or storage conditions will have on the shelf life of a given food. Unfortunately, the method is less useful in evaluating changes in expected shelf life because of major changes in food formulation. When a food is reformulated, the oxidation kinetics may have to be re-evaluated, and this part of the methodology is the most cumbersome and time consuming.

2. Mack et al. [58] and Singh et al. [59] at Michigan State University have worked on systems to simulate storage-related losses of oxidation-sensitive nutrients in liquid foods.

3. Flink and Goodhart [60] at M.I.T. are studying the role of oxygen diffusion in deteriorative reactions in frozen fish.

4. Hu [61] has developed an oxidation-sensitive indicator that may be useful in monitoring oxidation-related changes in products. Although the assumptions inherent in this system—as well as its suggested applications—may have only limited usefulness [62], the approach is of great potential value, and future developments in this field will bear watching.

VI. SUMMARY

In the preceding pages, I have attempted to present the current lines of research aimed at the prediction of nutrient retention during processing and storage on the basis of simple and rapid laboratory tests and the application of knowledge of nutrient retention to optimization of processing and storage conditions.

The industrial application of this research is still limited. The greatest immediate application potential is in optimization of thermal processing and in prediction of storage changes in so-called shelf-stable foods packaged in hermetically sealed containers. An excellent review of the scientific status of shelf life of foods was presented recently by the IFT Expert Panel on Food Safety and Nutrition [63]. This report presents a summary of scientific approaches to shelf-life analysis, industrial aids to monitoring food quality, as well as the consumer attitudes that will have a big impact on industrial practices of the future.

In my opinion, there are major stumbling blocks to rapid development of optimization procedures. These include:

1. There are still relatively few practicing food technologists with a thorough background in mathematics, physical chemistry, and

engineering, the disciplines required for solving the optimization problems.
2. We know relatively little about many important aspects of the foods we have to process. Information on physical properties, kinetic behavior of reacting nutrients, and interactions among food components is still scarce.

A recent NSF-sponsored conference on critical needs in food science and engineering highlighted the need for overcoming these stumbling blocks [64]. I believe this is one of the major tasks for the present generation of food scientists.

REFERENCES

1. R. O. Nesheim, Fed. Proc. 33:2267 (1974).
2. W. F. Kwolek and G. N. Bookwalter, Food Technol. 25:1025 (1971).
3. J. Olley and D. A. Ratkowsky, Food Technol. Austral. 25:66 (1973).
4. J. Kumamoto, J. K. Raison, and J. M. Lyons, J. Theor. Biol. 31:47 (1971).
5. S. R. Tannenbaum, in Food Chemistry (O. Fennema, ed.), Marcel Dekker, New York, 1975, pp. 347-381, 765-775.
6. M. Karel, O. R. Fennema, and D. B. Lund, Physical Principles of Food Processing, in Principles of Food Science (O. Fennema, ed.), Vol. II, Marcel Dekker, New York, 1975.
7. R. R. Legault, C. E. Hendel, W. F. Talburt, and M. F. Pool, Food Technol. 5:417 (1951).
8. A. F. Ross, Adv. Food Res. 1:257 (1948).
9. S. Mizrahi, T. P. Labuza, and M. Karel, J. Food Sci. 35:799 (1970).
10. D. G. Quast and M. Karel, J. Food Sci. 37:584 (1972).
11. G. Beetner, T. Tsao, A. Frey, and J. Harper, J. Food Sci. 39:207 (1974).
12. G. Beetner, T. Tsao, A. Frey, and K. Lorenz, J. Milk Food Technol. 39:244 (1976).
13. T. P. Labuza, CRC Crit. Rev. Food Technol. 3:217 (1972).
14. L. A. Wanninger, Food Technol. 26:42 (1972).
15. M. Karel and S. Mizrahi, Final Report on Contract Research Project #DAAK03-75-C-0038 with the U.S. Army Natick Development Center, Natick, Ma., 1977. Contract Title: Reaction Kinetics and Mass Transport Properties in Packaged Subsistence to Provide Parameters for Computer-Aided Calculations of Changes Occurring in Foods During Processing and Storage of Foods.
16. E. Dwobchak and M. Hegedus, Acta Aliment. 3:337 (1974).
17. R. Heiss and L. Robinson, Gordian 75:359 (1975).

18. W. Herlitze, R. Heiss, K. Becker, and K. Eichner, Chem. Ing. Technik 45:485 (1973).
19. M. Karel, Food Technol. 28:50 (1974).
20. F. Kiermeier and W. Waiblinger, Z. Lebensm. Unters. Forsch. 141:320 (1969).
21. R. Radtke, P. Smits, and R. Heiss, Fette Seifen Anstrichm. 72:497 (1970).
22. A. Sattar and J. M. deMan, J. Inst. Can. Sci. Technol. Aliment. 6: 170 (1973).
23. R. P. Singh, D. R. Heldman, and J. R. Kirk, J. Food Sci. 40:164 (1975).
24. J. N. Thompson and P. Erdody, J. Inst. Can. Sci. Technol. Aliment. 7:157 (1974).
25. G. Wildbrett, Fette Seifen Anstrichm. 69:781 (1967).
26. J. M. Aguilera, J. Chirife, J. M. Flink, and M. Karel, Lebensm.-Wiss. Technol. 8:128 (1975).
27. A. Kramer, Food Technol. 28:50 (1974).
28. S. E. Charm, R. J. Learson, L. R. Ronsivalli, and M. Schwartz, Food Technol. 26:65 (1972).
29. J. Herrmann, Nahrung 4:409 (1974).
30. D. B. Lund, Food Technol. 27:16 (1973).
31. A. A. Teixera, J. R. Dixon, J. W. Zahradnik, and G. E. Zinsmeister, J. Food Sci. 33:845 (1969).
32. D. G. Quast, M. Karel, and W. M. Rand, J. Food Sci. 37:673 (1972).
33. D. G. Quast and M. Karel, Mod. Packag. 46:50 (1973).
34. M. Karel and N. D. Heidelbaugh, in: Nutritional Evaluation of Food Processing, 2nd ed. (R. S. Harris and E. Karmas, eds.), Avi. Co., Westport, Conn., 1975, pp. 412-463.
35. M. Karel, AIChE Symp. Ser. 69:107 (1973).
36. D. Greiff and C. Greiff, Cryobiology 9:34 (1972).
37. A. I. Kay and T. H. Simon, J. Pharm. Sci. 60:205 (1971).
38. M. A. Zoglio, J. J. Windheuser, R. Vatti, H. V. Maulding, S. S. Kornblum, A. Jacobs, and H. Homot, J. Pharm. Sci. 57:2080 (1968).
39. A. R. Rogers, J. Pharm. Pharmacol. (Suppl.) 15:101T (1963).
40. H. V. Maulding and M. A. Zoglio, J. Pharm. Sci. 59:333 (1970).
41. G. Hoffman, Chem. Ind. (London) 729 (1970).
42. S. Mizrahi, T. P. Labuza, and M. Karel, J. Food Sci. 35:804 (1970).
43. S. Mizrahi and M. Karel, J. Food Sci. 42:958 (1977).
44. S. Mizrahi and M. Karel, J. Food Sci. 42:1575 (1977).
45. P. Nehring, Deut. Lebensm. Rundsch. 69:12 (1973).
46. K. Hayakawa, Can. Inst. Food Technol. J. 2:165 (1969).
47. Y. Y. Jen, J. E. Manson, C. R. Stumbo, and J. W. Zahradnik, J. Food Sci. 36:692 (1971).
48. E. A. Mulley, C. R. Stumbo, and W. M. Hunting, J. Food Sci. 40:993 (1975).

49. A. A. Teixera, C. R. Stumbo, and J. W. Zahradnik, J. Food Sci. 40:653 (1975).
50. A. A. Teixera, G. E. Zinsmeister, and J. W. Zahradnik, J. Food Sci. 40:656 (1975).
51. G. Kluge and R. Heiss, Verfahrenstechnik (Mainz) 1:251 (1967).
52. C. E. Hendel, V. G. Silveira, and W. O. Harrington, Food Technol. 9:433 (1955).
53. P. Görling, in Fundamental Aspects of the Dehydration of Foodstuffs, Society Chemical Industry, London, 1958, p. 42.
54. C. R. Oswin, Protective Wrappings, Cam Publishers, Ltd., London, 1954.
55. H. K. Purwadaria, D. R. Heldman, and J. R. Kirk, 1st Int. Conf. on Food Engineering, Collected Abstracts, Boston, August, 1976, p. 172.
56. D. Quast and M. Karel, J. Food Technol. 6:95 (1971).
57. I. B. Simon, T. P. Labuza, and M. Karel, J. Food Sci. 36:280 (1971).
58. T. E. Mack, D. R. Heldman, and R. P. Singh, J. Food Sci. 41:309 (1976).
59. R. P. Singh, D. R. Heldman, and J. R. Kirk, J. Food Sci. 41:304 (1976).
60. J. M. Flink and M. Goodhart, 1st Int. Conf. Food Engineering, Collected Abstracts, Boston, August, 1976, p. 215.
61. K. H. Hu, Food Technol. 26:56 (1972).
62. T. P. Labuza, Food Technol. 26:9 (1972).
63. IFT Expert Panel on Food Safety and Nutrition, J. Food Sci. 39:51 (1974).
64. M. Karel and D. I. C. Wang, A Workshop on Critical Needs in Food Science and Engineering, Vol. I, Document M.I.T.-NSF-76/01 (1976). Available from Input Branch, National Technical Information Service, U.S. Dept. Commerce, Springfield, Va.

CHAPTER 9

ANTINUTRITIONAL AND TOXIC SUBSTANCES: NATURALLY OCCURRING AND ACCIDENTAL CONTAMINANTS

Gerald N. Wogan

Department of Nutrition and Food Science
Massachusetts Institute of Technology
Cambridge, Massachusetts

I. INTRODUCTION

Nonnutrient chemicals in foods and food raw materials are of interest with respect to safety of the food supply. Chemically, these substances represent a wide range of compounds from simple inorganic substances to organic macromolecules. Their presence in food offers health risks of differing character and magnitude to the populations consuming them. The spectrum of possible risks extends from the danger of acute poisoning in a few instances to the undefined risk of prolonged exposure to compounds with mutagenic activity in experimental test systems. Evidence for real risks to human health also ranges in specific circumstances from well-documented cases of actual exposure and response to implications derived from purely experimental situations.

Under these circumstances, evidence that a given food might constitute a hazard to public health is often only presumptive. Much of the research effort devoted to identification of toxic substances in foods has been carried out on plant foodstuffs commonly used as food for farm animals. Because these are often consumed in large quantities over long periods of time, a toxic substance, if present, might produce adverse reactions that might otherwise be unapparent. Such observations have led to isolation and characterization of specific toxic factors present in important food commodities, even though evidence of their toxic properties has never been observed in human populations. In effect, therefore, the evidence for toxicity to man is often only the knowledge that a substance known to be toxic to animals under a given set of circumstances is present in food.

Information summarized in this chapter reveals the existence of a long list of such substances in foods regularly consumed by people. It is important to consider why, under these circumstances, their toxic effects are not more frequently manifest as outbreaks of recognizable toxic syndromes. It would appear that, through trial and error in the course of this evolution, man has learned to avoid those food materials that would produce immediate ill effects. He has also evolved processing methods to eliminate or reduce the level of toxicity. Cooking and other common means of food preparation effectively destroy many of the most prevalent toxic substances, particularly those found in important plant foodstuffs. It is also true that many of these toxicants occur in foods at sufficiently low levels so that, under conditions of normal consumption, their toxic effects never exceed capabilities for detoxification and are therefore not revealed. Abnormal patterns of consumption, however, as during periods of famine, sometimes reveal toxic manifestations that are not otherwise apparent.

The fact that some important foodstuffs are known only to cause toxic effects in animals is in itself of importance to the food supply. Dependency of much of the world's population on animal protein is already high and is likely to increase further. Were it not for the fact that such oilseeds as soybeans and cottonseed can be processed so as to inactivate their toxic constituents, these rich sources of plant protein could not occupy the position of importance they now do in the feeding of farm animals.

Given its multifaceted character, the general problem of toxicants in foods can logically be discussed by organizing available information around such themes as the chemistry of agents involved, characteristics of biological responses, or sources and routes of exposure. The latter organization, around sources of substances involved, will be used in this chapter because it is more in keeping with the organization of the book. Within this framework, the discussion comprises a general survey of currently known problem areas with more detailed presentation only of important examples of individual or classes of related substances in each major area. Available information is briefly summarized dealing with the chemical nature of compounds; their sources and routes of entry into the food supply;

their effects in biological systems pertinent to the question of food safety; evidence of human exposure and response; and the nature of control measures to reduce or eliminate hazards if they exist.

II. PRODUCTS OF MICROBIAL GROWTH

A. Bacterial Toxins

Botulism, associated with the growth of Clostridium botulinum, is probably the best known form of food poisoning because of the extraordinary potency of the toxin involved and the high fatality rate [1]. Poisoning by the enterotoxins produced by Staphylococcus aureus, though a much less serious illness, occurs far more frequently [2,3]. These two forms of bacterially induced food poisoning are the most clearly recognizable syndromes in this category, but a variety of less well-defined food-borne bacterial toxins are also known [4].

1. Botulism

Botulism is the disease resulting from the ingestion of foods contaminated with the preformed toxin produced by Clostridium botulinum, an anaerobic, spore-forming bacillus. Various aspects of the conditions for growth and toxin production as well as isolation and characterization of the toxin have been comprehensively reviewed [5].

Clostridia with the capacity to produce toxin are ubiquitously distributed with a natural habitat in the soil, where they are present as spores. Food contamination can readily occur. However, in spite of the wide distribution of various types of botulinum spores in nature, the prevalence of one or another kind of botulinum poisoning in a given locality is apparently attributable mainly to the dietary habits of the local inhabitants.

The six known serological types of toxin-producing Clostridium botulinum are designated A, B, (Cα, Cβ), D, E, and F. Of these types, only A, B, and E have been frequently associated with botulism in humans. The toxins produced by these organisms are proteins. The precise molecular weight of the active form of the molecule is unknown. However, crystalline toxin A as isolated from cultures of the bacterium has a molecular weight of 900,000. This molecule can be chromatographically separated into two fractions of molecular weight 128,000 (α), and 500,000 (β), both of which retain toxicity. Therefore the actual size of the toxic subunit is not clear.

The site of action of botulinum toxin has been established as those synapses in the peripheral nervous system which depend upon acetylcholine for transmission of nerve impulses. Death is a consequence of suffocation resulting from paralysis of the diaphragm and other muscles involved in respiration.

The potency of botulinum toxin is well known. In its purified form, 1 μg of the toxin contains about 200,000 minimum lethal doses for a mouse, and it is suspected that not much more than 1 μg of the toxin may be fatal for man. Botulinum toxins are thermolabile, losing their biological activity upon heating for 30 min at 80°C. This has great practical significance, since the toxins are inactivated by heat processing or most ordinary cooking conditions.

Today the most frequent cause of botulism in man is inadequately heated or cured food prepared in the home. Commercially prepared foods have been remarkably safe, with the exception of small-scale outbreaks over the last decade involving such products as smoked whitefish, and canned tuna, liver paste, and vichysoisse.

2. Enterotoxins Produced by *Staphylococcus Aureus*

Staphylococcal food poisoning is perhaps the most commonly experienced form of food-borne toxicity. The topic has recently been comprehensively reviewed [6,7], and only the main points relevant to this discussion are presented here.

Symptoms of poisoning generally appear 2-3 hr after eating, and consist of salivation followed by nausea, vomiting, abdominal cramps, and diarrhea. Most patients return to normal in 24-48 hr and death is rare. Because so few affected individuals seek medical treatment, the actual incidence of poisoning is unknown, but is thought to be extensive. Only outbreaks affecting large numbers of individuals usually come to the attention of public health officials.

The cause of the disease is the growth of and toxin production by *Staphylococcus aureus*, a common organism on the skin and external epithelial tissues of man and animal. Only a few of the various subtypes of *S. aureus* produce enterotoxin, and this is difficult to predict except by direct isolation of toxin from contaminated foods. At least four immunologically distinct enterotoxins are known, and they are designated types A, B, C, and D. Physicochemical studies of enterotoxins A, B, and C have shown them to be proteins of similar composition with molecular weights in the range of 30,000 to 35,000.

The pharmacological mode of action of the toxins is not clearly understood. They are very potent in producing their effects. The emetic dose of enterotoxin B in monkeys has been shown to be 0.9 mg per kg body weight, and it has been estimated that humans respond to as little as 1 μg of enterotoxin A.

Based on serological procedures, it has been found that enterotoxin A is most frequently encountered, followed by D, B, and C in that order. Emetic activity for monkeys is retained in crude culture extracts even after 1 hr of boiling. Purified toxins are somewhat more sensitive to heat inactivation, but still must be regarded as comparatively stable from the viewpoint of their toxicological activity.

Three conditions must be met for staphylococcal food poisoning to occur: (a) enterotoxin-producing organisms must be present; (b) the food must support toxin production (e.g., baked beans, roast fowl, potato salad, chicken salad, custards, and cream-filled bakery products are common vehicles); and (c) the food must remain at a suitable temperature for a sufficient time for toxin production (4 hr or more at ambient temperature).

B. Mycotoxins

Mold spores are ubiquitously distributed in nature, and the ease with which they germinate and grow on foods and feeds, especially if they become moist, is well known. The fact that moldiness generally results in unpleasant flavors or other undesirable changes in products has also been known for a long time. Another feature of mold spoilage was first recognized long ago, but its importance has come to be widely appreciated only within the last two decades. Some molds have the capacity to manufacture, during their growth period, chemical substances that are poisonous or produce toxic symptoms of various kinds when foods or feeds containing them are eaten by man or animals. These chemicals are referred to generically as "mycotoxins," and the toxicity symptoms produced by them as "mycotoxicoses."

Contamination of the food supply by mycotoxins gives rise to problems of several kinds. A direct hazard to human health can result when mycotoxin-contaminated foods are eaten by man. It is important to note that mycotoxins remain in the food long after the mold that produced them has died, and can therefore be present in foods that are not visibly moldy. Furthermore, many kinds of mycotoxins, but not all, are relatively stable substances that survive the usual conditions of cooking or processing. Problems of a somewhat different character can be created if livestock feed becomes contaminated by mycotoxins. In addition to the losses generated by toxicity syndromes that may occur in the animals themselves, mycotoxins or their metabolic products can remain as residues in meat or be passed into milk or eggs and thus eventually be consumed by man.

Historically, mass poisoning of human populations by mycotoxins has been recorded under two circumstances. Ergotism, a toxicosis resulting from eating grains contaminated with *Claviceps purpurea*, occurred in epidemic proportions during the Middle Ages, and small outbreaks have been documented as recently as 1951 in France. Alimentary toxic aleukia (ATA) is a mycotoxicosis caused by eating grain that became moldy as a result of overwintering in the field (see below). Both toxicoses are of acute onset and clearly associated with the consumption of large doses of the toxic substance responsible for the illnesses.

In contrast to the small number of documented mycotoxicoses in humans, there are literally hundreds of reports in the literature of toxicity syndromes in livestock that have been attributed to moldy feeds. In a few such instances, the causative fungi and toxic agents have been identified, but in most cases they remain unknown.

It is important to consider mycotoxins from the perspective of their real or possible significance to human health. In this context, based on present knowledge, the toxins and fungi producing them fall into one or more of the following general categories:

1. Mycotoxins known by direct evidence of exposure and response to have caused some form of toxicity in man.
2. Mycotoxins known to be toxic to animals and whose presence has been identified by chemical assay in human foods or foodstuffs, but without evidence of human exposure or response.
3. Fungi isolated from human foods or foodstuffs that produce mycotoxins when cultured under laboratory conditions but without evidence of actual occurrence of mycotoxins in foods.
4. Mycotoxins occurring in feeds or forage of domestic animals, causing toxicity syndromes in them, and presenting the possible risk of human exposure through residues in edible tissues or products.
5. Fungi isolated from animal feeds, litter, or various other sources that produce in laboratory cultures metabolites with toxic properties in experimental bioassays, but without evidence of occurrence in foods or feeds or of poisoning in animals or man.

1. Mycotoxins in Human Foods and Foodstuffs

Information in Table 1 is organized according to this framework and deals only with those problems that fall within the first three categories, i.e., where there is direct evidence or reasonable expectation of human exposure. For a detailed review of mycotoxins in these as well as in the remaining categories, the reader is referred to recent comprehensive reviews [10,17-20].

It is clear from the evidence summarized in Table 1 that mycotoxins present public health hazards of a wide range of types and severity. Aside from ergotism, as mentioned earlier, alimentary toxic aleukia is the only mycotoxicosis for which there is extensive direct evidnece of mass human poisoning. That syndrome, for which the exact chemical agents were never identified, occurred in various areas of the U.S.S.R. during the later years of World War II. Epidemics of the syndrome involving many thousands of persons were associated with the use of millet and other grains that could not be harvested at the usual time and were allowed to overwinter in the field. The grains became heavily molded with the spring thaws but were used because of food shortages, and toxicosis was the result.

The significance to man of the remaining mycotoxins listed in Table 1, except for the aflatoxins, must be assessed from inferences drawn from limited data on occurrence and biological activity of the various substances involved. It is not possible to make a meaningful generalization based on currently available evidence.

2. Aflatoxins

Because of the impact research on aflatoxins has had in stimulating other research in the mycotoxin field and also because of the comparatively large amount of information available, including their significance to man, this group of mycotoxins warrants further brief elaboration. Various aspects of the field have recently been comprehensively surveyed [9,10].

The aflatoxins are produced by a few strains of *Aspergillus flavus* or *A. parasiticus*, fungi whose spores are widely disseminated, especially in soil. Although toxin-producing fungi usually produce only two or three aflatoxins under a given set of conditions, a total of 14 chemically related toxins or derivatives have been identified. One of these, aflatoxin B_1, is most frequently found in foods and is also the most potent toxin of the group.

With respect to substrate, requirements for toxin production are relatively nonspecific, and the mold can produce the compounds on virtually any food (or indeed on synthetic media) that will support growth. Thus, any food material must be considered liable to aflatoxin contamination if it becomes moldy. However, experience has shown that the frequency and levels of aflatoxin found varies greatly among foods collected in a given region and in different regions.

With regard to their toxic and other biological effects, the aflatoxins are very interesting compounds. Acute or subacute poisoning can be produced in animals by feeding aflatoxin-contaminated diets or by dosing with purified preparations of the toxins. Although there are species differences in responsiveness to acute toxicity, no completely refractory species of animal is known. Symptoms of poisoning are produced in most domestic animals by aflatoxin levels in the feed to 10-100 ppm or less. Although cattle tolerate relatively high levels of the toxin, they secrete in milk aflatoxin M_1, a derivative that is also toxic.

Aflatoxin B_1 is among the most potent chemical carcinogens known, and it is this property that has provided an important stimulus for research on these mycotoxins. Carcinogenic activity of various aflatoxins has been demonstrated experimentally in the duck, rainbow trout, ferret, rat, mouse, and monkey.

A very important aspect of the aflatoxin problem is the emphasis that has been placed from the beginning upon the development of chemical assay methods for detection and quantitation of levels of the toxins in foods. Availability of such methods coupled with application of appropriate tech-

TABLE 1 Mycotoxins and Toxin-Producing Fungi from Human Foods or Foodstuffs

Toxin or syndrome	Fungal sources	Foods mainly affected	Chief pharmacological effects after ingestion	References
Aspergillus toxins				
Aflatoxins	*A. flavus* *A. parasiticus*	Peanuts, oil-seeds; grains; pulses; and others	Toxic to liver, carcinogenic to liver of several animals and possibly man	8-10
Sterigmatocystin	*A. nidulans* *A. versicolor*	Cereal grains	Toxic and carcinogenic to liver of rats	8
Ochratoxins	*A. ochraceous*	Cereal grains; green coffee	Toxic to kidney of rats	11
Penicillium toxins				
Luteoskyrin	*P. islandicum*	Rice	Toxic, possibly carcinogenic to liver of rats	12
Patulin	*P. urticae* *P. claviformi* and others	Apple products	Edema, toxic to kidney of rats	13

Fusarium toxins				
Zearalenone	*Gibberella zeae*	Corn	Hyperestrogenism in swine and laboratory animals	14
Alimentary toxic aleukia (ATA)	*F. poae* *F. sporotrichioides*	Millet and other cereal grains	Panleukocytopenia due to bone marrow damage. Mortality up to 60% in human epidemics	15
12,13 Epoxy-tricothecanes	*Fusarium* *Trichoderma* *Gliocladium* *Tricothecium*	Corn; other cereal grains	Cardiovascular collapse; increased clotting time; leukopenia. May have been involved in ATA in man	16

nology has served to minimize the risk of human exposure to aflatoxins in countries possessing the technological capabilities necessary for their successful implementation. It follows that the risk of contamination of food supplies would be greater in developing countries with less advanced agriculture and other technologies. It is indeed from these regions that the currently available information on implications of aflatoxins to human health has come.

This information derives from field studies designed on the basis of the following lines of evidence. With respect to both toxic and carcinogenic actions in animals, aflatoxins affect mainly the liver. It is reasonable to assume that man would respond similarly and liver disease, particularly liver cancer, is the principal illness presumed to be associated with aflatoxin exposure. It has been known for a long time that liver cancer, a relatively uncommon form of cancer in the United States and Europe, occurs at much higher frequency in some populations in other parts of the world, particularly in central and southern Africa and Asia. Several field studies have been conducted to determine whether elevated liver cancer incidence was associated with aflatoxin exposure [21].

Incidence patterns of primary liver cancer in Swaziland and Uganda were compared with frequency of contamination of dietary staples by aflatoxins. Geographical regions or tribal groups with elevated cancer incidence were associated with increased frequency of contamination. In further studies, aflatoxin ingestion was quantitatively measured in populations of Thailand, Kenya, Mozambique, and Swaziland. In every instance, elevated cancer incidence in subgroups of the general population was associated with highest levels of aflatoxin intake. These data suggest that aflatoxins are carcinogenic to man, and that regular ingestion of foods heavily contaminated with aflatoxins increases the risk of liver cancer.

The presence of these toxigenic molds in human foods presents an obvious potential risk to public health, which provides strong motivation for implementation of all available techniques for minimizing contamination of foods by mycotoxins.

III. NATURAL CONSTITUENTS OF FOODS

A. Plant Foodstuffs

For reasons that are not apparent, nature has provided plants with the capability of synthesizing a multitude of chemicals that cause toxic reactions when eaten by man or animals. In the course of his evolution, man must have learned by trial and error to avoid those plants that cause acute, easily recognizable poisoning [22], or to develop processing methods to reduce or eliminate this toxicity. Nonetheless, many foodstuffs that are still regularly consumed, including some of the major sources of plant protein of nutritional value, contain substances that are toxic if consumed in sufficient quantities. Cognizance must also be taken of the existence of these toxicants in formulation of novel products using plant protein sources.

Table 2 presents some essential features of most of the major toxicants in plant foodstuffs; other relevant points dealing with some of these problems are briefly summarized below.

1. Protease Inhibitors, Hemagglutinins, and Saponins

It is useful for this discussion to consider these three groups of substances together. Although they are not all related chemically or toxicologically, they are often present simultaneously in the same groups of pulses, legumes, and cereals. Indeed, much of what we now know about them has resulted from research stimulated by such early observations as the fact that the nutritive value of soybean meal is improved by heating, or that raw kidney beans cause weight loss and death when fed to rats.

The protease inhibitors [23] are proteins that have the property in vitro of inhibiting proteolytic enzymes by binding to the enzyme, apparently in a 1:1 molar ratio. Although a great deal of sophisticated physical biochemical research has been performed on the structure and mode of action of these inhibitors, their role in animal nutrition and toxicology remains largely undefined. Their ability to impair protein hydrolysis is possibly related to impaired nutritive value of raw products containing them, but this has not been conclusively proved. When fed to animals in purified form, the chief toxicological response is pancreatic hypertrophy, the significance of which is not clear. Since these inhibitors are inactivated by heating, their destruction may be related to the improved nutritive value of heated soybean meal.

Hemagglutinins [24,25] are also proteins that have in common the ability to cause agglutination of red blood cells in vitro. This effect, which is highly specific for each protein, results from binding to the erythrocyte plasma membrane, and the hemagglutinins have been referred to as "lectins" because of their specificity of binding. Lectins also have the ability to stimulate mitosis in cell cultures and have become useful tools in the study of membrane structure and function.

Although many hemagglutinins are known to exist, only a few have been isolated in pure form. Some purified proteins in this class are lethal when fed or injected into animals, the most toxic being castor bean ricin, which has an LD_{50} of 5 μg/kg in rats. By contrast, soybean and kidney bean lectins are less toxic by a factor of 1000, and those from lentils and peas are nontoxic. The toxicity of all hemagglutinins is destroyed by moist (but not dry) heat.

Saponins [26] are glycosides that occur in a wide variety of plants and are characterized by three properties, i.e., bitter taste; foaming in aqueous solution; and hemolysis of red blood cells. They are highly toxic to fish and other aquatic cold-blooded animals, but their effects on higher

TABLE 2 Toxic Constituents of Plant Foodstuffs

Toxins	Chemical nature	Main food sources	Major toxicity symptoms
Protease inhibitors	Proteins (mol wt 8,000-24,000)	Beans (soy, mung, kidney, navy, lima); chick pea; peas; potato (sweet, white); cereals	Impaired growth and food utilization; pancreatic hypertrophy
Hemagglutinins	Proteins (mol wt 36,000-132,000)	Beans (castor, soy, kidney, black, yellow jack); lentils; peas	Impaired growth and food utilization; agglutination of erythrocytes in vitro; mitogenic activity to cell cultures in vitro
Saponins	Glycosides	Soybeans, sugarbeets, peanuts, spinach, asparagus	Hemolysis of erythrocytes in vitro
Goitrogens	Thioglycosides	Cabbage and related species; turnips; rutabaga; radish; rapeseed; mustard	Hypothyroidism and thyroid enlargement
Cyanogens	Cyanogenic glucosides	Peas and beans; pulses; linseed; flax; fruit kernels; cassava	Cyanide poisoning
Gossypol pigments	Gossypol	Cottonseed	Liver damage; hemorrhage; edema

Lathyrogens	β-aminopropionitrile and derivatives	Chick pea; vetch	Osteolathyrism (skeletal deformities)
	β-N-oxalyl-L-α,β-diaminopropionic acid	Chick pea	Neurolathyrism (CNS damage)
Allergens	Proteins?	Practically all foods	Allergic responses in sensitive individuals
Cycasin	Methylazoxymethanol	Nuts of *Cycas* genus	Cancer of liver and other organs
Bracken fern Carcinogen	Unknown	Young fronds of fern ("fiddleheads")	Cancer of intestinal tract and other organs
Favism	Unknown (glycosides?)	Fava beans	Acute hemolytic anemia

animals vary. Chemically, they occur in two groups according to the nature of the sapogenin moiety conjugated with hexoses, pentoses, or uronic acids. The sapogenins are steroids (C_{27}) or triterpenoids (C_{30}). Interest in this group of substances was initiated mainly by their hemolytic activity, but this property seems to be unimportant with respect to in vivo toxicity.

2. Goitrogens

Goitrogens [27] are thioglucoside antithyroid agents that occur in plants of the family Cruciferae, especially the genus Brassica. These compounds are also responsible for the pungent nature of such plants. All natural thioglucosides, of which about 50 have been identified, occur in association with the enzyme(s) that hydrolyze them to yield glucose and bisulfate when wet, unheated tissue is crushed. Intramolecular arrangements may take place in the aglycone to yield in addition isothiocyanate, nitrile, or thiocyanate. Although the role of these antithyroid substances in the etiology of human endemic goiter is apparently minimal, their presence in commodities used as animal feeds is of considerable economic importance.

3. Cyanogens

Cyanide in trace amounts is widely distributed in plants and occurs mainly in the form of cyanogenetic glucosides [28]. Three glucosides have been identified in edible plants, namely amygdalin (benzaldehyde cyanohydrin glucoside), dhurrin (p-hydroxybenzaldehyde cyanohydrin glucoside), and linamarin (acetone cyanohydrin glucoside). Amygdalin is present in bitter almonds and other fruit kernels; dhurrin in sorghum and related grasses; and linamarin in pulses, linseed, and cassava. Yields of HCN as high as 245 mg/100 g from cassava and 800 mg/100 g from immature bamboo shoots have been reported. The lethal dose of HCN for man is on the order of 0.5-3.5 mg per kg body weight, and occasionally sufficient quantities of cyanogenic foods are consumed to cause fatal poisoning in humans. The possibility of chronic toxicity resulting from regular consumption of low levels has been suggested but not proved.

4. Gossypol

Gossypol [29] and several closely related pigments occur in pigment glands of cottonseeds at levels of 0.4-1.7%. It is a highly reactive substance and causes a variety of toxic symptoms in domestic and experimental animals. It also causes a reduction of nutritive value of cottonseed flour, a protein source of increasing importance for human feeding. Production of glandless, gossypol-free cottonseed by selective plant breeding is now being developed.

5. Other Plant Toxicants

The other plant toxicants listed in Table 2 create problems of different types, but tend to be of concern to more restricted populations by virtue of patterns of intake of sensitivity to the toxic agents. Human neurolathyrism [30], a crippling disease that results from degenerative lesions of the spinal chord, is known to occur only in India. Although it is associated with ingestion of certain varieties of Lathyrus sativus, the causative agent is unknown. The toxic amino acids listed in Table 2 induce lesions in animals reminiscent in some respects of those in humans, and occur naturally in the plant. They are therefore suspected, but not proved, to be involved in causing the disease.

Food allergens [31] are usually normal components of foods, and their undesirable properties are the result of altered reactivity (i.e., allergy) in individuals who respond to such otherwise innocuous substances. The range of food constituents known to cause allergic responses in sensitive individuals is very wide and incorporates virtually all kinds of foods.

Cycasin [32] represents still another kind of problem. This compound, the glucoside of methylazoxymethanol, is a normal component of a number of plants that serve as an emergency source of starch for some populations in the Pacific and Japan. Although it has very potent carcinogenic activity in animals, it appears that traditional methods of processing the starch effectively remove the toxic substance. Its importance to man is therefore uncertain.

The bracken fern, Pteridium aquilinium, is widely distributed throughout the world. It has been known for a long time that the plant is poisonous to livestock, particularly to cattle. Ingestion of fresh or dried bracken can result in two quite different toxicity syndromes, one clearly attributable to the presence of a thiaminase in the plant. This syndrome can effectively be reversed or prevented by adequate replacement of the vitamin. The other, occurring mainly in cattle and sheep grazing on bracken, comprises a syndrome whose main symptoms resemble those caused by ionizing radiation or by radiomimetic chemicals. The characteristics of the latter syndrome and attempts to identify the active agents have recently been reviewed [33]. Indications that bracken fern also contains carcinogenic substances have been provided through chronic feeding studies, which are discussed below. Components of bracken responsible for its radiomimetic and carcinogenic properties have not yet been chemically identified. It has not been rigorously proved that the same substances are responsible for these two properties, but attempts at isolation of active materials have involved the assumption that this is true. The present state of knowledge concerning the chemistry of the active material was summarized as follows by Leach et al. [34].

Isolation of the radiomimetic component has been hampered by lack of a convenient bioassay system and instability of the substance. A material which is mutagenic, carcinogenic, and acutely lethal to mice was extractable with hot ethanol. Subsequent purification procedures yielded a chromatographically homogeneous substance that is carcinogenic and toxic. Mass spectrometric data indicate a molecular weight of 156 and an empirical formula of $C_7H_8O_4$. The substance is nonaromatic and highly unstable with respect to its biological activity, and its chemical reactivity suggests the possible presence of a lactone in the molecular structure.

Published information on the carcinogenic properties of bracken fern has mainly derived from studies on fresh or dried plants or plant extracts, and little research has been done on highly purified material. Consequently, it is not possible to estimate accurately the potency of the active component.

Several general observations are warranted by the limited data available. Tumors induced in rats [35-37] and Japanese quail [38] by chronic administration of the plant or ethanol extracts consist primarily of intestinal adenocarcinomas, mainly of the ileum, with relatively fewer papillomatous lesions or carcinomas of the urinary bladder. The induction of intestinal tumors is noteworthy in view of the rarity with which they occur spontaneously in rodents and the small number of chemical carcinogens that induce them experimentally. On the other hand, Swiss mice [33] respond to similar exposure by development of lung adenomas, indicating a significant species difference in response.

The studies of Hirono et al. [36] indicate that bracken prepared according to Japanese tradition, i.e., immersion in hot water, retains its carcinogenic properties for rats although at an apparently reduced potency. The water used for processing was not carcinogenic under the conditions of the experiment.

Pamukcu et al. [39] reported that the incidence of urinary bladder tumors in rats fed bracken is increased by simultaneous administration of thiamine. Intestinal tumors appeared at a high incidence in bracken-treated rats whether thiamine was administered or not.

Published data also provide evidence relevant to the etiology of urinary bladder tumors in cattle. This disease, frequently observed in various parts of the world, has been associated epidemiologically with chronic bovine hematuria and also with the geographic distribution of bracken fern [40]. The fact that bracken contains a carcinogen for the urinary bladder of cattle is demonstrated by experiments in which high incidence of bladder carcinomas and papillomatous lesions was induced after feeding for periods of 9 months to more than 4 years [41]. Occurrence of tumors was invariably preceded by hematuria. The guinea pig responds to bracken feeding the same way as cattle, developing both hematuria and bladder tumors [33]. Relatively brief exposure and short latent period indicate a high sensitivity to the carcinogenic stimulus in these animals.

The existence of carcinogenic substances in bracken fern established by these experiments has several public health implications. Human exposure to them could take place by several routes. Direct consumption of bracken by humans is known to occur in Japan [42] and in other parts of the world as well. The implications of this fact are obvious, and the suggestion has been made that this practice might be associated with the relatively high incidence of stomach cancer that occurs in Japan and North Wales [43].

Indirect consumption might also occur. It has been established that the carcinogenic agent in bracken is readily transferred from mother to offspring in mice through placenta and milk. Passage of the toxic substance into milk of cattle fed the fern has also been demonstrated [44]. These observations have important implications for man, since they suggest the possibility of exposure through consumption of milk or possibly meat of animals grazing on bracken pastures. This kind of exposure could affect larger populations over longer periods than direct consumption restricted to a brief period during which the tender fronds ("fiddleheads") are available. The possibility that milk may become contaminated also has obvious implications with respect to early exposure of infants and children for whom milk represents a major proportion of the diet.

Favism [45] is a clinical syndrome in man consisting of acute hemolytic anemia and related symptoms resulting from the ingestion of fava beans (*Vicia faba*) or inhalation of pollen of the plant. The disease shows a remarkable localization in the insular and littoral regions of the Meditteranean area and has been attributed to an inborn error of metabolism with an ethnic distribution. Susceptible individuals have a deficiency of glucose-6-phosphate dehydrogenase in erythrocytes, which sensitizes them to the active agents in the bean resulting in acute hemolytic damage.

B. Animal Foodstuffs

Toxic substances also occur in foodstuffs of animal origin [46]. These toxicants fall into two categories, those that occur naturally and those which are deliberately or inadvertently introduced by man. The naturally occurring substances are limited in their distribution and seem to be confined to avian and fish eggs and to certain kinds of fish, shellfish, and amphibia. Examples of toxicity due to the introduction of man-made chemicals may be found in most meat and dairy products.

Contamination of meats and meat products by residues of food additives such as antibiotics or growth stimulants is a well-documented problem carefully regulated by legislation in many countries. Similarly, contamination of edible tissues by substances appearing by accident in feeds, such as pesticide residues, is a problem requiring continuing surveillance by regulatory agencies. Milk and milk products can also serve as vectors for a multitude of contaminants arising from animal feeds, including estrogens, toxic alkaloids and other plant substances, nitrates and nitrites, antibiotics, pesticides, mycotoxins, and radionuclides.

Among the naturally occurring toxicants in animal tissues, several proteins in avian egg white [47] have received a great deal of attention, mainly because of early observations that one of them, avidin, complexes biotin and can produce deficiency of that vitamin in laboratory animals. As a result of this research, a total of six egg white proteins that have capabilities of disrupting biochemical systems have been identified. These include:

1. Avidin, the protein responsible for "egg-white injury," which binds four moles of biotin per mole of protein
2. Ovomucoid, a mucoprotein which specifically inhibits bovine (but not human) trypsin
3. Ovoinhibitor, a protein which inhibits several proteolytic enzymes
4. Ovotransferrin, a metal-binding protein
5. Ovoflavoprotein, a riboflavin-binding protein
6. Lysozyme, a carbohydrase capable of lysing certain microbial cells

There have been few reports of adverse effects from consumption of egg white by man other than those attributable to microbial or chemical contamination or to allergic reactions. This is probably attributable to the fact that all of the activities listed above are destroyed by heating to usual cooking temperatures.

Poisoning caused by eating fish eggs containing naturally occurring toxins can be divided into three types [48]. By far the most important of these is tetrodotoxin poisoning caused by eating the roe of puffer fish ("fugu") which contain this toxin. Tetrodotoxin, which is also contained in other edible tissues of the fish, has been chemically characterized, and acts as a specific inhibitor of sodium transport across cell membranes. Thus it produces neuromuscular paralysis and causes death in about 50% of cases by respiratory failure. A completely different form of poisoning occurs after eating the roe of two species of marine fish, the blenny and the sculpin. The toxins have not been identified but appear to be lipoprotein in character.

Other naturally occurring toxicants of importance as food contaminants occur in various marine forms. Their presence among edible species of marine animals creates a problem that seems certain to become increasingly important as man is forced to turn to the oceans for additional sources of animal protein. The problem is especially difficult because present knowledge about the nature of the toxic agents and factors determining their occurrence is so limited that it is impossible to predict with any certainty when and where toxins are present.

More than 1000 species of marine organisms are known to be poisonous or venomous, and many of these are edible forms or otherwise enter the food chain [49-51]. Toxins causing them to be poisonous vary considerably in their chemistry and toxicology. Some appear to be proteins of large molecular weight, others are quaternary ammonium compounds of small size; most have not been isolated or purified.

The two main types of poisoning by marine animals are fish poisoning resulting from eating fish containing poisonous tissues, and shellfish poisoning resulting from ingestion of shellfish that have concentrated toxins from plankton constituting their food supply. These are known as "icthyotoxism" and "paralytic shellfish poisoning," respectively [52].

1. Icthyotoxism

About 500 species of marine fish are known to be poisonous and many of these are among edible varieties. Poisoning syndromes resulting from their ingestion are variable in character and are usually designated by the kind of fish involved, such as ciguatera, tetraodon, scombroid, clupeoid, cyclostome, or elasmobranch. The general character of the problem can be illustrated by selected examples.

Ciguatera poisoning is the most common form of fish poisoning. It can occur following ingestion of a wide variety of common food fish, such as grouper, sea basses, and snappers. This form of toxicity is associated with the food chain relationship of the fish. The toxic agent apparently arises in a blue-green algae and is then passed directly to herbivorous fish and indirectly to carnivorous species [54]. The toxic agent has been isolated in pure form and has an empirical formula of $C_{35}H_{65}NO_8$. Its LD_{50} in mice is 80 μg per kg body weight, but its precise mode of action is unknown. Death of poisoned individuals appears to be due to cardiovascular collapse.

Clupeoid poisoning sometimes occurs after eating of certain herring, anchovies, tarpons, and bonefish, and is particularly prevalent in the Caribbean. The situation may be related to ciguatera poisoning, but the source and character of the toxin are unknown. The clinical syndrome, however, is well characterized and has a high fatality rate.

Tetraodon (puffer fish) poisoning is probably the most widely known and studied fish poisoning [55]. Puffer fish are not widely used for food, but are consumed under special circumstances in Japan, where fatal poisonings are reported occasionally. Tetraodotoxin is probably the most lethal of all the fish poisons.

These few examples serve to emphasize the importance of additional research into the occurrence and nature of toxins in marine animals. Such information will be imperative in order to determine which marine animals can be safely harvested.

2. Paralytic Shellfish Poisoning

This syndrome is caused by eating clams or mussels that have ingested toxic dinoflagellates and effectively concentrated the toxic agents contained therein. Shellfish become toxic when local conditions favor growth ("blooms") of the dinoflagellates beyond their normal numbers; such circumstances are often referred to as "red tide." The organisms involved along the American coastlines are usually species of Gonyaulax, although other genera and species are also toxic [53].

The toxic agent saxitoxin has been isolated and purified from cultures of the dinoflagellate and from toxic shellfish. It has an empirical formula of $C_{10}H_{17}N_7O_4 \cdot 2HCl$, and its structure has recently been elucidated. It is stable to heat and thus not destroyed by cooking.

The purified toxin has an LD_{50} of 9 μg per kg body weight in mice and the estimated total lethal dose for man is thought to be 1-4 mg. The toxin depresses respiratory and cardiovascular regulatory centers in the brain, and death usually results from respiratory failure. The fatality rate of affected individuals is 1-10% in most outbreaks.

IV. INTENTIONAL FOOD ADDITIVES

Intentional food additives are chemicals added to foods to accomplish one or more general objectives, such as improvement of nutritive value; maintenance of freshness; creation of some desirable sensory property; or to aid in processing. At present, upwards of 3000 chemicals are added to foods for these purposes [56]. Their large number and the fact that relatively large quantities of some of them are ingested regularly over a lifetime emphasize the importance of establishing conditions that assure their safe use.

In most technologically developed countries, use of food additives is regulated through legal mechanisms that not only specify precisely the conditions under which the additive can be used, but also require evidence from studies on experimental animals that the compound does not induce adverse responses that would be detrimental to human health. During the last two decades, the amount and kinds of animal data required by regulatory agencies and other groups charged with the responsibility of safety evaluation has increased and become more sophisticated. Evidence is required from at least two species regarding acute and chronic toxicity, a variety of biochemical indices of toxicity, effect on reproduction, tests for carcinogenic, mutagenic, or teratogenic effects, and information on metabolic fate of the material.

Every newly proposed food additive must undergo this complete toxicological evaluation before a permit for use is issued. Under these conditions, such additives can be considered to be safe within the limits of the ability of animal testing to detect toxicological hazard. The great majority of food additives fall into this category, although they have not all undergone testing of the same intensive nature.

New evidence has become available, however, that indicates the existence of toxicological problems with certain food additives that were not previously recognized. Since it is not feasible to attempt a detailed discussion of all classes of food additives, one example of current interest will be discussed here to illustrate the kind of problems that can arise.

Sodium nitrite has a long history of use as a preservative and color stabilizer, particularly in meat and fish products. Regulations governing its maximum permitted use levels were based on no-effect levels with respect to acute toxicity in animals. Recently, however, it has been discovered that the use of nitrite may present another kind of hazard through its ability to interact with amines or amides, with the formation of N-nitroso derivatives of considerable toxicological interest.

Nitrosamines can form by the reaction of secondary or tertiary amines with N_2O_3, the active nitrosating reagent in most food products, through the following type-reactions:

$$R_2NH + N_2O_3 \longrightarrow R_2N{\cdot}NO + HNO_2$$

$$R_3N + N_2O_3 \longrightarrow R_2N{\cdot}NO + R$$

The kinetics of nitrosamine formation from secondary amines is a third-order reaction with a pH optimum of 3.4, the pK_a of nitrous acid:

$$v = k_n [HNO_2]^2 [amine]$$

These N-nitroso compounds are of toxicological interest because many of the known representatives of this class have potent carcinogenic activity in animals. Approximately 80% of more than 100 such compounds thus far tested are carcinogenic for one or more tissues of experimental animals [57]. Furthermore, formation of carcinogenic levels of nitrosamines in vivo by the simultaneous feeding of high levels of nitrite and amines to animals has been demonstrated [58, 59]. Nitrosamines can enter the diet through two principal routes. Early reports suggested that dimethylnitrosamine may be present in a variety of unprocessed food materials [60-62]. The origin of the nitrosamine in these cases is unknown. A second major route of entry into foods involves the reaction of nitrite with disubstituted amines to yield nitrosated products as outlined above. Because nitrite is widely used as a food additive and amines are widely distributed in food materials, the potential for nitrosamine formation seems great. Nitrosation of amines may take place in foods during storage or processing and nitrosamines would be ingested as such. Alternatively, it has been shown that nitrosation can also take place in vivo in the acid condition of the stomach, in which case ingestion of the precursors could result in local formation of nitrosamines and nitrosamides [63].

Detailed investigation of the prevalence of nitrosamine contamination of foods has been under way for several years. Quantitative determination of preformed nitrosamines has been limited by analytical technology to

TABLE 3 Occurrence of N-Nitrosamines in Foods

Source	N-Nitrosamine[a]	Concentration (ppb)
Meat products		
Salami, dry sausage	DMN	10-80
Fried bacon	DMN, DEN, NPYR, NPIP	1-40
Luncheon meat, salami, chopped pork	DMN, DEN	1-4
Uncooked and fried bacon	DMN, NPYR	2-30
Mettwurst sausage	NPYR, NPIP	13-105
Frankfurters	DMN	11-34
Fish		
Chinese marine salted fish	DMN	50-100
Raw and smoked salmon, shad, sable	DMN	0-26
Fresh, salted, fried cod; fried hake	DMN	1-9
Other foods		
Fruit (*Solanum incanum*)	DMN	ND[b]
Cheese	DMN	1-4
Soybean oil	DMN, DEN	ND

[a]DMN, dimethylnitrosamine; DEN, diethylnitrosamine; NPYR, nitrosopyrrolidine; NPIP, nitrosopiperidine.
[b]ND, not determined.

those representative compounds which are sufficiently volatile to permit analysis by gas-liquid chromatography. A summary of data published is presented in Table 3. Meats and fish preserved with nitrite are the main products involved, and dimethylnitrosamine the compound most frequently encountered, possibly because it is most readily detected by the analytical methods used.

A new analytical procedure using a thermal energy analyzer coupled with high-pressure liquid chromatography has recently been developed [64] and will permit quantitative measurement of nonvalatile N-nitroso compounds as well. This will permit accurate assessment of total levels of these compounds in foods.

In general, the levels of nitrosamines so far detected in foods have been far below the effective dose in animals. However, these observations have stimulated a reconsideration of the use of nitrite as an additive and additional research on conditions under which nitrosation can take place in vivo and in foods. The possible role of nitrosamines in the etiology of human cancers with widely different geographic distribution is also under investigation.

V. UNINTENTIONAL ADDITIVES

Unintentional food additives comprise chemicals that become part of the food supply inadvertently, through several different routes. For the most part, they are residues resulting from processing or other manipulation involved in food production and distribution, or that enter foods through purely accidental circumstances [65]. Major categories of important examples of this type of contaminant are summarized in the following sections.

A. Factors Arising from Processing

1. Fumigants

Ethylene oxide is commonly used as a fumigant to sterilize foods under conditions in which steam heat is impractical. This and other epoxides can produce adverse effects by destroying essential nutrients or by reacting chemically with food components to produce toxic products. Among other products that are of interest toxicologically, ethylene oxide can combine with inorganic chlorides to form the corresponding chlorhydrin. Ethylene chlorhydrin has been found in whole and ground spices fumigated commercially at concentrations up to 1000 ppm. The chlorhydrins are relatively toxic to animals, but effects of chronic exposure to low levels have not been evaluated and no tolerance limits have been set.

2. Solvent Extraction and the Production of Toxic Factors

Studies of trichloroethylene extracted oil-bearing seeds present an excellent example of the kind of problem that arises when there is an interaction between the substance being processed and the solvent, with the production of a highly toxic product, although the chemical used in processing is itself nontoxic.

Extraction of various oilseeds with trichloroethylene was formerly practiced in several countries, until the practice had to be abandoned when it was found that the extracted residue was toxic when fed to animals. Soybean oil meal, for example, when extracted in this fashion, invariably caused aplastic anemia when fed to cattle. Ultimately, it was shown that the toxic factor was S-(dichlorovinyl)-L-cysteine which was produced by the interaction of the solvent with cysteine in the proteins of the soybean meal.

3. Products of Lipid Oxidation

A large number of changes can be induced in lipids of foods during processing by commercial methods. Important among these from a toxicological point of view are some of the oxidative and polymerization reactions that take place, particularly after prolonged heating. For example, several investigators have shown that fatty acid monomers and dimers, differing from their natural counterparts, accumulate during deep-fat frying and similar heating conditions. Such heated oils when fed to rats result in depressed growth and food efficiency and also in liver enlargement. The mechanisms responsible for these changes are unknown, but the widespread use of these conditions in commercial practice makes further investigation of the problem worthwhile.

4. Carcinogens in Smoked Foods

The smoking of food for preservation and flavoring is one of the oldest forms of food processing. Despite its long use, surprisingly little is known about the toxicological connotations of the practice. For example, it is well known that products that are exposed directly to wood smoke become contaminated with polycyclic aromatic hydrocarbons, many of which are known to be carcinogenic for animals. It has, in fact, been suggested (but not proved) that the high incidence of stomach cancer in Iceland may be associated with the habitual consumption of heavily smoked meats and meat products. Woodsmoke condensate contains many other classes of compounds (phenolics, acids, carbonyls, and alcohols), most of which have not been investigated for toxicologic activity.

B. Accidental Contaminants

1. Heavy Metals

Heavy metals have been among substances receiving increased attention as widespread environmental contaminants and as accidental food contaminants. They enter the environment mainly as a result of industrial pollution and find their way into the food chain by a number of routes. The two metals of principal concern in this connection are mercury and cadmium [66,67].

Toxicological implications of mercury depend heavily on the chemical form involved. Exposure to organic mercurials, especially methyl mercury, is more dangerous than exposure to inorganic salts. The central nervous system (CNS) is the main site of toxic action of both forms, but exposure to methyl mercury is much more ominous, since lesions induced by it are irreversible. Human response data are available from a mass-poisoning episode in the Minimata Bay area of Japan, in which clinical signs of poisoning were evident when intakes of methyl mercury exceeded 4 μg per kg body weight. Taking into account total dietary sources of mercury, a "tolerable weekly intake" of 300 μg total mercury per person, of which not more than 200 μg should be methyl mercury, has been established.

Practically all of the methyl mercury in the diet comes from contaminated fish; other foods generally contain less than 100 ppb of total (inorganic and organic) mercury. In recent years, market-basket surveys reveal daily intakes of 1-20 μg of total mercury per individual in the United States and western Europe.

Cadmium is widely distributed in the environment, and is readily absorbed when ingested. All available information deals with inorganic salts; no toxicological information is available on organic compounds containing cadmium.

A small proportion of ingested cadmium is stored in the kidneys in the form of a metal-protein complex. Long-term exposure to excessive amounts results in renal tubular damage in animals and man. Other long-term effects include anemia, liver dysfunction, and testicular damage.

Human exposure to cadmium takes place mainly through foods, most of which contain less than 50 ppb of the metal. Representative intakes in various parts of the world are in the range of 40-60 μg per person per day, and a tolerance of 400-500 μg per week has been established.

2. Polychlorinated Biphenyls (PCBs)

In 1966 the discovery of PCBs in fish attracted attention among the scientific community, since it was thought that these chemicals were used only in closed, controlled systems. Subsequent research has revealed that they are in fact widespread environmental contaminants with many possible implications to public health [68].

With respect to PCBs in foods, current information suggests that they rarely appear in fresh fruits and vegetables; are frequently found in fish, poultry, milk, and eggs; and can enter foods through migration from packaging materials. Levels generally encountered are in the range of 1-40 ppm, with an average of less than 2 ppm.

The toxicological implication of these residues is not entirely clear. Although these chemicals do not have a high order of acute or chronic toxicity to animals, they accumulate in adipose tissues, and evidence of human poisoning has been reported at very high levels of intake.

3. Chlorinated Naphthalenes

These substances are of interest because they proved to be the causative agents in a previously unknown cattle disease, referred to as "bovine hyperkeratosis." This toxic syndrome is of great economic importance since it kills large numbers of cattle in affected herds. Chlorinated naphthalenes are commonly used in wood preservatives and lubricating oils, and processed feedstuffs become contaminated with inactive oil during processing, and are highly toxic to cattle eating them. Other domestic animal species are also susceptible to this poisoning, and the compounds are toxic to man. No evidence exists as to the possible presence of these contaminants in other portions of the food chain.

VI. SUMMARY AND CONCLUSIONS

This brief review has indicated the main features of the nature of real or potential problems created by the entry, by one route or another, of toxic substances into the human food supply. It is important that these problems be evaluated from the perspective of minimization of their possible impact on public health. Each of them offers interesting scientific and intellectual challenges of its own, and effective control measures require somewhat different scientific inputs.

In these terms, our present state of knowledge with regard to specific problem areas is highly variable. Some are well defined, and a large body of scientific information is at hand, as in the cases of certain of the natural constituents of plant foods or intentional additives (which is not to say that all problems in these areas have been solved). On the other hand, other important problems are still in their earliest stages of definition and obviously merit considerable further study, as in the instance, among others, of nitrosation of food constituents, producing compounds of possible great public health importance.

In all of these areas, progress toward problem recognition, definition, and evolution of control measures can be most effectively made by multidisciplinary research efforts, involving the joint participation of chemists, biologists, microbiologists, toxicologists, and epidemiologists. The need for this kind of approach is clearly manifest in such problems of current interest as the nitrosamines. In this, as in other areas, food chemistry plays a role of pivotal importance.

REFERENCES

1. C. E. Kimble, in Immunological Aspects of Foods (N. Catsimpoolas, ed.), Avi Publishing Company, Inc., Westport, Conn., 1977, pp. 233-259.
2. M. S. Bergdoll, in Immunological Aspects of Foods (N. Catsimpoolas, ed.), Avi Publishing Company, Inc., Westport, Conn., 1977, pp. 199-220.
3. T. E. Minor and E. H. Marth, Staphylococci and Their Significance in Foods, Elsevier Scientific Publishing Co., New York, 1976.
4. D. A. A. Mossel, in The Safety of Foods (J. C. Ayres, F. R. Blood, C. O. Chichester, H. D. Graham, R. S. McCutcheon, J. J. Powers, B. S. Schweigert, A. D. Stevens, and G. Zweig, eds.), Avi Publishing Company, Westport, Conn., 1968, pp. 168-182.
5. D. A. Boroff and B. R. DasGupta, in: Microbial Toxins (S. Kadis, C. Montie, and S. J. Ajl, eds.), Vol. IIA, Academic Press, New York, 1971, pp. 1-68.
6. M. S. Bergdoll, in Microbial Toxins (T. C. Montie, S. Kadis, and S. J. Ajl, eds.), Vol. III, Academic Press, New York, 1970, pp. 265-326.
7. E. M. Foster and M. S. Bergdoll, in The Safety of Foods (J. C. Ayres, F. R. Blood, C. O. Chichester, H. D. Graham, R. S. McCutcheon, J. J. Powers, B. S. Schweigert, A. D. Stevens, and G. Zweig, eds.), Avi Publishing Company, Westport, Conn., 1968, pp. 159-167.
8. R. W. Detroy, E. B. Lillehoj, and A. Ciegler, in: Microbial Toxins (A. Ciegler, S. Kadis, and S. J. Ajl, eds.), Vol. VI, Academic Press, New York, 1971, pp. 4-178.
9. L. A. Goldblatt, Aflatoxin: Scientific Background, Control and Implications, Academic Press, New York, 1969.
10. G. N. Wogan, in Methods in Cancer Research (H. Busch, ed.), Academic Press, New York, 1973, pp. 309-344.
11. P. S. Steyn, in Microbial Toxins (A. Ciegler, S. Kadis, and S. J. Ajl, eds.), Vol. VI, Academic Press, New York, 1971, pp. 179-205.
12. M. Saito, M. Enomoto, and T. Tatsuno, in Microbial Toxins (A. Ciegler, S. Kadis, and S. J. Ajl, eds.), Vol. VI, Academic Press, New York, 1971, pp. 299-380.

13. A. Ciegler, R. W. Detroy, and E. B. Lillehoj, in Microbial Toxins (A. Ciegler, S. Kadis, and S. J. Ajl, eds.), Vol. VI, Academic Press, New York, 1971, pp. 409-434.
14. C. J. Mirocha, C. M. Christensen, and G. H. Nelson, in Microbial Toxins (S. Kadis, A. Ciegler, and S. J. Ajl, eds.), Vol. VII, Academic Press, New York, 1971, pp. 106-138.
15. A. Z. Joffe, in Microbial Toxins (S. Kadis, A. Ciegler, and S. J. Ajl, eds.), Vol. VII, Academic Press, New York, 1971, pp. 139-189.
16. J. R. Bamburg and F. M. Strong, in Microbial Toxins (S. Kadis, A. Ciegler, and S. J. Ajl, eds.), Vol. VII, Academic Press, New York, 1971, pp. 207-292.
17. A. Ciegler, S. Kadis, and S. J. Ajl, eds., Microbial Toxins, Vol. VI, Academic Press, New York, 1971.
18. S. Kadis, A. Ciegler, and S. J. Ajl, eds., Microbial Toxins, Vol. VII, Academic Press, New York, 1971.
19. S. Kadis, A. Ciegler, and S. J. Ajl, eds., Microbial Toxins, Vol. VIII, Academic Press, New York, 1972.
20. G. N. Wogan, in Foodborne Infections and Intoxications (H. Riemann, ed.), Academic Press, New York, 1969, pp. 395-451.
21. G. N. Wogan, Cancer Res. 35:3499 (1975).
22. A. C. Leopold and R. Ardrey, Science, 176:512 (1972).
23. I. E. Liener and M. L. Kakade, in Toxic Constituents of Plant Foodstuffs (I. E. Liener, ed.), Academic Press, New York, 1969, pp. 7-68.
24. W. G. Jaffe, in The Safety of Foods (J. C. Ayres, F. R. Blood, C. O. Chichester, H. D. Graham, R. S. McCutcheon, J. J. Powers, B. S. Schweigert, A. D. Stevens, and G. Zweig, eds.), Avi Publishing Company, Westport, Conn., 1968, pp. 61-67.
25. W. G. Jaffé, in Toxic Constituents of Plant Foodstuffs (I. E. Liener, ed.), Academic Press, New York, 1969, pp. 69-101.
26. Y. Birk, in Toxic Constituents of Plant Foodstuffs (I. E. Liener, ed.), Academic Press, New York, 1969, pp. 169-210.
27. C. H. VanEtten, in Toxic Constituents of Plant Foodstuffs (I. E. Liener, ed.), Academic Press, New York, 1969, pp. 103-142.
28. R. C. Montgomery, in Toxic Constituents of Plant Foodstuffs, (I. E. Liener, ed.), Academic Press, New York, 1969, pp. 143-157.
29. L. C. Berardi and L. A. Goldblatt, in Toxic Constituents of Plant Foodstuffs (I. E. Liener, ed.), Academic Press, New York, 1969, pp. 211-266.
30. P. S. Sarma and G. Padmanaban, in Toxic Constituents of Plant Foodstuffs (I. E. Liener, ed.), Academic Press, New York, 1969, pp. 267-291.
31. F. Perlman, in Toxic Constituents of Plant Foodstuffs (I. E. Liener, ed.), Academic Press, New York, 1969, pp. 319-348.

32. G. L. Laqueur, Environmental Cancer (H. F. Kraybill and M. A. Mehlman, eds.), Hemisphere Publishing Corporation, Washington, 1977, pp. 231-262.
33. I. A. Evans, Cancer Res. 28:2252 (1968).
34. H. Leach, G. D. Barber, I. A. Evans, and W. C. Evans, Biochem. J. 124:13P (1971).
35. I. A. Evans and J. Mason, Nature, 208:913 (1965).
36. I. Hirono, C. Shibuya, K. Fushimi, and M. Haga, J. Natl. Cancer Inst. 45:179 (1970).
37. A. M. Pamukcu and J. M. Price, J. Natl. Cancer Inst. 43:275 (1969).
38. I. A. Evans, Naturally occurring chemical carcinogens: Bracken fern toxin, in Proc. 10th Int. Cancer Congress, Houston, 1972.
39. A. M. Pamukcu, S. Yalciner, J. M. Price, and G. T. Bryan, Cancer Res. 30:2671 (1970).
40. A. M. Pamukcu, Ann. N.Y. Acad. Sci. 108:938 (1963).
41. A. M. Pamukcu, S. K. Göksoy, and J. M. Price, Cancer Res. 27:917 (1967).
42. I. Hirono, C. Shibuya, M. Shimizu, and K. Fushimi, J. Natl. Cancer Inst. 48:1245 (1972).
43. I. A. Evans, B. Widdop, R. S. Jones, G. D. Barber, H. Leach, D. L. Jones, and R. Mainwaring-Burton, Biochem. J. 124:28P (1971).
44. I. A. Evans, R. S. Jones, and R. Mainwaring-Burton, Nature, 237: 107 (1972).
45. J. Mager, A. Razin, and A. Hershko, in Toxic Constituents of Plant Foodstuffs (I. E. Liener, ed.), Academic Press, New York, 1969, pp. 293-318.
46. I. E. Liener, ed., Toxic Constituents of Animal Foodstuffs, Academic Press, New York, 1974.
47. D. T. Osuga and R. E. Feeney, in Toxic Constituents of Animal Foodstuffs (I. E. Liener, ed.), Academic Press, New York, 1974, pp. 39-71.
48. F. A. Fuhrman, in Toxic Constituents of Animal Foodstuffs (I. E. Liener, ed.), Academic Press, New York, 1974, pp. 74-110.
49. B. W. Halstead, Poisonous and Venomous Marine Animals, Vol. I, United States Government Printing Office, Washington, 1965.
50. B. W. Halstead, Poisonous and Venomous Marine Animals, Vol. II, United States Government Printing Office, Washington, 1967.
51. B. W. Halstead, Poisonous and Venomous Marine Animals, Vol. III, United States Government Printing Office, Washington, 1970.
52. F. E. Russell, in The Safety of Foods (J. C. Ayres, F. R. Blood, C. O. Chichester, H. D. Graham, R. S. McCutcheon, J. J. Powers, B. S. Schweigert, A. D. Stevens, and G. Zweig, eds.), Avi Publishing Company, Westport, Conn., 1968, pp. 68-81.

53. E. J. Schantz, in Microbial Toxins (S. Kadis, A. Ciegler, and S. J. Ajl, eds.), Vol. VII, Academic Press, New York, 1971, pp. 3-26.
54. J. H. Gentile, in Microbial Toxins (S. Kadis, A. Ciegler, and S. J. Ajl, eds.), Vol. VII, Academic Press, New York, 1971, pp. 27-66.
55. F. E. Russell, in Advances in Marine Biology (F. E. Russell, ed.), Vol. 3, Academic Press, London, 1965, pp. 255-384.
56. T. E. Furia, Handbook of Food Additives, Chemical Rubber Company Press, Cleveland, Ohio, 1972.
57. P. M. Magee and J. M. Barnes, Adv. Cancer Res. 10:163 (1967).
58. S. S. Mirvish, M. Greenblatt, and V. R. C. Kommineni, J. Natl. Cancer Inst. 48:1311 (1972).
59. R. C. Shank and P. M. Newberne, Food Cosmet. Toxicol. 10:887 (1972).
60. I. S. DuPlessis, I. R. Nunne, and N. A. Roach, Nature, 222:1198 (1969).
61. L. Hedler and P. Marquardt, Food Cosmet. Toxicol. 6:314 (1968).
62. N. D. McGlashan, C. L. Walters, and A. E. M. McLean, Lancet ii/7576:1017 (1968).
63. J. Sander, Z. Physiol. Chem. 348:852 (1967).
64. D. H. Fine, R. Ross, D. P. Rounbehler, A. Silvergleid, and L. Song, Agr. Food Chem. 24:1069 (1976).
65. L. Friedman and S. I. Shibko, in Toxic Constituents of Plant Foodstuffs (I. E. Liener, ed.), Academic Press, New York, 1969, pp. 349-408.
66. L. Friberg, M. Piscator, and G. Nordberg, Cadmium in the Environment, Chemical Rubber Company Press, Cleveland, Ohio, 1971.
67. L. Friberg and J. Vostal, Mercury in the Environment, Chemical Rubber Company Press, Cleveland, Ohio, 1972.
68. D. H. K. Lee and H. L. Falk, Environmental Health Perspectives, Experimental Issue No. 1, April, 1972.

CHAPTER 10

AGRICULTURAL CHEMICALS

David J. Sissons
Geoffrey M. Telling

Unilever Research, Colworth Laboratory
Colworth House
Sharnbrook, Bedfordshire, U.K.

I. INTRODUCTION

With the almost exponential growth of world population in the last hundred years, the need to provide enough food to satisfy basic human nutritional requirements has grown on a parallel path. Today the production of food is as much an industry as the production of any other commodity.

The population of the world increases by about 6000 people every hour, and all these people will live longer and require much more food than their parents before them. It has been estimated that 15% of the world's population is permanently hungry and that a further 30-35% are undernourished. Although part of the solution to this problem is to develop new sources of food, a more important aspect is to protect and improve the supplies of food which are already available. Man has to compete for his food with a vast number of insects, animals, and plants. It has been estimated that over 50,000 different insect species can inflict damage on man and on his food and the number of different microorganisms and plants is no less. Melnikov [1] has stated that in agriculture no less than one-third of all crops are lost to pests and diseases and that, if they were not systematically controlled, losses would be considerably greater. World losses from pests, diseases, and weeds are estimated at 75 billion dollars per year. Among plant pests and causative agents of plant diseases are representatives of various living organisms including insects, mites, nematodes, fungi, bacteria, and parasitic plants. A large number of phytopathogenic microorganisms and insects infect potatoes, corn, cereals, beans, fruit, and vegetable crops. Similarly rodents and other small animals cause tremendous damage to crops which are stored after harvest. Insect pests, mites, and other parasites also represent a great danger to animal husbandry, since many of them are not only vectors of infectious disease but also cause discomfort to animals resulting in lower milk yields and loss of weight, etc.

Losses from weeds are equally great. Weeds deprive crop plants of the moisture and nutritive matter in the soil, shade the crop plants and hinder their normal growth, and contaminate harvested grain with seeds that are poisonous for man and for animals.

To control these competitors for our vital food, man is increasingly turning to the use of a wide range of chemicals. Until the discovery of the insecticidal properties of DDT in 1940, agricultural chemicals were generally restricted to inorganic compounds such as arsenic, copper, sulphur and mercury, and to a few natural insecticides such as nicotine, derris, pyrethrum, and coal tar oils. After World War II, agrochemicals became big business and today the list of synthetic chemicals used in agriculture, in the widest sense of the term, numbers almost 1000 [2]. This list includes insecticides, herbicides, fungicides, and chemicals used in the prophylactic treatment of animals as well as naturally occurring antibiotics. When chemicals used for the control of disease vectors, e.g., malarial mosquito, tsetse fly, etc., are added to this list, the total number is even greater. It has been estimated [1] that approximately 100,000 formulations of about 900 pesticides were being produced in amounts totalling 2×10^9 lb annually as of 1972.

In addition, a wide range of chemicals are also used to improve the yields of crops and of livestock. Fertilizers are extensively used in modern agriculture in a wide variety of formulations. A wide range of feed additives, e.g., vitamins, trace minerals, hormones, antibiotics, etc., are also used to improve the general well-being of animals, increase feed efficiency, and stimulate growth. Jukes [3] estimates the economic value of low levels of antibiotics in feed as 914 million dollars in 1970. Tranquilizers may be used to calm animals prior to slaughter. The fate of vitamins and minerals during processing is covered in Chaps. 2, 3, and 4 of this book and therefore will not be considered further.

Much of the early application of these chemicals was made without any realization or appreciation of the way in which background levels of these compounds could build up in the human food chain and in the general ecosystem. In the last 20-25 years a vast amount of work has been done to determine the fate of pesticides after application, which has revealed the widespread distribution of residues in the ecosystem. The levels of chemical residues in foods and their elimination during commercial and domestic processing, preparation and cooking has become a topic of great interest. Today it is recognized that standard commercial processing stages may considerably reduce, or even eliminate, levels of agricultural chemicals and that, by these means, foodstuffs which might contain unacceptable levels of such chemicals can be made acceptable, both legally and aesthetically, for human consumption.

II. GENERAL CLASSIFICATION

A. Pesticides

This is a generic term covering all classes of compounds used in plant protection and in the control of plant, insect, and animal populations. Pesticides may be divided into three main classes, namely insecticides, herbicides, and fungicides, but the term also covers plant-growth regulators, defoliants, desiccants, repellants, attractants, and sterilizing agents. Of these, the compounds most widely used in agriculture are herbicides, insecticides, and fungicides.

The range of chemicals covered by this classification is very wide and it is impossible to classify them into a few discrete groups. However, the main types of chemical class to be discussed in this chapter are shown in Table 1. A complete classification of compounds discussed in this chapter is given in the glossary, including explanation of abbreviations.

TABLE 1 Main Types of Chemical Used as Pesticides

Function	Chemical class	Examples
Insecticides, acaridicides, rodenticides	Organochlorine compounds	DDT, dieldrin, lindane
	Organophosphorus compounds	Malathion, parathion
	Carbamates	Carbaryl, aldicarb
	Nitro compounds	Binapacryl
	Rotenoids	Rotenone
	Pyrethroids	Pyrethrins
	Coumarones	Warfarin
Herbicides	Carbamates, ureas	Linuron, CIPC
	Inorganics	Sodium chlorate
	Nitrophenols	Dinoseb
	Chlorophenoxy acids	2,4-D, 2,4,5-T, MCPA
	Chloroaliphatic acids	TCA, dalapon
	Triazines	Simazine, prometryne
Fungicides	Inorganic	Sulfur, copper salts
	Dithiocarbamates	Zineb, thiram
	Trichloromercapto compounds	Captan, folpet
	Quinones	Chloranil, dichlone
	Nitrophenols	Dinocap
	Systemic compounds	Benlate, thiophanate
	Substituted diphenyls	Orthophenylphenol
	Antibiotics	Griseofulvin
Fumigants	Organophosphorus compounds	Sulfotep
	Aromatic compounds	Azobenzene
	Miscellaneous	Dichloropropane, dichloropropene, ethylene dibromide, hydrogen cyanide, methyl bromide, carbon disulfide
Plant-growth regulators	Auxin type	3-Indolebutyric acid
	Gibberellins	Gibberellic acid
	Miscellaneous	Maleic hydrazide, naphthalene acetic acid

TABLE 2 Drugs and Feed Additives

Function	Class	Examples
Growth stimulants	Hormone type	Hexoestrol, zeranol, diethylstilbesterol (DES)
	Antibiotics	Chloramphenicol, bacitracin, penicillins, tetracyclines
	Arsenicals	Arsanilic acid
	Goitrogens	Thiouracil
Prophylactics	Coccidiostats	Amprolium, ethopabate, decoquinate
	Antihistominals	Demitradizole, acinitrazole
Therapeutics	Corticosteroids	Cortisones, prednisone, tetracyclines
	Antibiotics	Penicillins
	Halogenated phenyls	Nitroxinil, bromophenophos
Tranquilizers		Azaperon

B. Drugs and Feed Additives

A range of chemicals is now commonly used as additives in animal feedstuffs to stimulate growth and increase feed efficiency and for the treatment or prevention of many pathological conditions (see Table 2). Additional limited uses involve tranquilizers to prevent cardiac arrest prior to slaughter of animals, and antibiotics for the preservation of a restricted range of foodstuffs, such as fish and poultry.

III. PROPERTIES

The large number of groups of pesticides in current use exhibit a wide range of properties that give them their unique control characteristics and significantly affect their subsequent role as residues and environmental contaminants.

Organochlorine insecticides are all low-molecular-weight (<500) solids with melting points typically below 200°C. They exhibit significant vapor pressures under normal environmental conditions which, together with their ability to codistill with water, partially accounts for their widespread global distribution. They are all nonpolar compounds with very low but definite water solubilities in the ppb range (μg/liter). This level of solubility is a further important factor in their widespread distribution, particularly in the marine environment. The mammalian toxicity of these compounds is generally moderate or low with the notable exception of endrin which has an acute oral LD_{50} for rats of 7-17 mg/kg.

The most significant properties of the organochlorine insecticides, however, are their extreme lipophilic nature and resistance to biodegradation, which results in their accumulation and concentration in fatty tissues and their extreme persistence in the environment. For example, DDT has a half-life of 10 years in soil. This stability has made them, in one form or another, the commonest residues in foodstuffs in particular and the ecosystem in general. As a class they are nonsystemic and invariably remain on the surface of plants from which they are readily washed off into the soil or penetrate into the waxy cuticle (see Sec. VI). They very slowly undergo dechlorination or dehydrochlorination reactions to yield insecticidal metabolites. The possible environmental degradation paths for pp'-DDT are shown in Fig. 1 [4].

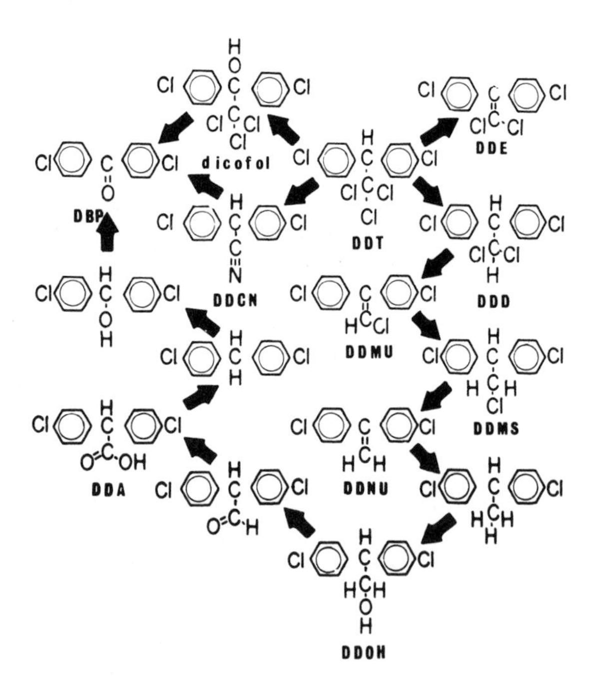

Fig. 1. Degradative pathways of pp'-DDT. (Reprinted with permission from Ref. 4, copyright by the American Chemical Society.)

Organophosphorus insecticides show a wider range of properties than their organochlorine counterparts. They are invariably more polar and can be divided conveniently into two groups, depending on their water solubilities. They are all liquids or low-melting solids and act as potent cholinesterase inhibitors, with many having acute oral toxicities within the LD_{50} range of 1-50 mg/kg. The majority of organophosphorus insecticides are derivatives of phosphoric, thiophosphoric, and dithiophosphoric acids with several being derivatives of phosphorus, phosphonic, and pyrophosphoric acids. In general, the phosphite esters and many of the mono- and diphosphate esters are preferentially soluble in organic solvents, having water solubilities on the order of mg/liter. However, they are sufficiently lipophilic to be readily absorbed by plants through the stem and leaves and, although poor systemic compounds, are much more readily translocated than are organochlorine insecticides. On the other hand, many phosphate and thiophosphate esters and most pyrophosphoric and phosphonic esters are very soluble in water. They are readily absorbed via roots of plants and rapidly translocated through the whole plant system.

Although the organophosphorus insecticides have significantly higher vapor pressures than organochlorine insecticides, they are very much more readily degraded both chemically and enzymatically, and no similar widespread distribution and build-up of residues has occurred. Degradation proceeds via oxidation and desulfurization initially to produce metabolites that are more polar and more active than the parent compounds. These are further degraded, ultimately by hydrolysis, to inactive esters of phosphoric acid and to the free acid. Typical metabolic pathways for the dithiophosphate insecticide dimethoate are shown in Fig. 2. For insecticides containing a thioether link in the side chain, degradation pathways are more complex because of sulfoxide and sulfone formation.

The degradation processes proceed at various rates, depending on the particular insecticide, climatic conditions, site of application, etc., and have been generally regarded as being complete within a period ranging from several days to weeks. Recent evidence suggests that, although disappearance of the parent species is generally rapid, the metabolites may well be more slowly degraded and residues may be detected several months after the original application.

A third group of insecticides are finding increasing use as they are much less persistent than the organochlorine insecticides and generally less toxic than the organophosphorus insecticides. Carbamate insecticides are all esters of alkylcarbamic acids. They are low-melting solids (30-150°C) and, although some are highly soluble in water, those most widely used have low systemic activity with water solubilities of 0.1% and less. Like the organophosphorous insecticides, they are cholinesterase inhibitors but acute oral LD_{50} values are generally 100 mg/kg and higher. A notable exception is aldicarb with an LD_{50} of 0.9 mg/kg for rats. Unlike other members of the group which are all aryl esters of alkyl carbamic acid,

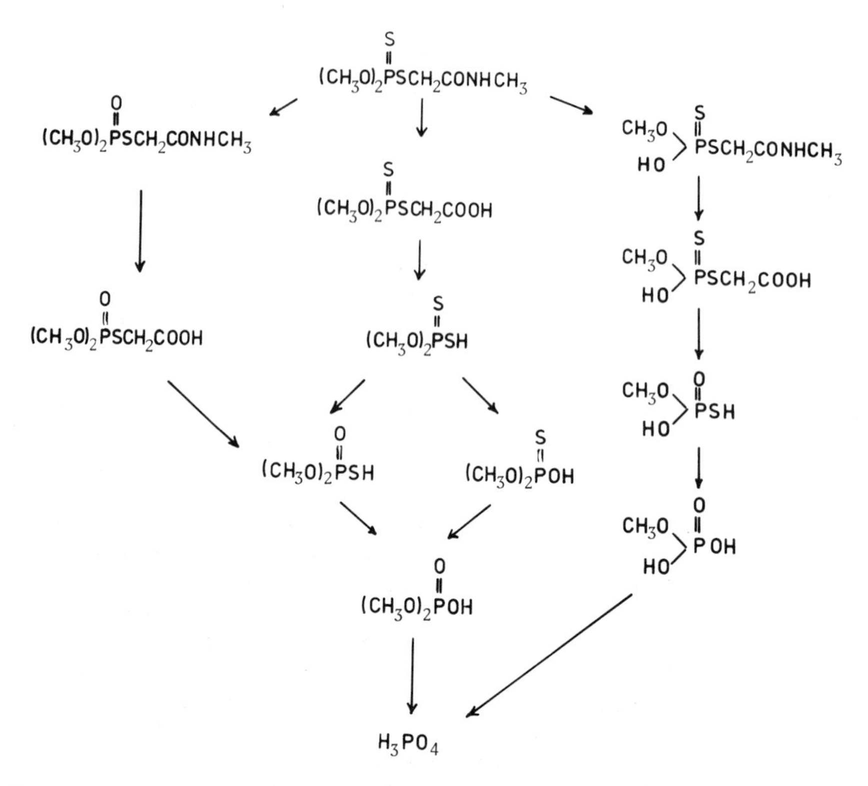

Fig. 2. Degradative pathways of dimethoate.

aldicarb is an alkyl oxime ester. Decomposition of carbamates occurs fairly readily and is primarily by oxidation and hydrolysis to inactive products as shown in Fig. 3.

Herbicides encompass a wide range of chemical types. Mono- and some dithiocarbamates, substituted ureas, and thioureas and esters of carbamic acids are predominantly liquids of low mammalian toxicity and low water solubility (mg/liter). They are lipophilic in nature, poorly absorbed and translocated and readily degraded to inactive products. The preemergent triazine herbicides are stable solids with melting points of up to 225°C. They lack systemic activity and are more resistant to degradation than most herbicides, and simazine is known to persist in soils following high application rates of up to one year. On the other hand, chlorosubstituted phenoxy carboxylic acids and the bipyridylium herbicides are readily soluble in water and efficiently translocated. Metabolism of these latter compounds in plants and animals is minimal, and degradation is predominantly by soil-borne organisms and photodecomposition. However, under normal

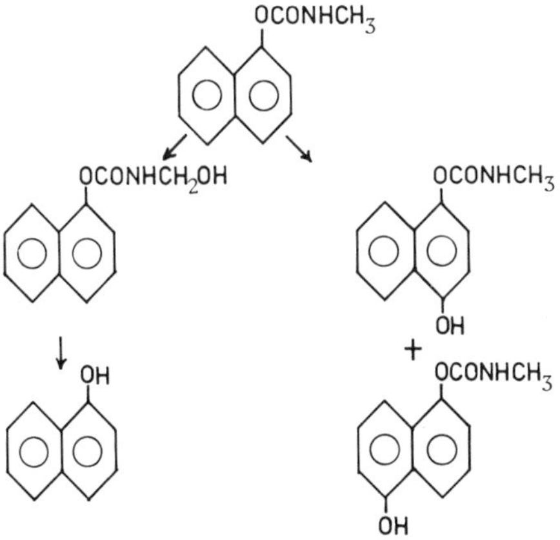

Fig. 3. Degradative pathway of carbaryl in plants and animals.

conditions of usage residues of the vast majority of herbicides are degraded to insignificant levels within a period of weeks or months.

Fungicides as a group are poor systemic compounds of low water solubility and low mammalian toxicity. They are effectively metabolized in plants and are unstable in soils with the result that residues are seldom detected several weeks after application at normal rates of usage. Several novel systemic fungicides, e.g., benomyl and the thiophanates, have recently been introduced. These compounds are of extremely low mammalian toxicity and are readily hydrolyzed in plants to inactive products. Alkyl mercuric salts are widely used as fungicidal seed dressings. These compounds are solid with melting points in the 150-250°C range and are highly soluble in water. They are resistant to degradation and most of the mercury fungicides are leached out of soils following sowing, to contribute to the reservoir of mercury in the oceans.

The peculiar properties of the various groups of pesticides dictate whether or not, after performing the function for which they were designed, they remain to cause problems as contaminants. As shown in Table 3, the major areas of concern are the insecticides, such as organochlorines because of their extreme persistence and organophosphates because of their high toxicity.

Under normal conditions of use, carbamate insecticides, herbicides, and fungicides are degraded prior to harvest and present very few residue problems by comparison. This is further borne out by data on freqency of detection of pesticide residues in foodstuffs. Figure 4 gives the distribution of residues in total diet samples in the United States from 1967 to 1970 [5], and clearly shows that over 90% of all residues detected derived

TABLE 3 Variations in Properties of Various Types of Pesticides

Class	Toxicity	Persistence (time)	Systemic property
Insecticides			
Organochlorine	Low to moderate	Years	Poor
Organophosphorus	Moderate to high	Days to weeks (months)	Good
Carbamate	Low	Weeks	Poor to good
Fungicides	Low	Weeks	Poor
Herbicides	Low to moderate	Weeks to months	Poor to good

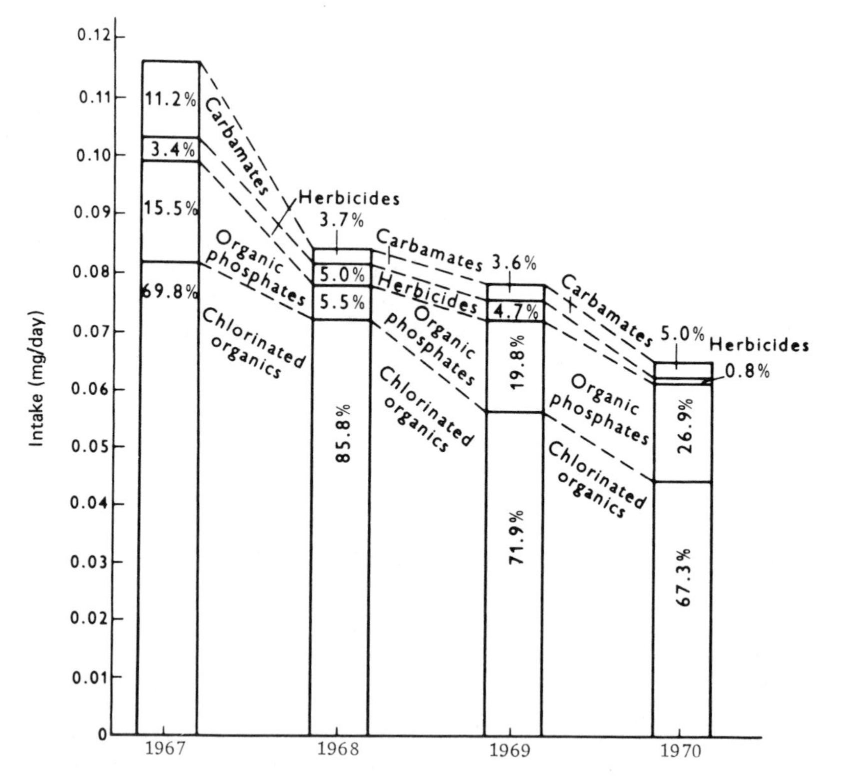

Fig. 4. Distribution of residues in total diet samples by chemical class, 1967-1970. (Reprinted from Ref. 5 by permission of the authors.)

from the application of insecticides and at least 65% from organochlorine insecticides.

The increase in the proportion of organophosphorus residues detected during this period reflects the increasing use of this class of chemical, whereas the fairly constant level of organochlorine residues, in spite of decreasing application, reflects their extreme persistence and stability. The decreasing proportion of herbicides detected is mainly due to a significant reduction in the use of 2,4-D and pentachlorophenol.

Since the 1950s a large number of different types of drugs have been used for a variety of reasons in animal husbandry (see Table 2). They exhibit a wide range of properties, depending on their particular function. The antibiotics are all complex, water-soluble compounds that are readily absorbed and rapidly distributed throughout the body tissues. Initial buildup occurs in the liver and particularly the kidney, through which excretion readily occurs. The sulfonamides as a group are composed of simpler molecules. Sulfanilamide is water soluble, readily absorbed and distributed, and excreted through the kidneys as the parent species in many animals. Sulfaquinoxaline, on the other hand, is only slightly water soluble, is less readily absorbed and distributed, and is excreted more slowly, mainly as a conjugate. Synthetic and natural hormones and antiinflammatory steroids are lipid-soluble compounds that tend to become concentrated in the fat of animals in addition to liver and kidneys. They are, however, readily excreted in conjugated form.

In general, therefore, the rapid excretion of most hormones and antibiotics added at low and carefully controlled levels to feeds, results in no significant residue problems in either foodstuffs or the environment.

The major source of residues in foodstuffs of animal origin derives from the therapeutic use of drugs, especially when they are used prophylactically. This is particularly so in the case of certain endemic diseases such as mastitis in cows. In this instance high levels of antibiotics are employed, either systemically or by intermammary infusion, together with corticosteroids. Failure to employ the correct dose or, more usually, failure to discard milk for a specified period following medication, is known to result in significant residue levels in milk. The cause for concern with these residual levels is not primarily one of toxicity, but is due to their physiological action in humans. This is especially so for antibiotics where, in addition to their ability to act as sensitizing agents for allergic response, the possible development of resistant strains of microorganisms presents a distinct hazard.

IV. LEGISLATION

Legislation controlling the use of agricultural chemicals is continuously increasing, and regulatory agencies, both national and international, have established a wide range of control systems to limit the human and animal intake of these chemicals. By far the most comprehensive legislation is that established to control levels of pesticide residues and hence this will be discussed in some detail.

A. Pesticide Residues

The degrees of control of pesticide usage and of levels present in foodstuffs at the moment of sale to the general public vary widely from country to country, but there are three main types of control which are applied.

1. Trade Legislation

This is designed to control the import, manufacture, and sale of pesticides in a particular country. It may vary from a mandatory control where no new pesticide may be marketed or used without prior approval by the government, e.g., in eastern European countries, to the voluntary system which operates in the United Kingdom, where a list of Ministry Approved Chemicals is issued annually, but which has no legal standing.

2. Codes Governing Application

Essentially these codes may specify (a) which pesticides are to be used on which crops; (b) when and in what quantities they should be applied; and (c) safety intervals governing the minimum time which must elapse between the last application of a pesticide and harvesting of the crop. These codes show the same variation between mandatory regulations and voluntary agreements as above.

3. Legal Tolerances

A legal tolerance is the maximum concentration of a particular pesticide allowed on or in a particular crop or food product. Its value is generally no higher than the residue likely to arise from "good agricultural practice." For many years the United States was almost alone in defining legal tolerance but gradually the concept has been adopted by the majority of countries. A notable exception is the United Kingdom where, apart from a few recommended maximum levels, e.g., for dieldrin, no formal legislation governing levels of pesticide residues in foodstuffs exists.

Unfortunately, "good agricultural practice" varies widely in different parts of the world, depending on climate, insect infestation, etc. Hence the level of pesticides used and that allowed on a particular crop may vary widely, according to the area of the world in which it is grown. In the so-called "developed world" there are undoubtedly various examples of the excessive use of pesticides which could be eliminated without leading to increased crop losses, but in other parts of the world there is a need for more, not less, use of pesticides. Since national legislation on permitted levels of pesticide residues must be realistic and take account of these varying needs for pesticide usage, a comparison of legislation between different countries will show wide variations. In extreme cases, products which satisfy legislation in a producer country would be legally unacceptable in a consumer country. In an attempt to overcome this problem WHO/FAO sponsored the Codex Commission on Pesticide Residues (CCPR), which has spent many years drawing up a series of pesticide residue tolerances to be applied to products moving in international trade. It recognized that these proposed tolerances are generally higher than those laid down in corresponding national legislation governing products on sale to the general public, but implies that pesticide residues will be further reduced during storage and normal commercial processing.

The Draft Report of the Eighth Session of the Codex Committee on Pesticide Residues (March 1975) (CX/PR75) contains a recommendation that the Committee should examine the problem of pesticide residue tolerances for processed or semiprocessed foods such as dried and frozen fruits and vegetables and for fruit juice concentrates. It is requested that specific indications should be given where maximum limits are applied to products other than the raw commodity.

Current proposals already recognize that, in certain instances, processing leads to decreased levels of pesticide. Thus the limits for paraquat in rice with husk and in polished rice are 10 and 0.5 mg/kg, respectively. Similarly, limits for fumigants on cereals are 5-to-10-fold higher on raw cereals than on the milled products.

National legislation generally controls limits in foodstuffs on sale to the general public. The most common form of such legislation is typified by that of West Germany [6] for residues in fruits and vegetables. This legislation lists some 140 pesticides and specifies the maximum levels permitted on particular crops. In addition, it specifies a list of 22 compounds which must not be present at levels greater than an arbitary limit of analytical detection, generally 10 μg/kg. Permitted levels of particular pesticides found on a crop which is not listed in the regulations are 10% of the lowest legal limit. This latter criterion can lead to considerable problems, as will be discussed below.

In the West German legislation on fruits and vegetables, which came into force in 1972, a section is included stating that levels of residues in raw materials may be higher than the specified limit if the processor-manufacturer can guarantee that, after processing, levels will comply with this limit. Hence legislation has begun to reflect a recognition of the fact that normal commercial processing of foodstuffs may, and often does, lower the level of pesticides present in foodstuffs at harvesting. Most of the early legislation was concerned with residue tolerances for fruits and vegetables, where use and uptake of pesticides may be effectively controlled. A comparison of tolerances over the last 10-15 years clearly reflects the changing pattern of pesticide usage. The list of permitted compounds has increased steadily but tolerances for organochlorine insecticides have decreased rapidly and in the West German legislation of 1972 levels of chlorinated cyclodiene compounds such as dieldrin and heptachlor are restricted to a maximum of 10 μg/kg.

Legislation on residue levels in animal fats, milk, fish, etc., has appeared only relatively recently [7]. Tolerances for residues in such foodstuffs are generally much higher than those for fruit and vegetables, reflecting the lack of control of the uptake of pesticides in such systems and the cumulative effect of the more persistent compounds such as organochlorine insecticides.

B. Drugs and Feed Additives

Much stricter control of the use of antibiotics and hormones has been instituted from the start and the same problems as have arisen with residues of pesticides have not been encountered. Legislation is confined mainly to a strict control of the products permitted for use and to safety intervals controlling the time between the last treatment of an animal and the sale of the product. All new uses are strictly controlled and products and treatments are withdrawn if they are shown subsequently to give rise to problems.

As in the case of pesticides, control measures for antibiotics and hormones vary widely from country to country. For example:

1. In the United Kingdom, veterinary drugs for therapeutic use are only available on prescription.
2. In the United States, compounds of special importance in the treatment of human illness are not permitted for animal use, including chloramphenicol, semisynthetic penicillins, or gentamicin. This has led to a more widespread use of compounds such as bacitracin and virginiamycin.
3. The EEC additives directive does not permit the use of hormones in animal feeds, although under certain circumstances they may be used as growth promoters when used as implants.

V. ANALYTICAL METHODOLOGY

The level at which legislation to control organic residues in foodstuffs can be set effectively, the ability of government agencies to enforce such legislation, and the means of following the fate of chemicals in a wide variety of situations is directly linked to the availability and sensitivity of suitable analytical methods. Thus, the earliest limit for DDT residues in fruit and vegetables was set at 7 ppm to coincide with a detection limit of 3 ppm obtained with the then current colorimetric procedure. With the development of the highly sensitive electron-capture detector in the early 1960s, the way was open to determine residues of all organochlorine insecticides below mg/kg levels. This detector is, however, not specific and required the development of adequate cleanup steps before the current procedures capable of detecting residues in a wide range of foodstuffs at levels as low as 1 μg/kg could be used routinely. The increasing use of organophosphorus insecticides led to the initial development of insensitive and nonqualitative total phosphorus colorimetric screening procedures. These methods were replaced in the late 1960s by more sensitive qualitative procedures based on the alkali flame-ionization detector (AFID or thermionic detector) and, more recently, on the more specific and stable flame-photometric detector, which are both capable of detecting residues down to the 10 μg/kg level. The increasing use of carbamate insecticides has led to the development of gas chromatographic procedures based on a modified AFID that is sensitive to nitrogen-containing compounds and most recently on sensitive electrolytic conductivity detectors such as those developed by Coulson [8] and by Hall[9], which can also be used for the specific determination of chlorine- and sulfur-containing residues.

The use of gas-liquid chromatography (GLC) with sensitive and specific detectors has revolutionized insecticide residue analysis in the past decade. Although some herbicides and fungicides are amenable to analysis in this way, many are insufficiently volatile, decompose at the elevated temperatures required in GLC analysis, or are insensitive to the available detector systems. Derivative formation followed by GLC has been used with success in the low-level analysis of many of these compounds such as carbamates, phenols, and phenoxy carboxylic acids. For the remainder, the majority can be analyzed satisfactorily at the 0.1 mg/kg level by spectroscopic techniques, either directly as in the case of the bipyridillium herbicides, or indirectly as in the case of dithiocarbamate fungicides, which are hydrolyzed and the liberated carbon disulfide either complexed and determined colorimetrically or determined directly using head space analysis and gas chromatography.

Determination of the range of drugs added to medicated animal feeding stuffs at levels > mg/kg is readily accomplished using a variety of procedures. Antibiotics are determined by microbiological methods, whereas colorimetric or fluorimetric procedures are generally employed for other drugs following extensive cleanup of sample extracts. The analysis of drug residues in tissues, organs such as liver and kidney, and products such as eggs and milk is much more difficult to achieve at concentrations down to the μg/kg level. Direct spectroscopic methods lack the required sensitivity, and a stringent cleanup of extracts is essential to remove the bulk of interfering coextracted material. The analysis can be further complicated by the presence of conjugates that require preliminary hydrolysis prior to extraction. Although microbiological assay methods are used in some instances, analysis is usually achieved by a final TLC separation (thin-layer chromatographic), either of the parent species or a derivative, followed by spectroscopic or spectrofluorimetric scanning of the developed plates. More recently, the increased specificity and sensitivity of GLC has been employed. Diethylstilboestrol, for example, can be detected at the 1 μg/kg level, following formation of the bisdichloroacetate derivative and analysis by GLC with an electron-capture detector [10].

The application of newer techniques, such as gas chromatography-mass spectrometry, high performance liquid chromatography, and gel permeation, offers the possibility of further improvements in analytical methodology of both pesticides and drugs in the future.

There are, however, legislative limits for both pesticides and drugs which are almost impossible to detect even with these modern techniques. It also follows that much of the early work on losses during processing where "zero" levels were found has only limited value, since detection limits of the methods used were much higher than those possible today. Furthermore, the significance of metabolites and their relative stability was not appreciated in those early days.

VI. SOURCES AND LOCATION OF RESIDUES IN FOODSTUFFS

Contamination of foodstuffs arises either from deliberate treatment or by accident in a variety of ways, as shown in Fig. 5 [11]. The exact pathways by which residues get into foodstuffs and where they finally end up is governed by a large number of factors. Their importance varies with different classes of foodstuffs, which will be considered separately.

A. Fruit, Vegetables, and Grains

Residues in or on plant products are generally but not always the result of deliberate application. They may arise in the following ways:

1. Insecticides and fungicides as seed dressings.
2. Herbicides, both pre- and postemergent.
3. Insecticides, applied directly or to the surrounding soil.
4. Fungicides, applied directly.
5. Postharvest treatment by (a) fumigants, (b) desiccants, (c) defoliants, and (d) antisprouting agents.

In addition, unintentional contamination can arise in a variety of ways:

1. Cross-contamination, particularly during harvesting, such as contamination of peas during separation from haulms during mechanical vining.
2. From rodent and pest control in factories and foodstores.

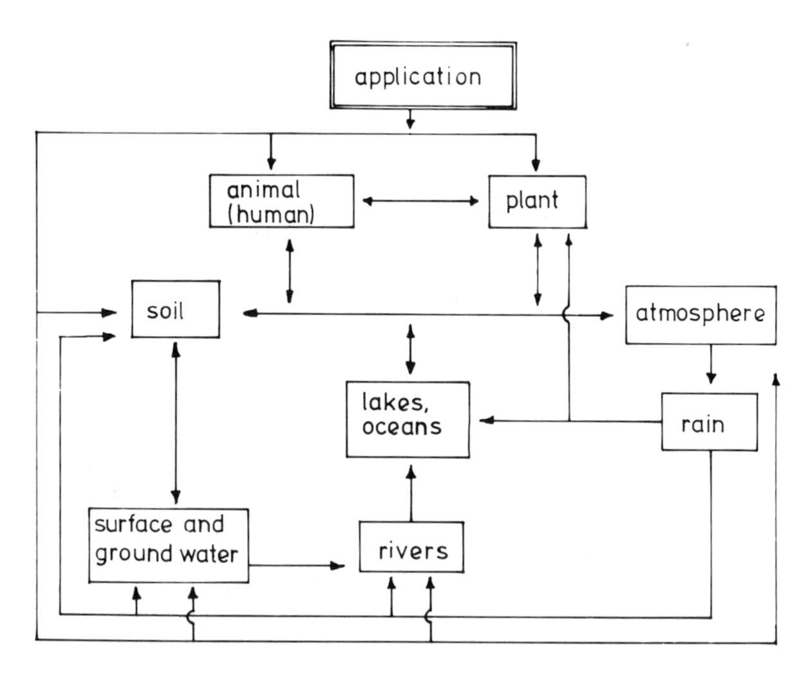

Fig. 5. Basic pathways of pesticides in the food chain. (Reprinted from Ref. 11 by permission of Springer Verlag.)

3. Because of misuse of pesticides, including (a) drift of spray onto adjoining crops, (b) harvesting too soon after treatment, and (c) too heavy an application.
4. From residues present in soil from a previous crop.

The way in which pesticides derived from any of these sources are absorbed into the plant depends on a variety of factors, including volatility, solubility, resistance to chemical degradation, etc. In general nonpolar lipophilic pesticides, such as organochlorine insecticides and carbamates, readily penetrate the outer regions of the plant cuticle which contains a high proportion of waxes and other highly lipophilic material. Because of the very low water solubility of these species they tend to concentrate in the cuticles of stem and leaves and are only very slowly and inefficiently translocated within the plant. Similar absorption and retention occurs in fruit. Systemic insecticides on the other hand are retarded by the lypophilic nature of the cuticular barrier and are much more effectively absorbed through the root system, partly because of the much greater surface area and partly because of the much lower level of lipophilic constituents. The efficiency of this method of absorption of systemic insecticides has resulted in the increasing use of granular applications. Translocation from the roots occurs through the xylem. It is a rapid process and results in a fairly uniform distribution of pesticide throughout the plant and leaves. However, translocation is governed by the same principles that control the movement of nutrients, with the result that higher concentrations tend to accumulate in sites of highest metabolic activity such as younger leaves and seeds. Many systemic insecticides are also absorbed through leaves and efficiently transported via the phloem, e.g., the growth retardant chlormequate is absorbed in this way and effectively translocated and concentrated in the roots.

A significant quantity of the nonsystemic pesticides that are applied to foliage as sprays end up on the soil either by run-off or wash-off. Absorption of these compounds also proceeds through the roots although less effectively than that of the systemic ones. This leads to the accumulation of residues in the root, as these pesticides are not translocated to any significant extent. Soil type has a significant effect on the rate of uptake via roots. It is now well established that the depth of penetration of nonsystemic pesticides such as the cyclodienes is dependent on the lipophilic content of the root cells, and the greater the content the greater the penetration and level of accumulation. In about 80% of studies carried out on potatoes, residues were firmly retained in the skin and had not migrated into the tuber [12]. On the other hand, penetration into carrots tends to be greater because of the higher lipid content of the inner cells. These generalizations are, however, strongly influenced by varietal differences. In a study of the uptake of aldrin and heptachlor by carrots [13], four varieties retained 70% of the absorbed insecticides in the outer layer. A fifth (white) variety, however, was found to contain half of the absorbed residues in the internal part of the root. In experiments with DDT and malathion the depth of penetration has also been shown to depend on formulation, the degree of weathering of the pesticide, and also on the length of the period of storage after harvesting [14].

Once in the plant, loss of residues proceeds by metabolic processes. In general, these have only a minimal effect on the nonsystemic pesticides, whereas systemic compounds are more readily degraded to inactive species. The process resulting in the greatest reduction in residues is quite often, however, the diluting effect of plant growth, which also provides an increasing volume for metabolic degradation to occur. In the case of a fast growing species, such as spinach, the effect of increased plant weight can be highly significant as shown in Fig. 6.

The majority of fungicides, such as dithiocarbamates, captan, and sulfur, show little if any penetration and residues are only found on those parts of the plants to which the spray is applied. Therefore subsequent growth is free from contamination. Residual levels depend on weathering conditions,

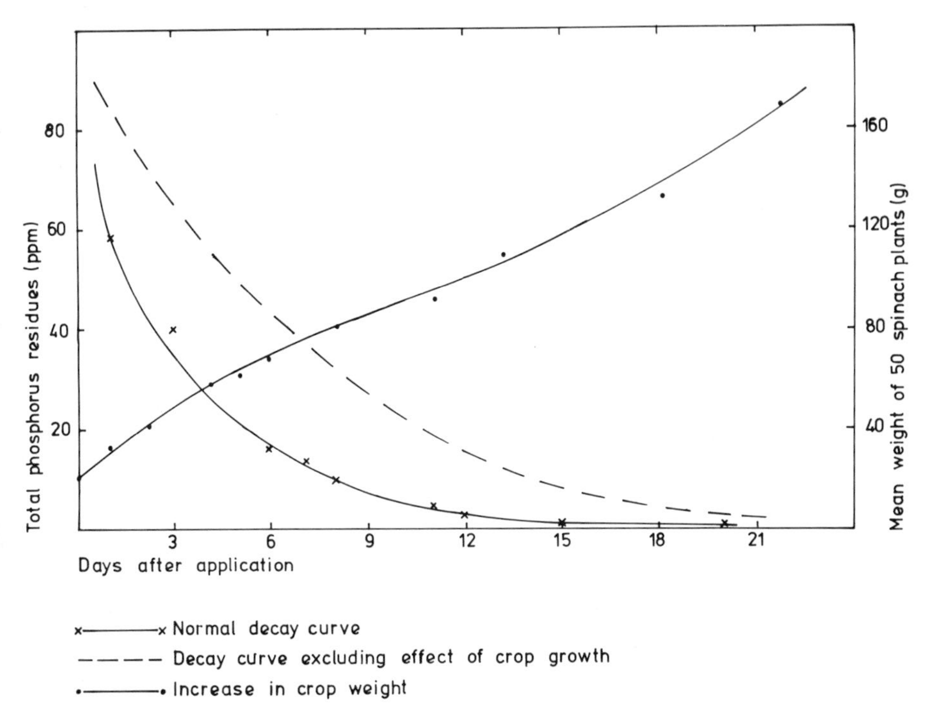

Fig. 6. Decay curve of demeton-S-methyl residues on spinach.

such as wind, rain, and temperature, rates of volatilization from the surface, and photodecomposition, which is dependent on the wavelength of light and duration of daylight. Other fungicides, particularly benomyl and the thiophanates, are systemic to varying extents and are degraded mainly by metabolic processes.

Treatment of seeds with systemic compounds, particularly insecticides, has proved a valuable protection to young seedlings. Coating seeds with impregnated powder results in the diffusion of the active ingredient into the soil from where it is absorped by the developing roots. Soaking seeds in pesticidal solution prior to planting results in the direct absorption and translocation of the pesticide along with the food reserves.

The postharvest application of pesticides is, in contrast to crop spraying, a deliberate attempt to contaminate a product with a pesticidal residue. Weathering and soil effects are not available to reduce the level of residues, and volatility and metabolism are the principal agents. Malathion, lindane, and synergized pyrethrins, together with several other organophosphorus compounds are most commonly used for postharvest application of insecticides. Grains are by far the most widely treated product with malathion most frequently used. Abbott et al. [15] reported in 1970 finding residues of malathion in 12 out of 66 samples of cereals analyzed during a total diet survey. Both polar and nonpolar species are rapidly absorbed especially via the attachment region of the pericarp. Organophosphorus residues are effectively metabolized as in plants although at reduced rates to give inactive phosphate esters. Pyrethroids are readily degraded by hydrolysis, whereas lindane only slowly undergoes dechlorination. The rates of transportation and degradation are dependent on moisture content but proceed at levels as low as 7%. Fumigants such as methyl bromide and ethylene dibromide are frequently used to disinfect stored grains. The former is absorbed and rapidly decomposed to inorganic bromide and a series of methylated derivatives, principally of the gluten fraction. Ethylene dibromide is readily adsorbed into the grain and is mainly lost by volatilization during aeration.

Sprout inhibitors such as IPC, CIPC, and MENA are widely used on potatoes stored for extended periods. The vast majority of the material applied remains on the surface of the tuber, with a small quantity being absorbed into the outer layer of skin. No movement of these compounds into the tuber itself occurs. Growth-regulating compounds on the other hand are readily absorbed and translocated in plants. Most are removed by metabolic degradation but for some compounds, e.g., 2,3,6-TBA, exudation of the unchanged species via the roots assumes significant proportions.

B. Meat and Animal Products

The majority of residues detected in meat and animal products derive from unintentional sources including:

1. Residues present in animal feedstuffs, forage, etc.
2. Cross-contamination following pesticidal treatment of barns, stalls, chicken houses, etc.
3. Contamination of drinking water and air.
4. Intake from mammalian milk during weaning.

In addition residues may arise from the deliberate use of pesticides and drugs for prophylactic or therapeutic purposes. Because of the persistence of organochlorine compounds, organophosphorus esters are finding widespread application in this area. Kaemmerer and Buntenkotter [16] have extensively reviewed their application and list 25 organophosphorus insecticides used in various ways in connection with animal husbandry, in the form of dips, dusts, and sprays against a wide range of ectoparasites and insects, and in the form of drenches, injections, doses, and feed additives against worms and other endoparasites. The available data suggests that the majority of organophosphorus compounds used for prophylactic purposes are rapidly metabolized and excreted, and no toxic species are detected within a week of treatment. Various degradation routes occur, with hydrolysis resulting in the formation of inactive dialkyl phosphate being the most common. However, certain compounds, such as chlorfenvinphos, fenchlorphos, dioxathion, or malathion, are degraded more slowly and finite levels of toxic residues may be detected in a variety of organs. One of the most persistent is fenchlorphos which has a zero tolerance in and on meat products in the United States and specified delay periods between application and slaughtering of up to 60 days, depending on the specific treatment. Levels of fenchlorphos in various tissues and organs 7 days after oral administration of a single dose of 100 mg per kg body weight are shown in Table 4. The distribution of fenchlorphos in the various organs and tissues is typical of the majority of organophosphorus insecticides and demonstrates the accumulation of residues, principally in those organs containing the greatest degradative capacity.

As discussed in Sec. IV, strict control of the addition of drugs to animal feeding stuffs ensures zero residues at slaughter. However, the therapeutic use of drugs can result in residues in milk, for example, unless delay periods are strictly adhered to.

Residues in meat and animal products that derive from unintentional sources are predominantly organochlorine insecticides. The most frequently detected residues are DDT (dichlorodiphenyltrichloroethane), as the metabolite DDE, and dieldrin, although others including lindane, heptachlor epoxide, and the fungicide hexachlorobenzene are often encountered. This is due to their extreme persistence in crop waste used directly, or after processing, as animal feeding stuffs, and in domestic and commerical waste such as vegetable and fruit peelings which contain the majority of the residues in these products. Also the volatility and finite water solubility, allied to the persistence of these compounds, ensures that they are distributed widely from the sites of original application and are absorbed from soil into grass and forage crops. Once ingested, the organochlorine insecticide residues are efficiently absorbed and deposited in the body fat of animals. The stored deposits are in equilibrium with a low level of residues in the bloodstream, which results in the passage of residues into the milk. Following ingestion of organochlorine insecticides, the concentration of residues in milk has been shown to rise within hours or days before leveling off at a plateau characteristic of the particular dietary level. The level present in the milk is dependent on the particular pesticide, and it has been demonstrated that heptachlor epoxide is excreted into milk twice as readily as dieldrin, seven times as readily as endrin, approximately 14 times as readily as lindane or DDT, and 2200 times as readily as methoxychlor [17,18].

Birds accumulate organochlorine insecticide residues in a way similar to animals, and DDE and dieldrin are again most frequently detected in poultry. The concentration of residues stored in the fat is roughly proportional to the dietary intake, and the level in abdominal fat is at the same

TABLE 4 Fenchlorphos and Derivatives in Cow Tissues 7 Days after Oral Administration

Tissue	Residue (mg/kg)			
	Total	Fenchlorphos	Water soluble	Skellysolve-B soluble
Subcutaneous fat	44.3 ± 7.7	8.2 ± 5.1	19.1 ± 1.0	18.0 ± 5.0
Mesenteric fat	23.1 ± 3.5	1.0 ± 0.2	13.2 ± 2.7	8.9 ± 2.5
Loin muscle	18.2 ± 1.1	3.2	10.8	4.2
Neck muscle	7.5 ± 1.0	3.6 ± 0.8	2.0 ± 1.0	1.9 ± 1.0
Liver	31.9 ± 6.7	6.8 ± 1.1	23.3 ± 1.1	1.8 ± 0.4
Spleen	11.6 ± 2.1	0.3 ± 0.5	10.3 ± 0.5	1.0 ± 0.4
Kidney	80.0 ± 16.5	21.9 ± 3.9	48.7 ± 7.4	9.4 ± 3.9
Lung	45.6 ± 6.7	8.0 ± 1.5	28.4 ± 2.8	9.2 ± 1.5
Heart	25.4 ± 2.8	5.5 ± 2.0	18.2 ± 1.6	1.7 ± 1.2
Brain	13.8 ± 1.2	1.0	10.9	1.9
Spinal cord	6.2 ± 0.8	0.3	3.7	2.2
Tongue	9.7 ± 0.1	3.2 ± 0.9	5.8 ± 1.0	0.7 ± 0.1
Hide	9.6 ± 1.5	3.7 ± 0.6	3.9 ± 2.1	2.0 ± 1.5
Diaphragm	11.3 ± 2.2	4.8 ± 0.3	1.6 ± 0.8	4.9 ± 0.7
Rumen wall	8.0 ± 0.7	1.6 ± 0.2	3.6 ± 1.5	2.8 ± 1.3
Rumen contents	8.3 ± 2.8	0.9 ± 0.4	5.8 ± 0.9	1.6 ± 0.8
Small intestine and contents	32.6 ± 7.7	0.8 ± 0.2	28.9 ± 1.1	2.9 ± 1.0
Bone (rib)	14.4 ± 0.9	-	-	-
Ovary	27.7 ± 1.4	-	-	-

Source: From Ref. 16, reproduced by permission of Springer-Verlag.

concentration as that in the fat of eggs. Heptachlor epoxide and dieldrin are more efficiently retained in the fat than DDT or endrin and much more so than lindane [19].

C. Fish and Marine Products

With the possible exception of prophylactics used in fish farming, residues are invariably due to unintentional contamination arising from a variety of sources.

1. The vapor pressure of many persistent pesticides is sufficient for them to enter the atmosphere and provide a constant and far-reaching contamination of the aquatic habitat.
2. Run-off from agricultural land either by solubilization or adsorption onto sediment, the latter providing a continuous reservoir to replenish losses from water.
3. Direct contamination as effluent from industrial processes and dumping of sewage sludge. This route is of importance in connection with organomercury fungicides that are widely used in paper pulp mills.
4. Contamination of fish food used in fish farms.

Source 1 results in a general contamination of the environment, whereas sources 2 and 3 result in local contamination of rivers, lakes, estauries, coastal areas, and seas, and lead to wide variations in residue levels in fish taken from different areas. In addition, pesticide concentrations in different layers of water can vary significantly and over depths of a few centimeters. This results in significant residue differences in different species. Fish feeding in layers of higher contamination show higher residue levels than those feeding in adjacent layers. Bottom feeders that ingest sediment can again show distinctly higher residue levels than those feeding in the adjacent water layer.

Uptake of residues by fish occurs in two ways. In certain species significant amounts of pesticides are absorbed directly from the water during its passage through the gills. In other species, especially larger fish, the major source of residues is via contaminated food. Phytoplankton are the primary source of food for marine animals, and have the ability to concentrate pesticides to over 100 times the level in their immediate surroundings without adverse toxic effect. Further accumulation of residues occurs at each higher trophic level, resulting in a final level of residues in species toward the top of the food chain that can be several thousands of times the environmental concentration. Once ingested, residues of the persistent insecticides are deposited and stored in the fatty tissues. The distribution of these residues therefore depends primarily on the location of fat within different species. In fatty fish such as herring, stored fat is distributed throughout the fish, and the edible portions therefore reflect the total body burden. Seasonal variations, such as reproductive activity, that significantly affect the total fat reserves of these species have a marked effect on tissue residue contents. On the other hand many species store fat in speci-

TABLE 5 Residue Levels in Fish from Various Waters, mg per kg whole muscle.

Fish	Source	Number	PCB		Total DDT		Dieldrin		Ref.
			Mean	Range	Mean	Range	Mean	Range	
Atlantic Salmon	New Brunswick	4	0.45	-	0.35	-		-	21
	Baltic Sea	11	0.3	0.014 - 0.54	3.4	0.26 - 7.1		-	22
Coho Salmon	Lake Erie	30	2.1	1.0 - 4.3	0.58	0.21 - 1.7	0.06	0.03 - 0.15	23
Herring	Newfoundland	30	0.4	0.19 - 0.73	0.25	0.02 - 0.53		-	21
	Baltic Sea	18	0.27	0.009 - 1.0	0.68	0.09 - 2.3		-	22
	Gulf of Bothnia	4	0.065	0.026 - 0.091	0.26	0.15 - 0.42		-	22
	North Sea	6	1.0	0.8 - 1.15	0.09	0.07 - 0.13	0.06	0.04 - 0.075	24
	Firth of Clyde	5	1.6	0.9 - 3.2	0.28	0.15 - 0.33	0.08	0.06 - 0.1	24
Mackerel	Newfoundland	4	0.35	-	0.16	-		-	21

Cod	Nova Scotia Banks	3	0.02	-	0.01	-			21
	Baltic Sea	5	0.033	0.012 - 0.057	0.063	0.027 - 0.11			22
	Swedish West Coast	4	0.02	0.006 - 0.03	0.005	0.001 - 0.006			22
	North Sea	6	0.02	0.01 - 0.03	0.002	0.001 - 0.003	< 0.002	-.002	24
	Icelandic Waters	6	0.01	<0.01 - 0.04	0.003	0.001 - 0.006	<0.002	-	24
Plaice	Nova Scotia Banks	1	0.03		0.01	-		-	21
	Baltic Sea	6	0.017	0.01 - 0.032	0.018	0.006 - 0.036		-	22
	Swedish West Coast	3	0.021	0.002 - 0.056	0.006	0.003 - 0.009		-	22
	North Sea	8	0.13	0.03 - 0.55	0.02	0.005 - 0.037	0.004	<0.002 - 0.013	24

fic locations, such as the liver in cod and in pockets along the backbone in salmonids. The edible muscle from these fish contains only very low levels of residues, whereas extracted oil from cod liver, for example, contains maximum levels of residues. Fish offal and waste that is processed into fish meal and fish feed can also contain significant levels of residues.

The most frequently detected residues are those of DDT and its metabolites, particularly DDE, and, to a lesser extent, dieldrin. Residues of other persistent insecticides are less generally encountered and invariably reflect local contamination such as the use of camphechlor in cotton-producing areas of the United States. Mercury is absorbed by fish and crustacea as methyl mercury, following conversion by microorganisms. It is a natural "contaminant" of seawater and results in a general background level of up to 0.1 ppm in the flesh of most fish species. Local contamination arising from the industrial use of mercury compounds results in a significant increase in tissue levels. Analysis of methyl mercury in muscle from pike from an uncontaminated area showed a natural level of 0.06 mg/kg, from an agricultural area 0.1 - 0.3 mg/kg, from a contaminated area 0.2 - 1.2 mg/kg, and downstream from a paper mill 0.5 - 6.0 mg/kg [20]. Polychlorinated biphenyls (PCBs) are also a contaminant of certain marine areas. These compounds are widely used in industry because of their chemical and biological stability and have, during the last 25 years, become a major pollutant of coastal waters adjacent to highly industrialized areas. They are similar in structure and properties to DDT and, as such, are frequently present, along with DDT and its metabolites, in samples taken from areas such as the Baltic and North Sea, the Pacific coastal waters of the United States, and around Japan. Typical levels of PCBs, total DDT and dieldrin residues in different species of fish taken from various locations are shown in Table 5.

VII. COMMERCIAL PROCESSING STAGES AND THEIR EFFECTS ON RESIDUE LEVELS

A. Fruits and Vegetables

It has long been recognized that the widespread use of the whole spectrum of pesticides is essential for efficient production of fruits and vegetables. It has been estimated that without pesticides, losses due to depredation would run into billions of dollars per year. Although attempts are now being made to cut back on use of pesticides by use of integrated pest-control methods, and the use of certain persistent groups is now being phased out, the total world pesticide usage is still increasing rapidly.

As discussed earlier in this chapter, residues of herbicides in fruits and vegetables at harvest are minimal and so this section will be concerned mainly with removal of insecticide residues and, to a lesser extent, fungicide residues.

It has long been recognized that the normal commercial processes used in the production of frozen, canned, and dried fruits and vegetables plays an important part in lowering levels of pesticide present in the final products sold to the general public. Many of these pesticide losses are adventitious as they arise as a by-product of commercial stages necessary to give a suitable product but, in certain cases, modifications to normal processes have been made with the express intention of removing pesticide residues.

Various groups of workers have reported studies on commercial and domestic processing losses mainly during the period 1965-1971; the scope of these studies is shown by the examples given in Table 6. The pattern of findings fits together very well and allows fairly accurate predictions as to the losses which will occur during various stages of commercial processing.

Processing of fruit and vegetables may consist of a number of stages, all or some of which may be applied to a particular crop.

1. Cold-Water Washing

This consists of washing at room temperature by flotation techniques or rotary washing. The basic technique has been modified in various ways, such as the use of detergent, increased temperature, and addition of potassium hydroxide or hydrogen peroxide solutions. The extent to which this step removes pesticides depends on a number of factors.

1. The character of the surface—smooth or rough, waxy or nonwaxy. Thus parathion residues on celery and spinach are virtually untouched by water washing ($< 10\%$ removal) and use of detergents only increases to $\sim 25\%$.
2. The surface-to-volume ratio of the material—the bigger the product, the more effective the washing.
3. The level of residue initially present—the higher the initial level, the bigger the percentage removal.
4. The chemical/physical properties of the pesticide. Thus when no surface penetration occurs (e.g., carbaryl, malathion, and a large part of DDT), washing with cold water removes 75-95% of any pesticides present. That part of DDT which penetrates the surface can only be removed by peeling; pp'DDT is more easily removed than op'DDT.
5. The time that the residue has been in contact with the food.
6. The formulation in which the pesticide is applied.

Fungicides such as dithiocarbamates also appear to show some surface penetration as shown by the following results obtained in the authors' own

TABLE 6 Studies on Removal of Pesticide Residues from Fruits and Vegetables during Normal Commercial Processing

Crop	Pesticide	Processing steps									Ref.
		Wash	Peel	Blanche	Juice	Can	Store	Jam	Freeze	Cook	
Apples	CIPC, IPC						x				25
	DDT	x	x							x	26
	Captan	x				x			x		27
Apricots	Wide range of insecticides					x	x				28
Beans, green	DDT	x	x							x	29
	DDT, malathion, carbaryl	x	x								30
	Azodrin, Gardona	x		x		x					31
Broccoli	Parathion, carbaryl	x		x						x	32
	Malathion								x	x	33
Cherries	Gardona					x			x		34
Gooseberries	CIPC, IPC					x		x			25
Lemons	Tedion	x									35
Oranges	Guthion	x			x						36
Papaya	Morestan				x						37
Peaches	Gardona		x			x					34

Pears	Gardona		x			x				34
Peppers, green	DDT	x				x				38
Plums	CIPC, IPC						x			25
Potatoes	DDT	x	x						x	39
Snap beans	Diazinon	x	x	x					x	40
	DDT	x		x		x		x		41
Spinach	DDT, parathion, carbaryl	x		x		x				42
	Diazinon	x	x	x						40
	Wide range of insecticides					x	x			28
Sweet corn	Gardona					x				31
Tomatoes	CIPC, IPC						x			25
	Azodrin	x	x		x	x				43
	Gardona	x			x	x				31
	Aldrin	x	x		x					44
	Diazinon	x	x	x						40
	DDT, malathion, carbaryl	x	x	x		x				45
Turnip greens	DDT		x							46

laboratory by Usher and Denton [47]. A series of spinach samples were heavily oversprayed with zineb and analyzed before and after washing:

Sample	Trial 1			Trial 2		
	min	mean	max	min	mean	max
Before washing	8.8	10.3	12.9	242	327	484
After washing		2.5		3.1	4.1	6.2

2. Blanching

Some form of blanching is usually essential to inactivate enzyme systems, especially when products are to be frozen. Blanching may be carried out in either hot water or steam. Hot-water blanching is more effective in removing pesticide residues than is steam. The same variations in efficiency of pesticide removal, depending on the nature of pesticide and crop, is found as with washing. Thus blanching of green beans containing DDT gives 50% removal, whereas blanching of spinach gives only 10% removal of DDT.

Hot-water blanching hydrolyzes many organophosphorus residues but has little effect on levels of parathion on celery and broccoli. This indicates that waxy surfaces are only moderately affected by blanching.

3. Trimming and Peeling

Nearly all fruits and root vegetables are subjected to chemical and/or mechanical peeling operations prior to canning. One of the most common techniques is that of lye peeling where vegetables are treated with 1% alkali at about 200°F (93°C). Vegetables with loosened skins are then conveyed under high velocity jets of water which wash away lye and any residual skin. It is almost universally true that when peeling operations are used removal of nonsystemic pesticides is virtually complete. Workers at NCA [38] studied the removal of DDT, malathion, and carbaryl from potatoes and tomatoes and found removal ranged from 90 to 99%. Other workers have shown that organochlorine insecticides concentrated in the outer layers of carrots are almost totally removed by lye peeling [48].

4. Freezing

Losses during this step would be expected to be minimal, if any.

5. Canning

Commercially canned foods supply an important part of the diet and Farrow et al. [49] reported that when spinach containing DDT was canned the DDT was dechlorinated to TDE(DDD) and other compounds during the autoclave stage used to sterilize the contents of the can. As a follow up to this work, Elkins et al. [28] surveyed the effects of thermal processing and storage on 15 representative pesticides. Selected compounds covering the range of organochlorine, organophosphorus, and carbamate pesticides were added to samples of spinach and apricots. The results obtained by these workers are given in Table 7, indicating that losses of particular compounds during thermal processing varied according to the type of chemical and the type of substrate. Thus malathion was more stable in apricots at pH 3.4 than in spinach at pH 5.4, and similar conclusions could be drawn for methyl parathion.

6. Dehydration

It has been shown by several workers that pesticides such as DDT codistill with water [50, 51], and therefore it is to be expected that commercial dehydration processes should result in the removal of a significant part of any residues present. Very little data have been published on losses during dehydration of vegetables; this type of process will be discussed below in more detail.

7. Juicing

In the juicing of products, the process follows the normal pattern of washing and peeling before the juicing stage. This will generally remove over 90% of any pesticide residues present, and canning sterilization will remove most of the remainder.

Care must be taken in the production of citrus juices. Pesticides in citrus fruits have been shown to be concentrated in the rinds and therein mostly in the oils. Although the oils are mainly excluded from commercial juice products, the completeness of separation of peel and flesh is an important consideration in limiting residues in the final product.

8. Storage

The work of Elkins et al. [28] indicates that storage of canned products can lead to significant losses in levels of pesticides present at time of initial processing and that storage at elevated temperatures leads to still greater losses. Losses from frozen products would be expected to be minimal during storage. Storage losses from fresh fruit dipped with malathion or captan, for example, are significant when samples are held at ambient temperature for a period of months. These losses are probably due to enzymic processes.

It must be appreciated that most of these reported findings apply to pesticides found on or near the surface of the fruit or vegetable. When pesticidal compounds are systemic they penetrate the whole plant and often form more polar metabolites. In such cases criteria governing losses may be completely different.

TABLE 7 Effect of Heat Processing and Storage on Chlorinated Hydrocarbon, Organophosphate, and Carbamate Pesticides in Spinach and Apricots

Pesticide	Percent reduction in residue level							
	Spinach				Apricots			
	Initial level (ppm)	Processed	Ambient 1 year	100°F (38°C) 1 year	Initial level (ppm)	Processed	Ambient 1 year	100°F (38°C) 1 year
Chlorinated Hydrocarbons								
Captan	35.7	93	100	100	88.5	97	99	100
Lindane	10.1	33	49	99+	6.8	13	56	100
TDE	6.82	8	32	49	6.81	9	38	64
Thiodan	1.84	19	19	85	1.79	13	22	85
Toxaphene	6.5	27	60	95	6.8	7	35	92
Methoxychlor	12.6	21	65	100	12.5	0	82	100
Organophosphate								
Diazinon	0.74	58	100	100	0.45	100	100	100
Guthion	1.2	100	100	100	1.00	61	100	100
Malathion	7.74	96	99+	100	7.63	32	84	100
Methyl parathion	0.88	100	100	100	0.85	54	100	100
Trithion	0.76	17	71	83	0.76	35	84	84
Carbamate								
Zineb	6.8	100	100	100	6.7	20	42	48
Ziram	4.2	100	100	100	5.0	40	88	94
Maneb	9.9	100	100	100	9.8	40	53	54
Carbaryl	10.5	44	46	67	11.4	12	16	17

Source: Reprinted with permission from Ref. 28. Copyright by the American Chemical Society.

Inefficient storage of spinach prior to processing can lead to the reduction of nitrate to nitrite under both anaerobic and aerobic conditions [52].

9. Domestic Processing

Studies by various workers on domestic processing of foodstuffs indicates a pattern of losses similar to those obtained during commercial processing. However, the domestic equivalents, e.g., washing, are generally less efficient; although when potatoes are washed and peeled, residue reductions are similar by both processes.

An important aspect of domestic processing is the cooking of spinach. The high nitrate level of spinach can give rise to significant levels of nitrite if spinach is cooked according to traditional methods of Western Europe. Sinios and Wodsak [53] state that cases of methaemoglobinemia have been reported from West Germany in infants aged 2-10 months as a result of eating spinach which had been bought fresh and cooked at home. In these cases the spinach became toxic after 24-48 hr postcooking storage, due to conversion of nitrate to nitrite.

B. Dairy Products

The presence of pesticides, and especially of organochlorine compounds, in dairy products has been a subject of considerable concern in recent years. Since the introduction of the first synthetic insecticide (DDT), a wide range of chlorinated hydrocarbon and organophosphate pesticides has been used in the control of cattle insects, particularly grubs, flies, and lice. If cattle ingest forage contaminated with chlorinated hydrocarbon pesticides, the pesticide passes into the bloodstream and is eventually stored in the depot fat. An equilibrium is established between the pesticide in the blood lipids and the depot fat. Some of that in blood lipids is finally secreted into milk. There is, however, a lesser problem with the ingestion of organophosphate pesticides since these are generally metabolized by mammals and hence do not contaminate milk. In 1959, Marth and Ellickson [54] reviewed the status of pesticides in milk and milk products and reported that surveys of milk market supplies indicated that 25-62% of samples contained measurable amounts of chlorinated pesticides. More recently, Richou-Bac [55] reported on levels of organochlorine insecticides in milk and dairy products produced in France and included the levels of particular compounds during the period 1967-1971 (see Table 8).

The pattern and levels of organochlorine insecticide residues in milk also vary considerably from country to country. In 1972, we have found in our own laboratory [56] the variations listed in Table 9 in mean levels of organochlorine insecticides in samples of dairy products from Belgium, Italy, Germany, and France.

Hence experimental results confirm that levels of organochlorine insecticides in milk and dairy products are much higher than in fruit and vege-

Table 8 Residues of Organochlorine Pesticides in French Milk, Arithmetic Means in mg/kg on Fat Basis.

Pesticide	1967	1968	1969	1970	1971
α-BHC	0.73	0.64	0.46	0.21	0.16
β-BHC	0.16	0.11	0.10	0.08	0.09
γ-BHC	0.03	0.03	0.03	0.03	0.04
Total BHC (benezenehexa-chloride)	0.92	0.78	0.59	0.32	0.29
Heptachlor and heptachlor epoxide	0.02	0.025	0.03	0.06	0.10
Dieldrin and aldrin	0.02	0.03	0.04	0.05	0.09
Total DDT	Trace[a]	0.02	Trace	Trace	Trace

Source: Ref. 55.
[a]Trace ≤0.001 mg/kg.

tables, and that milk and dairy products contribute a significant part of the human intake of these compounds.

Some efforts have been given to the removal or reduction of residues in dairy products through normal or modified processing procedures but these have met with only limited success. The normal processes which may be

TABLE 9 Mean Levels of Organochlorine Insecticide Residues in Dairy Products, mg/kg on Fat Basis.

Country	No. of samples	γ-BHC	Heptachlor epoxide	Dieldrin	DDE	TDE	DDT
Belgium	60	0.084	0.107	0.104	0.078	0.153	0.231
Italy	6	0.049	0.038	0.066	0.042	0.042	0.077
Germany	31	0.035	0.017	0.033	0.031	0.066	0.108
France	36	0.047	0.079	0.081	0.044	0.133	0.158

used commercially in the production of milk products can be summarized as follows:

Whole Milk

Pasteurized Sterilized Homogenized Milk	Condensed Evaporated Dried Milk	Separation	
		Skim Milk Buttermilk Whey	Cream Butter Butter oil Butter serum

Few of these processes involve a stage during which organochlorine compounds might be expected to be removed. The lipophilic organochlorines remain in the fat portion by preferential adsorption on the surface layer of fat globules, and most dairy processes merely involve a redistribution of pesticides with the butter fat. Cream, butter, and cheese will generally have the same level of chlorinated hydrocarbons as the milk fat from which they are produced. However, there is some evidence that residues of DDT and DDE may be degraded if they are present in certain types of surface-ripened cheese [57]. Levels in whey, skim milk, and buttermilk tend to be higher (on a fat basis) than in the original milk [58].

Pasteurization and sterilization conditions range from 150° F (65°C) for 30 min to 270° F (132°C) for 1 sec, but these would generally not lead to any elimination of chlorinated hydrocarbons although it has been reported [59] that DDT may be converted to DDE during sterilization. Some milk products, such as condensed or dried milk, may lose a proporation of their initial level of chlorinated hydrocarbons in the condensing process by forced codistillation with water vapor at elevated temperatures. In such cases the loss will vary widely, depending on the physical properties of the particular pesticide [58].

A number of papers have been published on the use of techniques such as UV irradiation, molecular distillation, and high-temperature deodorization of milk and butter fat. The latter techniques, as expected, effectively removed chlorinated hydrocarbons from butter oil, but they can scarcely be considered to be practicable commercial techniques for dairy products.

In general, therefore, whatever chlorinated hydrocarbons are present in raw milk will be present in all fat-based products derived from that milk. This fact is reflected in the levels of chlorinated hydrocarbons permitted in milk and milk products by the WHO/FAO practical residue limits and the West German proposed limits (see Table 10).

C. Oils and Fats

The present world consumption of oils and fats, excluding butter and "unseen" fats, i.e., those associated with meat, is about 46 million tons. The major source is plant seeds which provide vegetable oils, with the commercially important fish oils and animal fats being produced in smaller quantity.

TABLE 10 Residue Limits of Chlorinated Hydrocarbons in Milk and Milk Products, mg/kg on Fat Basis.

Chlorinated hydrocarbon	West Germany	WHO/FAO
Total DDT	1.0	1.25
Aldrin and dieldrin	0.15	0.15
Chlordane	0.05	0.05
Endrin	0.02	0.02
Heptachlor and heptachlor epoxide	0.15	0.15
Hexachlorbenzene	0.3	0.3
Lindane	0.2	0.1
α- and β-BHC	0.1	-

Strict legislation has evolved in the last decade in the western world to regulate the level of residues in products such as these, both in international trade and in the products as sold to the public. For reasons stated before, this legislation is primarily concerned with levels of organochlorine insecticide residues. However, whereas limits have been specifically set for oils and fats of marine and animal origin, no specific limits have been set for vegetable oils, and the inferred limits which have to be used are very much lower than for the former. The effect of processing on levels of organochlorine residues in vegetable oils is therefore of major significance relative to these lower limits.

1. Vegetable Oils

Over 40 different species of trees and plants are used for the commercial production of oils. The principal seeds used to produce edible vegetable oils are listed below and account for over 90% of the annual production.

soya bean	cottonseed	olive
sunflower	rapeseed	palm
groundnut	coconut	palm kernel
		sesame

A significant quantity of these oil seeds are grown in the so-called developing countries for refining and consumption in western countries. The consuming countries have only limited control over pesticide usage in the producing countries. Widespread use of pesticides, both specifically on oil seed crops and in general agriculture and pest control programs, is undoubtedly made in these countries. The effect of processing on the level of residues in vegetable oils is therefore important and has been studied extensively over the last 10 years.

Crude oils are separated from the seed husk either by a single application of pressure or by a combination of pressure and solvent extraction. In either case, fat-soluble pesticides are efficiently removed with the crude oil. These residues will be present in the so-called "untreated oils" preferred by certain diet cults, which received no further processing after cold pressing. This fact was recognized in the mid-1950s by the Italian Ministries of Agriculture and Health who instigated a research program to find a suitable alternative to DDT and the various other organochlorine insecticides that were widely used at that time in controlling the olive fly on the economically very important olive crop. The conclusions from this work were that, although adequate control could be obtained using alternative organophosphorus insecticides, only dimethoate (Rogor) gave a residue-free oil. Its hydrophilic nature ensured that it was retained in the cellular tissues during pressing. This insecticide has subsequently been used in the Mediterranean area on olive groves with minimal residue problems.

Levels of persistent residues in crude extracted oils vary widely depending on the type of oil, its source, pattern of pesticide usage, etc. Typical residue levels obtained by monitoring crude oil intake to a refining plant over an 18 month period in 1972 - 1973 are shown in Table 11 [60]. By far the most frequently detected residues are of organochlorine insecticides, with DDT and metabolites predominating, although lindane, dieldrin, heptachlor epoxide, and endrin are frequently detected. Mean residue levels approximate to 0.1 mg/kg or less in most crude oils, although values in excess of 1 mg/kg have been widely reported.

Crude oils are of poor taste and keeping quality, and the vast majority are further processed in a series of operations designed to give a product that has neither color, taste, or smell. In practice this situation is unattainable and refining is a compromise between elimination of all interfering substances and the formation of physiologically hazardous substances. The term "refining" will be used in this Chapter in its broadest sense to cover all processes used to improve the quality of crude oils rather than in its original sense (which is still used in the United States) to define processes designed to remove free fatty acids, phosphatides, mucilagenous material, and other gross impurities from oils. In practice, six or seven different refining processes can be employed, but only the neutralization, bleaching and deodorization steps are essential to all oils. The effect of these processes on the fate of pesticide residues is therefore of prime interest and has been studied in some depth.

a. Neutralization. The removal of free fatty acids from oils is achieved in a variety of ways. Treatment with strong alkali, alkali carbonates, or sodium silicate has no significant effect on pesticide residues although some

TABLE 11 Typical Pesticide Residue Levels in Some Crude Vegetable Oils, μg/kg oil

	Soybean		Arachis		Sunflower		Palm		Coconut		Rapeseed	
	Range	Mean	Range	Mean	Range	Mean	Range	Mean	Range	Mean	Range	Mean
Aldrin	0–5	3	2–395	130	0–50	5	0–5	2	0–5	2	0–5	5
Dieldrin	5–120	80	0–75	15	0–5	3	0–5	2	0–25	6	0–15	9
Eldrin	5–240	85	0–80	23	0–20	3	0–5	1	0–5	5	trace	
Heptachlor epoxide	10–45	20	0–20	5	0–5	2	trace		0–5	2	0–5	2
Total DDT	10–400	110	60–405	150	5–100	30	5–55	25	20–70	50	5–675	10
Total endosulfan	15–185	65	5–175	45	5–30	8	0–20	10	–		20–95	25
HCB (hexachloro-benzene)	0–5	1	0–5	2	trace		trace		–		–	
α-BHC	10–150	50	5–60	40	0–65	13	trace		–		–	
δ-BHC	0–15	3	5–15	3	trace		trace		–		–	
γ-BHC	5–170	25	2–165[a]	70	0–10	5	0–5	2	0–30	7	30–75	45
Methoxychlor	0–10	1	0–50	15	trace		–		–		100–240	145

[a]Excluding one sample containing 8.9 mg/kg γ-BHC.

dehydrohalogenation of organochlorine insecticides probably occurs. Alternative processes involving reaction with glycerine, extraction with ethanol, furfural, etc., addition of urea or the use of ion-exchange columns has no effect on pesticide residue levels. The removal of volatile fatty acids under vacuum or at temperatures in excess of 200°F with steam or other convenient carrier gas is a process that is gaining popularity, particularly for palm and coconut oils. Although no data are available, these procedures should result in some loss of pesticide residues either by degradation or volatilization.

b. Bleaching. The removal of carotenes, xanthophylls, chlorophylls, gossypols, and condensation products formed during extraction and storage is accomplished by adsorption onto activated earths, carbon, silica, or alumina at about 180°F (82°C) followed by filtration through a bed of similar material; chemical bleaching is rarely used in modern refineries. No removal of pesticidal compounds has been noted during this process as they are readily desorbed from most adsorbents and eluted by the oil.

c. Deodorization. Compounds such as saturated hydrocarbons, aldehydes, and ketones which impair the taste of oils at levels down to mg/kg are removed at elevated temperatures under reduced pressure. This vacuum steam distillation stage has been shown to result in significant removal of pesticides due to their general volatility under the conditions employed. It has been reported by Parsons [61] that at 190°C and 4 mm Hg for 4-1/2 hr the process is only partially successful in removing pesticides (20%) but that 70% removal is obtained during 5 hr at 250°C. Smith et al. [62] working with cottonseed and soybean oils at 240-250°C and 6 mm Hg for 1 hr obtained removal of pesticides, including the less volatile endrin, dieldrin, and pp'-DDT, to below the limit of the analytical procedure employed (0.03 mg/kg). Saha et al. [63] used simulated commercial refining techniques to follow the rate of radioactively labeled lindane and pp'-DDT added to rapeseed flakes at the 0.3-ppm level. No significant reduction in the levels of radioactivity occurred during the solvent extraction, solvent removal, neutralization, or bleaching stages. Altogether, 95% of the lindane and 98% of pp'DDT were removed during deodorization at 230-260°C and 6 mm Hg pressure for 4 hr. Although the deodorization stage accounts for the largest single removal of pesticides during refining, the extent of removal is dependent on the conditions of time, temperature, and pressure employed. Traditional batch deodorization is carried out at 6 mm Hg and 180-190°C and will result in incomplete elimination of pesticides. Continuous processes are increasingly operated at higher temperatures, 220-280°C, and lower pressures, 1-3 mm Hg, and result in more complete removal of pesticides. However, high temperature deodorization has, until recently, been suspected of forming obnoxious and physiologically hazardous compounds such as dimers and polymers in oils. It has now been established that this formation is more a function of time than temperature, and the

German Society of Fat Research is of the opinion that temperatures of up to 270°C can safely be used for periods of up to 20 min. Under these conditions the majority of any residues present should be eliminated and levels in the majority of oils reduced to acceptable amounts.

In addition to the essential steps outlined above, further processes are required with certain oils or to achieve certain specific effects. Desliming is an important step in the refining of soybean and, to a lesser extent, groundnut and rapeseed oils, and is designed to remove gums, resins, phosphatides, and similar materials. Various methods are employed including water or acid leaching at temperatures below 100°C, heat desliming at temperatures between 240 and 280°C, and less frequently by use of adsorbents, flocculating agents, and special "degumming agents." An important by-product of this process are the commercially valuable phosphatides used in lecithin production. Depending on the particular process used, levels of pesticides of up to 1 mg/kg have been detected in lecithins, but the overall effect of the desliming process in reducing residue levels in bulk oils is minimal.

Hardening of oils by hydrogenation of unsaturated fatty acid esters is achieved catalytically. This process has been shown to result in significant reduction in residue levels in certain instances. The mechanism of removal was thought to be by adsorption of pesticides onto the catalyst system and to be strongly influenced by the catalyst support employed. Significant reduction in residue levels has been obtained when carbon was used to support the catalyst but no effect was noted with a silica support. Efforts to recover pesticides from the support by desorption have been unsuccessful and removal is thought now to be by dechlorination of pesticides rather than adsorption. Interesterification processes by which the character of an oil, or more usually a mixture of oils, is changed by altering its glyceride structure have no effect on residue levels.

In summary, organochlorine insecticides have low but significant vapor pressures and are mainly removed during refining by volatilization and codistillation with steam during the deodorization process. The more volatile pesticides, such as lindane and hexachlorobenzene, are removed at the lower temperatures employed in batch deodorization but less volatile insecticides, such as endrin, dieldrin, and pp'-DDT, are only readily volatilized at the higher temperatures used in the continuous process. Degradation products such as DDE and TDE (DDD) are more volatile than the parent DDT, and possible thermal rearrangement products of endrin are at least as volatile as the parent species. Hydrogenation may also contribute significantly to pesticide removal, depending on the precise conditions employed. Small losses may occur by dehydrohalogenation or volatilization during other refining processes, but these are generally insignificant compared to those occurring in the deodorization and, to a lesser extent, hydrogenation processes. High-temperature vacuum-steam deodorization is therefore primarily responsible for reducing the level of pesticide residues in the majority of vegetable oils to below the levels included in the stringent current

TABLE 12 Typical Pesticide Residue Levels in Margarines Prepared from Vegetable Oils, μg/kg.

Pesticide	Range	Mean level
Aldrin	0-5	trace
Dieldrin	0-10	4
Endrin	0-10	trace
Heptachlor epoxide	0-5	trace
Total DDT	0-60	25
Endosulfans	0-30	4
Lindane	0-10	3

legislation. The data in Table 12 give typical residue levels in 157 samples representing the output from a margarine factory over an 18 month period from 1974 [64].

2. Confectionary Fats

The unique properties demanded for this product are met by only a small number of fats derived from vegetable seeds, notably cacao butter, Borneo tallow, and illipe butter. The commercial and economic importance of cacao butter in the production of chocolate is so great that much effort has gone into the development of synthetic substitutes. Various fractions of other vegetable oils rich in palmitic, stearic, and oleic acids have been employed directly and after chemical modification in this synthesis.

The bulk of fats used for confectionary purposes are not given the extensive refining used in the preparation of vegetable oils. Cacao beans are dried prior to transportation for processing. After cleaning, the beans are roasted at temperatures up to 120°C and the shells cracked and removed. The resulting cacao nibs are then milled to produce chocolate liquor which is further refined to the desired smoothness prior to molding. None of these processes has any significant effect on levels of pesticide residues that are invariably present in the initial seed, with the result that this type of product is generally more contaminated than typical refined vegetable oils. The range and mean levels of residues determined in 20 samples of cocoa butter taken from various reported sources are shown in Table 13.

3. Marine Oils

The most significant contamination of oils occurs with those of marine origin. Most species of marine animals used in oil production are high up

TABLE 13 Pesticide Residues in 20 Random Cocoa Butter Samples, μg/kg of Fat.

Pesticide	Range	Mean
α-BHC	trace - 960	78
γ-BHC	16 - 870	101
Total DDT	trace - 2410	156
Dieldrin	0 - 21	6
Heptachlor epoxide	0 - 5	0

in food chains and can concentrate both pesticides and PCBs to levels very much higher than those found in crude vegetable oils. Wide variations exist depending on the source, with highest levels present in fish caught off the coast of industrial areas of Northern Europe, the United States, and Japan. Typical residue levels reported by various authors are shown in Table 14.

These oils are refined using processes similar to those employed for vegetable oils, which results in a significant reduction in the level of pesticides present in the final product. Because of the higher initial levels of pesticides, the levels of residues present in the fully refined oils are invariably higher than those found in refined vegetable oils. Typical levels range from 0.01 - 0.05 mg/kg in fully refined oils to 0.001 - 0.01 mg/kg in vegetable oils, and consist principally of DDT and metabolites. The fate of PCBs during the refining of fish oils is less clear and very sparsely reported. Addison and Ackman [65], working with normal production runs of Canadian fish and seal oils containing 3-13 ppm of PCBs, corresponding to Aroclor 1254, obtained a reduction in PCB content to below 0.3 ppm (limit of detection of the analytical procedure used) on refining in each instance. Not only the lower chlorinated PCBs, but the higher chlorinated species, which have vapor pressures that are almost certainly lower than that of pp'-DDT, were removed during refining. Removal of lower chlorinated PCBs would occur along with DDT during deodorization which was carried out at 250°C and 7-10 mm Hg pressure for 2-1/2 hr. The mechanism by which the less volatile, higher chlorinated PCBs were eliminated was not investigated. The authors' laboratory analysis of herring oils withdrawn at different stages in refining showed an overall reduction of 60% in PCB content from an initial level of 1.7 ppm. The largest single loss occurred during deodorization with smaller losses at other stages in the process. The oil was deodorized in a batch process at 200°C and residual PCBs were composed mainly of the higher chlorinated species.

TABLE 14 Residue Levels in Crude Fish Oils, mg/kg Oil

Oil	Source	DDE	DDD	DDT	Dieldrin	PCB
Herring	Gulf of St. Lawrence	3.9 - 9.8	0.9 - 2.6	1.8 -5.1	0.0 -0.12	1.7-10.9
Seal	Nova Scotia	0.6 - 4.2	0.4 - 3.1	0.0 -1.5	nd[a]	0.0- 3.7
Whale	Antarctic	0.06-24.25	0.01-10.7	0.0 -0.14	nd	0.0- 1
	North Atlantic	2.5 -24.8	0.0 -16.1	nd	nd	0.0- 7.0
Pilchard	Portugal	0.01- 0.25	0.06- 0.25	0.07-0.34	0.01-0.06	-
Various fish	East Canadian Coast	0.7 - 3.8	0.0 - 1.0	0.2 -2.8	0.0 -0.03	3-12

[a]Not detected.

4. Animal Fats

As has been previously stated, the ingestion of pesticide residues by farm animals via animal feeding stuffs, forage crops, and water, or following their application in the farm situation to control a variety of lice, mites, ticks, etc., is generally accepted as inevitable. Whereas the systemic compounds are readily metabolized and excreted, the highly lipophilic species, almost exclusively organochlorine insecticides, are stored in the body fat of the animal for extended periods at levels related to the dietary intake. The raw materials used in the production of animal fats therefore contain generally low but finite levels of persistent organochlorine insecticides.

The rendering of fatty tissues is achieved by the application of heat, which denatures the proteins of the fat cell walls and renders them permeable to the fat and water contained in them. Wet rendering, in which substantial amounts of water are added to the chopped fatty tissue at minimum temperatures, results in the production of the blandest fats. The process is, however, time-consuming and most rendered fats are obtained by autoclaving in live steam at temperatures of up to 150°C. Some fats are also dry rendered in closed vessels under partial vacuum at temperatures up to 110°C. This latter process produces a distinctive cooked flavor which is preferred in some areas. No information is available on the effect of these rendering processes on residue levels but the relatively high pressures, low temperatures, and neutral pH would suggest little, if any, losses. However, it is increasingly common for off-flavors produced in the high-temperature rendering processes to be removed by deodorization. As seen with vegetable oils, this process will significantly reduce or eliminate residues depending on the precise conditions employed. Rendered products such as lard (originally derived from omental fat but now produced from any fresh hog fat), drippings (from sheep and oxen), suet and tallows or cooking fats prepared from mixtures of any of these fats with fish or hydrogenated vegetable oils, would therefore be expected to contain variable levels of residues, depending on the source of the fat and the exact processing conditions employed. This is well illustrated by the data for drippings and, to a lesser extent, lard, as presented in Table 15 [66]. Samples of both beef and mutton drippings were analyzed.

The very high dieldrin levels were entirely due to its use in sheep dips in the UK, a practice that was subsequently banned. High levels of BHC (average 1.1 mg/kg) in mutton drippings from Argentina were responsible for the relatively high value obtained for imported drippings in 1965.

A small but significant portion of animal fat output is used as raw material in the oleo-chemical industry in the production of fatty acids and glycerol. Following hydrolysis, the fatty acids are distilled under vacuum at temperatures up to 260°C and organochlorine residues are codistilled and collected in the first distillate fraction. Residual fat from this process, used in the production of compound animal feedstuffs, is therefore pesticide free.

TABLE 15 Residue Levels in Certain Animal Fats, mg/kg.

Product	Year	No. of samples	Source	BHC isomers		Dieldrin		Total DDT	
				Range	Mean	Range	Mean	Range	Mean
Lard	1964	6	UK	0.0-0.95	0.02	0.05		0.1-1.1	0.45
	1967	51	UK	<0.02-0.91	0.04	<0.02-0.12	0.007	<0.05-2.6	0.18
	1968	57	UK	<0.02-0.13	0.02	<0.02-0.05	0.004	<0.05-1.6	0.14
	1969	8	UK	0.0-0.01		nd[a]		0.0-0.003	
Drippings	1965	66	Imported	0.0-15.5	0.34	0.0-0.45	0.04	0.0-3.2	0.13
		166	UK	0.0-5.1	0.29	0.0-8.2	0.73	0.0-0.74	0.06
	1968	48	UK	<0.02-0.67	0.03	0.02-0.48	0.02	<0.05-0.45	0.02

Source: From Ref. 66.
[a]Not detected.

Fractional distillation of the initial distillate to produce pure fatty acid results in the concentration of organochlorine compounds in the residues which are discarded.

D. Poultry, Meat, and Fish

1. Poultry

Due to the efficient feed conversion ratio, there has been a tremendous upsurge in poultry production in recent years. A 4-lb broiler, for example, can be produced in 8 weeks from a chick and just 8-1/2 lb of feed. The processing operations required to produce the saleable article consist of the following:

1. Slaughter and bleeding.
2. Scalding—conditions range from 125°F (51°C) for 2 min. to 140°F (60°C) for 1 min.
3. Defeathering—rapidly by mechanical means.
4. Eviscerating—by hand in a cool room.
5. Chilling—rapidly to 35°F (1.6°C) to minimize bacteriological spoilage.
6. Packing and storage—at temperatures below 35°F (1.6°C).

The only stage which results in a reduction of pesticide residue levels is the evisceration step when removal of organs and associated fat reduces the total level of lipophilic residues. The processed product as sold contains the major part of any residues present in the tissue at slaughter in an unchanged form.

Cooking, however, whether carried out domestically or as a factory processing stage, has an effect on levels of lipophilic residues. Ritchey et al. [67] fed chickens a diet containing low levels of lindane and DDT and studied the effects of baking and frying on residue levels in the carcass. Losses of lindane were generally greater following frying than baking, whereas DDT losses were lower and less dependent on the cooking process, as shown in Table 16.

In subsequent studies involving higher tissue residue levels, Ritchey et al. [68] extended the range of insecticides and cooking procedures and obtained the data shown in Table 17. Pressure-cooking for 15 min at 10 lb pressure resulted in larger losses of both lindane and DDT than either oven-baking for 60 min at 350°F (176°C) or frying in oil for 40 min. Losses of other residues were generally independent of the cooking procedure. In addition, samples of ground tissue were also oven-heated in sealed containers for varying periods at 350°F (176°C) to determine losses that occurred by heat destruction as opposed to rendering of fat. Little or no destruction of endrin, aldrin, or dieldrin occurred. A significant proportion of the heptachlor (as heptachlor epoxide) was rapidly destroyed, whereas destructive losses of lindane and, to a lesser extent, total DDT increased slowly

TABLE 16 Effect of Cooking on DDT and Lindane Levels in Chicken.

Level in raw carcass, mg/kg		Cooking losses, %			
		Lindane		Total DDT	
Lindane	Total DDT	Baking	Frying	Baking	Frying
0	0.5	-	-	30	45
0	1.68	-	-	11	5
0.037	0	39	61	-	-
0.65	0	8	24	-	-
0.03	0.56	50	50	16	32
0.54	0.47	46	66	27	0
0.06	1.84	50	40	55	42
0.56	2.34	46	57	39	41

Source: Table compiled from data presented in Ref. 67.

with time. Destruction of pp'-DDT was shown to occur via TDE(DDD) with little change in the level of DDE originally present in the raw tissue.

In a further study, involving lower tissue residue levels than those used by Ritchey et al., Liska et al. [69] obtained up to 90% loss of lindane and DDT during simmering of carcasses at 190-200°F (88-93°C) for 3 hr. Losses of heptachlor and dieldrin were lower and only averaged 50%. However, following high-temperature pressure-cooking at 250°C (121°C) for 3 hr, almost complete removal of residues was obtained at higher initial levels. Heptachlor was again the most persistent residue, but 90% was removed under these conditions. A significant difference in the initial residue content and rate of loss of residues from dark and light meat was also noted by Liska et al. White meat contained lower initial levels and lost proportionately more during cooking than the darker meat.

Cooking is, therefore, the only means by which residues in chicken tissue are significantly reduced during processing. The extent of any loss depends on the initial residue level present and the cooking procedure, with high-temperature processing giving near complete removal of most residues. Heating results in some destruction of DDT and lindane, but has little effect on cyclodienes. The major loss of residues therefore occurs by rendering of fat, and subseuqunt use of this fat reintroduces the extracted residues.

Prophylactics pose a possible residue problem, especially in intensively reared animals such as broiler chickens. Provided the recommended

TABLE 17 Cooking Losses of Organochlorine Pesticide Residues in Chicken.

Pesticide	Level in raw carcass, mg/kg	Cooking losses, %					
		Baking	Frying	Pressure-cooking	Oven-Cooking,[a] min		
					30	60	90
Total DDT	11.9	5	28	38	16	20	25
Lindane	2.3	0	25	45	29	19	79
Endrin	9.6	26	20	31	0	0	0
Heptachlor	8.7	20	0	21	43	30	35
Dieldrin	15.2	22	42	37	18	7	13
Aldrin	14.2	24	30	22	20	0	10

Source: Table compiled from data given in Ref. 68.
[a]Cooked in sealed containers to prevent loss of fat and moisture.

interval between withdrawal of the drug and slaughter is maintained, tissue levels should be insignificant because of rapid metabolism and excretion [70]. If detectable levels of these drugs are present at slaughter, processing, including cooking, will have no effect on muscle tissue or liver levels. Significant quantities of the intact drugs can be detected in rendered fat and juices after cooking of contaminated birds but derive almost entirely from adipose tissue.

2. Meat

Meat packaging constitutes an important industry in developed countries. Total production of meat and meat products in the United States doubled between 1942 and 1970 to 18 million tons per year. The industry is centered on cattle, calves, pigs, sheep, and lambs. After slaughter, the animals are skinned and dressed, various organs are removed, and excess fat is trimmed, before chilling. After chilling the carcasses of cattle are distributed whole to the retail meat trade or cut and boned and used in the production of a variety of meat products. Cutting of hogs is dictated by the classification of the carcass. Loin hogs are usually cut for the retail meat trade, whereas other types are primarily cut for bacon, ham, and lard production. Lambs may be cut up, but calves are generally distributed to the retail trade as the whole carcass. None of the procedures involved in the preparation of carcasses for the retail trade has any effect on residue levels. Trimming of the carcass removes a significant amount of the total body burden of residues, and the fate of these on rendering is discussed separately. The most significant effect on residue levels in lean meat will therefore occur during cooking. Very little work has been reported, however, on losses specifically from meat. Carter [71] and Carter et al. [72] cooked beef cuts, which contained between 15 and 27 ppm of DDT, by broiling, frying, roasting, braising, and pressure-cooking. Reduction in DDT levels increased from zero during broiling (drippings included in analyses) to approximately 50% during pressure-cooking. These results are generally similar to those obtained with poultry, and indicate that residue losses are roughly proportional to the severity of the cooking procedures employed.

Whereas high-quality meat is invariably sold as fresh meat, significant quantities of lower-grade meat are processed into a wide variety of products before retailing. Cured products are prepared in several ways. Injection and pickle curing are widely used in the production of bacon, hams, tongues, and corned beef, etc. Dry curing is employed to prepare certain bacon cuts and types of hams and, following addition of the curing ingredients to comminuted meat, in the production of a wide range of sausages. Little loss of lipophilic residues would be expected to occur in these processes. Yadrick et al. [73] studied the effect of curing and heat drying on dieldrin residues in pickled bellies. Results were inconclusive; in one sample dieldrin content apparently rose by 12%, whereas in two other samples reductions of 10 and 62% were noted. More consistent results were obtained when bacon slices were subsequently cooked. Approximately half of the dieldrin was lost during either pan-frying or baking. However, 45-75% of the dieldrin removed during frying was recovered in the drip, whereas only 10-30% was recovered in the drip following baking.

Many meat products including bacon, hams, and sausages are smoked, originally to assist preservation but now almost exclusively to impart particular flavors to the product. Smoking is carried out at temperatures of 100-170°F (37°C) for periods of up to 24 hr. Higher temperatures are required for uncooked products. Products smoked for shorter periods at lower temperatures are cooked either before or after smoking, by immersion in hot water for sufficient time to reach an internal temperature of around 150°F (65°C). No leaching of constituents occurs during these processes and the duration and temperatures reached are too low for any significant destruction of residues to occur.

Canning of meats on the other hand is a far more drastic process. Exact conditions depend on the type of product and the size and shape of the container. Temperatures of up to 250°F (121°C) for times extending to 5 hr may be employed. Under these conditions no losses of residues can occur by leaching, but destruction of residues by dechlorination occurs to an extent, depending on the time and temperature employed.

3. Fish

No loss of residues occurs from fresh or frozen whole fish from the time they are caught until they are purchased from the retailer. Cutting to produce fillets, sticks, and steaks from fish such as cod, haddock, and whiting effectively removes all lipophilic residues which are retained in the viscera and residual carcass. However, fish such as herring, pilchard, mackerel, tuna, and salmon retain significant quantities of residues in the fat which is associated with the edible portions of these species. These fish do not store as well as low-fat fish and consequently are not retailed fresh or frozen on any scale, but are generally processed by canning, smoking or, or a lesser extent, salting.

In the canning of salmon, the fish are mechanically cleaned, dressed, cut to the required size, compacted, and automatically filled into cans containing salt. After vacuum sealing, the cans are sterilized in retorts at 240°F (115°C) for the required time, approximately 90 min for a 500-g can. A reduction in the total residue content of the salmon results during autoclaving by dechlorination processes. Tuna, on the other hand, are precooked prior to canning to reduce the oil content of the product and facilitate separation of meat from the bone. Up to 30% weight loss of water and oil occurs at this stage accompanied by a significant reduction in residue content by leaching. In addition to salt, vegetable or fish oil is added to the can before filling and sterilizing under conditions similar to those for salmon. Reduction of residue levels again occurs by dechlorination, and in addition redistribution of lipophilic residues occurs during autoclaving and on subsequent storage between the native tuna oil and the added oil. If this free oil is discarded when the can is opened, the residue content of the meat will be many times lower than that of the fresh fish. In addition, pilchards, brisling, sardines, and various other species, notably herring, are also canned in oil or oil-based sauces. There is an increasing tendency to replace added vegetable oil with fish oil in these products which could result in an increase rather than a reduction in residue levels in the canned fish product.

Salting, smoking, and kippering of fish are carried out on a limited scale. None of these processes involves stages that result in a significant reduction of residue levels.

A further major industry is centered around fish waste leading to the production of fish meal, fish body oils, and a range of minor products including gelatin, glue, etc. Waste and fish of high oil content are precooked in a continuous process in live steam at up to 240°F (115°C); low-fat fish are generally batch dried before or during cooking. Liberation of oil occurs and this is removed under pressure to give a residual cake containing about 3% oil and up to 50% water. This cake is dried either in direct flame-heating rotary drums or at lower temperatures in steam tube driers.

The final fish meal contains relatively little of any residues present in the original fish. The major part will be removed along with the oil and further losses by volatilization and co-distillation will occur during drying, particularly at the higher temperatures reached in the rotary drier. The low to very low level of residues normally present in fish meals is significant, with this product finding widespread use as animal feed, particularly for poultry, where residues are efficiently concentrated in the tissues and eggs. Recently, much effort has gone into the complete extraction of oils and fatty substances from fresh tissue to produce an odorless and tasteless product, fish flour or fish protein concentrate, which is expected to find increasing use as a protein supplement in human diets. Residues in this product are virtually nonexistent because of the almost complete removal of oil and fat during processing.

Fish liver oil prepared from cod, halibut, tuna, and a variety of shark is a prime source of vitamin A. Livers rich in the vitamin are either rendered with steam or digested with acid or alkali to release the oil. Complete saponification of the oil results in the production of a vitamin A concentrate that can be further concentrated by molecular distillation. Lipophilic residues will be extracted and concentrated along with the vitamins. Cyclodiene insecticides and polychlorinated biphenyls are unaffected by alkali treatment but BHC isomers are completely degraded, pp'- and op'-DDT residues are dehydrochlorinated to the corresponding DDE compound and TDE is dehydrochlorinated to DDDE (DDMU).

E. Cereals

Cereals such as wheat, barley, maize, and rice make a very important contribution to human food consumption and they are, at the same time, one of the major areas of pesticide usage. This usage starts with the application of seed dressings such as mercury, or γ-BHC, continues with the wide use of herbicides such as barban, 2-4 D, Suffix, and di-allate to remove competing species such as wild oats, and also includes a wide range of insecticides used during normal growth. In certain countries the fungicide hexachlorobenzene is commonly used as protection against botrytis. In addition to these uses, which are similar to those on fruits and vegetables, a further widespread use is in postharvest treatment of stored grains. Postharvest uses cover two main groups of compounds:

1. Contact insecticides, which prevent infestation occurring during storage, e.g., pyrethrins, malathion, γ-BHC, and dichlorvos.
2. Fumigants, which are applied to control infestation which is already present at harvest. These fumigants may be further subdivided into (a) gaseous compounds such as hydrogen cyanide, methyl bromide, ethylene oxide, and phosphine, and (b) liquid fumigants such as the halogenated hydrocarbons, carbon tetrachloride, 1,2-dichloroethane, and 1,2-dibromoethane. These fumigants may be applied directly to grain stored in bulk, to storage sacks, and for the fumigation of bins and processing machinery.

The distribution of residues of these chemicals follows the pattern discussed elsewhere in this chapter in that the major part of the residues will generally be found in the outer husk of the grain, although significant penetration does occur in certain cases. Any process which involves removal of the outer layers of the grain will therefore result in a major reduction in residue levels present. Fumigants generally enter grain by a sorption process which is usually reversible. Methyl bromide tends to break down partially in the grain giving methylation of the gluten or protein factor of the grain, and leaving a residue of inorganic bromide. A survey of bromide contents of 108 samples of imported cereals was reported by Hill and Thompson (see Table 18) [74].

Removal of fumigants is best achieved by aeration to allow desorption to occur and stringent controls on aeration periods have been laid down by various authorities.

1. Wheat

The production of flour from wheat (milling) consists of the following main stages: First, a dry cleaning removes dirt, cellulose coverings, light grain, and other cereals such as oats, barley, and rye. These products removed in preparing the wheat for milling are known as "screenings."

TABLE 18 Bromide Residues in Cereals and Cereal Products Sampled during the period July 1968 to June 1971.

Country of origin	No. of samples	Shipments represented	Range distribution of samples: Range (mg/kg)	Less than 50 mg/kg	50-100 mg/kg
Maize					
Brazil	23	4	1-57	22	1
Bulgaria	2	1	8-10	2	0
East Africa	8	1	60-91	0	8
Kenya	55[a]	3	45-95	10	45
Malawi	2	2	22-32	2	0
South Africa	11[a]	1	20-57	8	3
United States	7	1	1-4	7	0
Wheat					
Australia	1	1	9	1	0
France	5	1	6-100	2	3
United States	1	1	12	1	0
Wheat flour					
Australia	68[b]	6	< 1-79	56	12
Wheat pollards					
Guyana	1	1	58	0	1
Oats					
Australia	34[b]	4	6-62	27	7
Barley					
Australia	8	3	8-31	8	0
Rice					
Australia	15[b]	6	1-35	15	0
Burma	3	1	1-42	3	0
United States	2	1	5-9	2	0
Rice bran					
Burma	11	2	24-76	6	5
India	22	15	9-199	16	2[c]
Kenya	1	1	62	0	1
Tanzania	3	2	10-167	1	1[d]
Bird seeds[e]					
Argentina	1	1	10	1	0
Australia	16[b]	7	1-28	16	0

Source: From Ref. 74, reprinted by permission of Society of Chemical Industry.

[a]Most of these samples were taken after fumigation in port silo.
[b]Samples taken from shipments in freight containers.
[c]In addition, four samples contained more than 100 mg/kg.
[d]In addition, 1 sample contained more than 100 mg/kg.
[e]Millet, panicum, canary seed.

Second, the prepared wheat then goes through a series of rollers which break up the grains and allow for the gradual separation of the embryo and endosperm from the husk. End products vary widely according to the process used but, for the purposes of this chapter, they can be classified as flour, fine offal, and bran.

A paper from FMBIRA [75]gives data on the distribution of malathion in the various milling products obtained from a sample of wheat which, after dry cleaning, contained 13 mg/kg malathion. Reported findings are given in Table 19. In the same paper it was reported that when a flour containing 2.6 mg malathion per kg was baked into bread by various processes, 46-49% of original malathion was still present in the bread.

Thompson et al. [76] determined organochlorine residues at various stages in the processing of wheat into flour and the results are given in Table 20.

2. Barley

Barley is used in the production of malt for alcoholic beverages. In the malting process the barley is first dried to about 12% water, then stored for a period of 6-8 weeks. The barley is then steeped in cold water for periods which vary from 48 to 85 hr, the water is drained away, and the barley is spread out in thin layers where it remains for 6 to 9 days. During this period germination occurs under controlled humidity conditions. The newly made malt is then transferred to kilns where it is dried, first at a moderately low temperature, and later at a temperature which is sufficiently high to bring about a cessation of enzymic activity without actually destroying the enzymes. When this curing process is complete, the barley is removed and allowed to cool. The barley has now been converted into malt. Under these conditions of successive drying, washing, and drying it is to be expected that at least some of any pesticides present will be removed by codistillation.

TABLE 19 Distribution of Malathion in Milled Wheat Products.

Product	Yield (%)	Malathion content (mg/kg)	Distribution (%)
Flour	70.8	2.6	17
Fine offal	9.1	34.0	30
Bran	20.2	27.0	53

Source: Table prepared from Ref. 75.

TABLE 20 Insecticide Residues in a Consignment of USA Dark Northern Spring (DNS) Wheat at Various Stages between the Ship's Hold and the Completion of the Milling Process in 1968.

Month of sampling	Position of grain in ship	Nature of sample	Location from which sample was taken	Insecticide content (mg/kg)	
				BHC	DDT
January	Hold 1	Grain	Ship's hold	8.7	nd
February	Hold 3	Grain	Silo 1	nd	nd
	Hold 4	Grain	Silo 1	0.13, 0.30, 0.94, 1.5, 1.6	nd
	Hold 5	Grain	Silo 1	nd	nd
February	Hold 1	Grain	Silo 2	0.20	nd
	Hold 2	Grain	Silo 2	0.21, 0.55	nd
	Hold 3	Grain	Silo 2	nd, nd, 0.04, 0.11	nd
February	Holds 1 and 5	Grain	Floor storage granary	0.05, 0.80, 1.3, 1.6, 1.7, 2.6, 2.7, 2.9	nd
		Dust	Floor storage granary	2.5, 3.1, 27	nd
July	Holds 1 and 5	Dust	Floor storage granary	1.5, 28	tr, 0.20
July	Holds 1 and 5	Grain	Lorries loaded from granary	1.7, 3.1, 6.1	tr, 0.05, 0.19
March		Wheat[a]	Mill	0.06	0.05[c]
		Grist	Mill	0.02, 0.51	tr, 0.25[d]
		Offal	Mill	0.05	tr[c]
		Flour	Mill	0.02	nd
June		Screenings	Mill	0.06, 0.07	0.07, nd

TABLE 20 (continued)

September	Wheat[b]	Mill	0.73	0.06
	Mixed wheat containing 15% DNS	Mill	0.33	tr
	Fine screenings	Mill	1.6	0.08[e]
	Coarse screenings	Mill	0.48	tr
	Bran	Mill	1.1	tr
	Flour	Mill	0.04	nd

Source: From Ref. 76, reprinted by permission of Society for Chemical Industry.
[a]From silo.
[b]From granary.
[c]Sample contained a trace (tr) of DDE.
[d]One sample contained a trace and the other 0.03 ppm DDE.
[e]Sample contained 0.02 ppm DDE.

Apart from brewing and whisky distilling, malt is used for the production of vinegar, malt extract, and certain milk beverages and breakfast cereals.

Barley is also used in a flaked form in the brewing industry, and pot or pearl barleys, prepared by removal of the outer husk by abrasion, are used as cereal for the preparation of invalid and infant foods and as the processed grain for breakfast cereals.

The bran or dust resulting from the abrasion process is incorporated with other components into animal feedstuffs.

3. Maize

Maize is used for human food and animal feeding and maize starch is the principal raw material for the manufacture of glucose syrups. The by-products from the production of starch and glucose also provide animal feedstuffs, e.g., maize gluten feed.

Maize has a high oil content and, for the prevention of rancidity during storage, the germ which contains the oil is generally removed by decortication. Maize oil is used as a salad oil, as the oil basis of mayonnaise, and, more recently, as a basis for margarine.

Various surveys of levels of organochlorine insecticides have been reported [76] and typical results are given in the Table 21.

4. Rice

Rice, when threshed, has a thick fibrous husk and in this state is known as "paddy" or "rough" rice. When commercially processed, rice is ground to remove the skins of the grain by sifting and winnowing and then the product is polished. The by-products from this process are again used as animal feeds. In the case of these grains, as with wheat, commercial processing removes the outer layers of the product and most of any pesticide will be removed.

5. Rye

Rye is generally converted to flour, although this is likely to contain more of the husk than is found in wheat flour. Consequently residue levels tend to be higher than in other flours.

Current legislation and recommendations on levels of pesticides in grains and milled products reflect the losses likely to occur in various processing steps, as shown in Table 22.

TABLE 21 Organochlorine Residues in Maize on Arrival in Britain Sampled during the Period January 1967–June 1969.

Country of origin	No. of samples examined	Shipments represented	No. of samples containing BHC, mg/kg				Max BHC found, mg/kg	No. of samples containing DDT, mg/kg			Max DDT found, mg/kg
			<0.01	0.01–0.10	0.10–1.0	> 1.0		<0.05	0.05–1.0	>1.0	
Argentina	4	3	1	1	2	–	0.52[a]	2	–	–	0.02[a]
France	6	4	–	6	–	–	0.09	6	–	–	–
Kenya	11	5	–	3	8	–	0.56	8[b]	3	–	0.52
Malawi	3	2	–	–	3	–	0.8	–	3	–	0.5
Romania	1	1	1	–	–	–	tr	1[c]	–	–	–
South Africa	13	8	1	9	3	–	0.38	13	–	–	tr
United States	4	4	–	3	1	–	0.16	5	1	–	0.13
Total	42	27	3	22	17	–	–	35	7	–	–

Source: From Ref. 76, reprinted by permission of Society of Chemical Industry.

[a]The two highest concentrations (which were in samples from the same shipment) consisted predominantly of α-BHC. Examination by TLC failed to detect any hexachlorobenzene. One of the other samples contained a trace of DDE.

[b]One sample contained 0.15 ppm DDE and another traces of this compound.

[c]Sample contained a trace of DDE.

F. Compound Animal Feeds

The introduction of compounded animal feeds has been instrumental in dramatically increasing the feed efficiency ratio of farm livestock during the last 50 years. Although hay, forage, and other farm-produced feedstuffs are still widely used, the production of sophisticated, nutritionally balanced, animal feeds is now a major industry which has enabled increasing quantities of meat, milk, and eggs to be produced with maximum efficiency.

In addition to vitamins, minerals, and nonnutrient additives, a wide range of raw materials from an even larger number of different sources are employed in the production of animal feeds to meet the various demands of agriculture.

1. Grains

Cereal products, such as maize (corn), wheat, barley, oats, grain sorghums, and, to a lesser extent, rye and rice, form the largest single component of most animal feeds. They constitute upward of 50% of pig

TABLE 22 FAO/WHO Guidelines on Limits of Fumigant Residues, mg/kg.

Compound	Raw cereals at point of entry	Milled cereal products which will be subjected to baking or cooking	Bread and other cooked cereal products[a]
Carbon disulfide	10	2	0.5
Carbon tetrachloride	50	10	0.05
1,2-Dibromoethane	20	5	0.1
1,2-Dichloroethane	50	10	0.1
Phosphine	0.1	0.01	-
Methyl bromide[b]	50	10	0.5

[a]The limits quoted here generally reflect current limits of analytical detection.
[b]As inorganic bromide.

and poultry feeds, whereas in cattle feeds levels of whole grain are decreasing from 25% with a corresponding increase in the content of grain by-products such as middlings, husks, and screenings. Pesticide residue levels in these feedstuffs are similar to those previously discussed in Sec. VII.E, with significantly higher levels associated with the grain by-products. Depending on price and availability, various molasses, by-products of sugar refining, have found increasing use in recent years as a substitute carbohydrate.

2. Protein Supplements

A wide variety of raw materials is used in the production of high-protein supplements. Animal products, including meat and bone meals, fish and shrimp meals, feathermeal, and oilseed meals prepared from soybeans, peanut, coconut, cacao, cottonseed, sunflower, palm kernel, and linseed, etc., are extensively used for this purpose. Both types of product contain small amounts of residual fat or oil including low levels of any fat-soluble residues present in the original product. The majority of such persistent residues will, however, have been previously eliminated during the rendering of meat and fish, or the pressing or solvent extraction of oilseeds. Further losses of residues occur, principally by volatilization and codistillation, during subsequent drying. Temperatures used to dry oilseed meals rarely exceed 250°F (121°C) but animal products may be subjected to higher temperatures and pressures. Typical residue levels found in these foodstuffs are given in Table 23 [77].

3. Fats

In addition to residual fats and oils present in protein supplements, a variety of refining by-products, including rendered animal fats, tallows, and acid oils are utilized in animal feeds as an additional source of energy. Such products are invariably contaminated with residues, particularly those from organochlorine insecticides, and receive no additional processing prior to incorporation into compounded feeds. Typical residue levels in such products are included in Table 23.

4. Vegetable Waste

Increasing use as feedstuffs is currently being made of various waste vegetable products from farms, packing sheds, and processing plants. These include vines, leaves, tops, roots, skins, and peelings of such crops as peas, tomatoes, beet, spinach, broccoli, etc., as well as citrus peels, fruit hulls, apple pulp, coffee grounds, seed husks, cereal straw, etc. Many of these products are known to contain the majority of any pesticide residues present in the original material before processing for human consumption. The extent to which such products are used is dictated

TABLE 23 Organochlorine Residue Levels in Constituent Animal Feedstuffs, μg/kg.

Feedstuff	HCB			BHC isomers			Heptachlor[b]			Dieldrin[c]			Total DDT		
	%[a]	max	mean	%[a]	max	mean	%[a]	max	mean	%[a]	max	mean	%[a]	max	mean
Wheat feed		nd		100	58	13		nd		12	6	<1		nd	
Wheat brans	100	22	6	90	41	3	60	39	7	40	3	1	60	18	5
Rice brans	36	2	<1	94	71	30	25	6	1		nd		78	519	154
Cereal screenings		nd		100	22	16		nd		15	23	3	15	27	5
Molasses		nd			nd			nd			nd			nd	
Seed husks	88	2	<1	88	59	26	175	5	2	37	3	1	75	63	27
Various oilseed meals	31	6	1	58	84	7	40	9	<1	21	19	2	71	519	26
Meat and bone meals		nd		94	20	10	52	5	3		nd		55	18	10
Fish meals	60	4	2	30	10	2		nd		60	7	5	75	83	34
Feather meals		nd		100	64	22	10	3	<1		nd			nd	
Various rendered fats	99	1300	138	64	354	62	62	190	25	53	142	31	96	6275	384
Tallows	33	36	4	65	160	27	89	59	25	17	39	3	100	269	239
Palm acid oil	55	100	23	82	640	188	9	15	1	100	99	31	100	1200	197
Fish acid oil	100	92	86	100	111	101	100	61	57	100	24	23	100	187	176

[a]Percentage of total samples containing detectable levels of residue.
[b]Including heptachlor epoxide.
[c]Including aldrin.

by quantity, availability, and price, and most are utilized by small local compounders. Little published information is available of the fate of pesticide residues present in such products during their preparation as feedstuffs.

Gunther [78] investigated the fate of various pesticides in the production of citrus pulp cattle feeds. The basic process involved treatment of the ground peel (remaining after separation of the juice and oil) with lime, pressing to produce a pulp of about 40% moisture content, and oven-drying overnight at 60-65°C to a final moisture content of approximately 9%. His results showed that significant (90%) loss of all residues investigated occurred through this process, with the exception of dioxathion. Major losses of azinphos-methyl (Guthion), carbaryl (partially hydrolyzed to α-naphthol), dicofol, malathion, Morestan, and tetradifon (Tedion) occurred during the liming and pressing stages, whereas significant losses of Bidrin, dimethoate, and Omite occurred during drying.

Archer and Crosby [79] examined the suitability of a number of laboratory-scale techniques for the removal of organochlorine insecticides from a range of raw materials including alfalfa hay, field hay, almond hulls, and seed crop screenings. Washing with cold or hot water was ineffective in eliminating organochlorine residues, although as expected, complete elimination of malathion was achieved. When benzene was used instead of water, almost complete extraction of organochlorine insecticides was obtained from the waxy cuticular layers. Vapor washing with water and other solvents and various chemical wash treatments were equally effective in removing both types of insecticides. Although these results demonstrate that residues can be eliminated from plant feedstuffs by exploiting certain of their properties, these procedures are not economically feasible as yet in commercial production. Little adverse effect on the nutritional value of the feedstuffs was demonstrated.

The effects of drying on the pesticide residue contents of feedstuffs was also studied by Archer and Crosby [79]. Results showed that when hay or almond hulls were dried at 100°C for 12 hr, approximately one-third of the organochlorine residues present were lost by volatilization. Short-time high-temperature commercial dehydration was found to be no more effective than oven-drying. When feedstuffs were saturated with water prior to oven-drying, elimination of residues was doubled (60-87%) due to enhanced removal by codistillation. Both wet and dry heating procedures were equally effective in eliminating malathion residues.

The production of animal feeds from the variety of available feedstuffs does not involve any further steps that will significantly reduce residue levels. Steaming and drying stages in the production of pelleted feeds will have a marginal effect in reducing levels by volatilization and codistillation of the more volatile insecticides, but in general levels present in the products used in formulating the feed will persist through compounding.

TABLE 24 Residue Levels Determined in Various US Poultry Feedstuffs, μg/kg.

Pesticide	Soybean meal			Corn meal			Alfalfa meal			Fish meal			Fats		
	%[a]	Max	Mean	%[a]	Max	Mean	%[a]	Max	Mean	%[a]	Max	Mean	%[a]	Max	Mean
Heptachlor	16	24	1.2	17	29	1.4	30	130	5.5	24	64	2.5	16	734	11.3
Heptachlor epoxide	8	7	0.2	12	36	0.6	27	138	4.2	18	116	2.0	25	62	2.4
Aldrin	35	278	6.0	32	1909	15.6	34	322	8.0	43	78	4.4	15	8395	39.3
Dieldrin	26	28	1.3	19	203	2.5	46	183	7.2	43	2450	23.0	71	389	19.2
pp'-DDT	43	47	3.4	48	131	3.9	78	532	35.2	78	266	30.8	46	448	9.7
DDE	6	68	0.5	4	4	0.2	20	114	14.8	13	1627	134	22	455	83.4
TDE(DDD)	12	41	0.6	7	603	4.2	32	2155	16.5	68	156	20.1	86	701	54.0
Lindane	15	13	0.4	15	33	0.7	12	109	1.8	40	347	5.7	36	172	4.2
Methoxychlor	6	53	1.0	6	151	5.1	16	1947	57.5	3	232	2.6	1	151	5.0

Source: Compiled from data presented in Ref. 80.

[a]Percentage of total samples containing a detectable residue.

Waldron and Naber [80] reported on the incidence of pesticide residues in 1162 samples of feedstuffs used to prepare poultry feeds in the United States. Detectable levels of at least one residue were found in 67% of soybean and corn meals, 94% of alfalfa meals, and 98% of fish meals and fat samples. Data relating to the incidence and levels of residues determined are given in Table 24. These results once again illustrate the higher level of residues associated with fatty materials. Since the fat content of marketed feeds is generally below 5%, significant dilution of these relatively high residue levels occurs during compounding. Typical residue levels in 175 samples of compound animal feeds produced over the period 1972-1975 are given in Table 25 [77]. These levels are very similar to those that can be calculated from the data of Waldron and Naber [80], and demonstrate that low but detectable levels of residues are normally present in compound feedstuffs. Such residues can therefore be recycled via a wide range of animal products into the human food chain. An increasing awareness of this situation over the past decade has resulted in the introduction of strict legislation to control residue levels in animal feeds in order to restrict the accumulation of persistent residues in animal products. In many instances these permitted levels are more stringent than the requirements for some human foods. The current West German limits [81] for organochlorines residues in animal feeds are given in Table 26.

TABLE 25 Typical Residue Levels Determined in Compound Animal Feeds, μg/kg.

Pesticide	%[a]	Max	Mean
HCB	97	35	3
α-BHC	91	57	4
β-BHC	2	13	<1
Lindane	92	68	6
Heptachlor	67	37	2
Heptachlor epoxide	25	11	1
Aldrin	15	6	<1
Dieldrin	43	13	2
pp'-DDT	78	140	13
op'-DDT	63	40	4
pp'-DDE	76	30	3
pp'-TDE(DDD)	53	26	4
op'-TDE(DDD)	25	30	4
Endrin	15	17	<1
PCBs	26	400	15

[a]Percentage of total sample containing a detectable residue.

TABLE 26 West German Limits for Residues in Animal Feeds

Pesticide	Maximum permitted content (mg/kg) in		
	Poultry feed	Pig feed	Calf feed
Aldrin and dieldrin	0.02	0.03	0.05
Endrin	0.02	0.02	0.02
Heptachlor and epoxide	0.03	0.03	0.05
HCB	0.025	0.02	0.06
Lindane	0.03	0.02	0.05
Chlordane	0.05	0.05	0.05
Total DDT	0.2	0.2	0.3

VIII. CONCLUSIONS

Many of the standard commercial techniques applied in the processing of foodstuffs can lead to significant reductions in the level of pesticide residues present in the starting material.

Up to 10 years ago such losses were considered to be incidental to the main function of the processing stages, but increasing awareness of the importance of controlling pesticide intake in human diet has led to a new approach. In many products the fate of pesticides during processing has received careful study, resulting in optimized conditions giving maximum removal of pesticides. A knowledge of the chemical and physical properties of pesticides, linked with an understanding of the physical structure of the foodstuff to be processed, allows an acceptable estimate of optimum conditions for the removal of pesticides. In the processing of oils and fats, commercially acceptable changes in deodorization conditions give almost complete removal of organochlorine insecticide residues. For other products, such as milk and compound animal feedstuffs, conditions necessary for the elimination of residues have been determined but are currently not economically viable.

The routes for major reductions in many products are those concerned with the physical removal of residues by peeling or trimming of such foods as vegetables, fruits, meat, fish, and poultry. This has a significant effect in reducing the level of residues directly entering the human food chain. However, increasing quantities of these waste products are being utilized as animal feeds, which results in the reintroduction of residues into the human food chain via a wide range of animal products. The findings of Duggan and Corneliussen [5] and Manske and Corneliussen [82] clearly illustrate this point: that is, 50% of dietary residues are derived from dairy products, meat, fish, and poultry, which contribute less than 30% of total diet.

Although many residues, such as organophosphates, carbamates, and certain organochlorine compounds, may be degraded to nontoxic species during processing, many others are removed as intact species. Washing removes a wide range of residues, and volatilization or codistillation of organochlorine insecticides is a major pathway for removal during many processes.

The fate of these compounds appears to have received little attention but it could well be that satisfactory elimination of residues, especially organochlorine compounds, from many foodstuffs results in little more than the redistribution of these persistent residues within the ecosystem and eventually the human food chain.

GLOSSARY*

acinatrazole	D	2-acetylamino-5-nitrothiazole
aldicarb	I	2-methyl-2-(methylthio)propionaldehyde 0-(methylcarbamoyl)oxime
aldrin	I	1,2,3,4,10,10-hexachloro-1,4,4a,5,8,8a-hexahydro-1,4,endo,exo-5,8-dimethanonaphthalene
amprolium	D	1-[(4-amino-2-n-propyl-5-pyrimidinyl)methyl]-2-picolinium chloride
Aroclor 1254		chlorinated biphenyl containing 54% chlorine (a PCB)
arsanilic acid	D	p-aminobenzenearsonic acid
azaperon	D	4-fluoro-4-[4-(2-pyridyl)-1-piperazinyl]butyrophenone
azobenzene	I	diphenyl diimide
azodrin	I	cis-3-(dimethoxyphosphinyloxy)-N-methylcrotonamide
bacitracin	A	polypeptide antibiotic derived from certain strains of _B. licheniformis_ and _B. subtilis_ var. Tracy
barban	H	4-chloro-2-butynyl-N-(3-chlorophenyl)carbamate

*Compounds are categorized into the following general types: antibiotics, A; drugs, D; fungicides, F; plant growth regulators, G; herbicides, H; insecticides, I; other pesticidal compounds, P.

Benlate	F	methyl-1-(butylcarbamoyl)-2-benzimidazole carbamate
BHC isomers	I	isomers of 1,2,3,4,5,6-hexachlorocyclohexane
binapacryl	I	2-sec-butyl-4,6-dinitrophenyl 3-methyl-2-butenoate
bipyridylium compounds	H	see diquat and paraquat
Bidrin	I	cis-3-(dimethoxyphosphinyloxy)-N,N-dimethyl-crotonamide
bromophos	I	0-(4-bromo-2,5-dichlorophenyl)-0,0-dimethyl phosphorothioate
camphechlor	I	see toxaphene
captan	F	N-trichloromethylmercapto-4-cyclohexene-1,2-dicarboximide
carbaryl	I	1-naphthyl N-methylcarbamate
chloramphenicol	A	D(-) threo-2-dichloracetamido-1-p-nitrophenyl-propane-1-3-diol
chloranil	F	2,3,5,6-tetrachloro-1,4-benzoquinone
chlordane	I	1,2,4,5,6,7,8,8-octachloro-3a,4,7,7a-tetra-hydro-4,7-methanoindane
chlorfenvinphos	I	2-chloro-1-(2,4-dichlorophenyl)vinyl diethyl phosphate
chloromequate	G	2-chloroethyltrimethyl ammonium ion (usually as chloride)
CIPC	H	isopropyl N-(3-chlorophenyl)-carbamate
coat tar oils	I	heavy creosote and anthracene oil ranges
copper salts	F	carbonates, basic chlorides, sulfates, naph-thenates
corticosteroids	D	group of steroid hormones originally derived from the cortex of the adrenal gland
2,4-D	H	2,4-dichlorophenoxy acetic acid
dalapon	H	α,α-dichloropropionic acid
DDE		1,1-dichloro-2,2-bis(p-chlorophenyl)ethylene, metabolite of DDT
pp'-DDT	I	1,1,1-trichloro-2,2-bis(p-chlorophenyl) ethane
decoquinate	D	ethyl 6-(n-decycloxy)-7-ethoxy-4-hydroxy-3-quinoline carboxylate
demitradizole	D	1,2-dimethyl-5-nitroimidazole
demeton-S-methyl	I	S-2-(ethylthio)ethyldimethylphosphorothiolate
derris	I	extract of Derris roots, main constituent rotenone
di-allate	H	5-(2,3-dichloroallyl)-N,N-diisopropylthio-carbamate

diazinon	I	0,0-diethyl 0-2-isopropyl-4-methyl-6-pyrimidyl phosphorothioate
dichlone	F	2,3-dichloro-1,4-naphthoquinone
dichloropropane	I	1,2-dichloropropane
dichloropropene	I	1,3-dichloropropene
dichlorvos	I	2,2-dichlorovinyl dimethyl phosphate
dieldrin	I	1,2,3,4,10,10-hexachloro-6,7-epoxy-1,4,4a,5,6,7,8,8a-octahydro-1,4-endo,exo-5,8-dimethanonaphthalene
diethylstilbestrol	D	3,4-bis(p-hydroxyphenyl)-3-hexene
dimethoate	I	0,0-dimethyl S-(N-methylcarbamoylmethyl) phosphorodithioate
dinocap	F	mixture of mainly 2,4-dinitro(6-methylheptyl) and 2,6-dinitro(4-methylhelptyl)phenyl crotonates
dinoseb	H	2-(1-methylpropyl)-4,6-dinitrophenol
dioxathion	I	2,3-p-dioxanedithiol S,S-bis(0,0-diethyl phosphorodithioate)
diquat	H	1,1'-ethylene-2,2'-dipyridylium cation (generally as dibromide)
dicofol	I	2,2,2-trichloro-1,1-bis(-4-chlorophenyl)ethanol
endosulfans	I	6,7,8,9,10,10-hexachloro-1,5,5a,6,9,9a-hexahydro-6,9-methano-2,4,3-benzodioxathiepin 3-oxide; A and B isomers
endrin	I	1,2,3,4,10,10-hexachloro-6,7-epoxy-1,4,4a,5,6,7,8,8a-octahydro-1,4-endo,endo-5,8-dimethanonaphthalene
ethopabate	D	methyl 4-acetamidole-2-ethoxybenzoate
ethylene dibromide	I	1,2-dibromoethane
ethylene oxide	I	1,2-epoxyethane
fenchlorphos	I	0,0-dimethyl 0-(2,4,5-trichlorophenyl) phosphorothioate
folpet	F	N-(trichloromethylthio)phthalimide
Gardona	I	2-chloro-1-(2,4,5-trichlorophenyl)vinyl dimethyl phosphate
gentamicin	A	mixture of aminoglycosides produced by the growth of species of Micromonospora, particularly M. purpurea, normally as the sulfate
gibberellins	G	growth promoting substances derived from certain strains of the fungus Fusarium
griseofulvin	F	7-chloro-4,6-dimethoxycoumaran-3-one-2-spiro-1'-(2'-methoxy-6'-methylcyclohex-2'-en-4'-one)

Guthion	I	0,0-dimethyl S-[(4-oxo-1,2,3-benzotriazin-3(4H)-yl)methyl] phosphorodithioate
heptachlor	I	1,4,5,6,7,8,8-heptachloro-3a,4,7,7a-tetrahydro-4,7-methanoindene
heptachlor epoxide		metabolite of heptachlor
hexachlorobenzene	F	hexachlorobenzene
hexoestrol	D	3,4-bis(p-hydroxyphenyl)-n-hexane
hydrogen cyanide	I	hydrocyanic acid
indolebutyric acid	G	indole-3-butyric acid (β-indolyl butyric acid)
IPC	H	0-isopropyl N-phenyl carbamate
lindane	I	γ-BHC
linuron	F	3'-(3,4-dichlorophenyl)-1-methoxy-1-methylurea
malathion	I	0,0-dimethyl S-diethyl mercaptosuccinate phosphorodithioate
maleic hydrazide	G	6-hydroxy-3(2H) pyridazinone
maneb	F	manganese ethylenebis(dithiocarbamate)
MCPA	H	(4-chloro-2-methylphenoxy) acetic acid
MENA	G	methyl ester of 1-naphthyl acetic acid
mercury salts	P	mercuric and mercurous chloride
methoxychlor	I	1,1,1-trichloro-2,2-bis(p-methoxyphenyl)ethane
methyl bromide	I	bromomethane
methyl parathion	I	0,0-dimethyl 0-p-nitrophenyl phosphorothioate
Morestan	F	6 methyl-2,3-quinoxalinedithiol cyclic S,S-dithiocarbonate
Naphthalene acetic acid	H	1-naphthalene acetic acid
nicotine	I	S-3-(1-methyl-2-pyrrolidinyl)pyridine
nitroxynil	D	2-nitro-4-cyano-6-iodo phenol
Omite	I	2-(p-tert-butylphenoxy)cyclohexyl 2-propynyl sulfite
orthophenyl phenol	F	2-phenyl phenol
paraquat	H	1,1-dimethyl-4,4-bipyridylium ion (as the dibromide or dimethylsulfate)
parathion	I	0,0-diethyl 0-p-nitrophenyl phosphorothioate
PCBs		mixed isomeric chlorinated biphenyls, commercially marketed for various industrial purposes with a range of chlorine contents
penicillins	A	N-acyl derivatives of 6-amino penicillanic acid
PCP	F	Pentachlorophenol

prednisone	D	21-acetoxy-17α-hydroxypregna-1,4-diene-3-11-20 triene
Prometryne	H	4,6-bis(isopropylamino)-2-methylthio-1,3,5-triazine
pyrethrins	I	insecticidal compounds derived from the flowers of Pyrethrum cineraefolium
rotenone	I	major insecticidal component of derris root
simazine	H	2-chloro-4,6-bisethylamino-1,3,5 triazine
suffix	H	ethyl ($\pm$)-2-(N-benzoyl-N-3,4-dichloro analino)propionate
sulfotep	I	0,0,0,0-tetraethyl dithiopyrophosphate
sulphaquinoxaline	D	2-p-aminobenzenesulphonamidoquinoxaline
sulphanilamide	D	p-aminobenzenesulphonamide
2,4,5 T	H	2,4,5-trichlorophenoxy acetic acid
2,3,6 TBA	H	2,3,6-trichlorobenzoic acid
TCA	H	trichloroacetic acid
TDE(DDD)	I	1,1 dichloro-2,2-bis (p chlorophenyl) ethane
Tedion®	I	S-p-chlorophenyl,2,4,5-trichlorophenylsulphone
tetracyclines	A	derivatives of 4-dimethylamino-1,4,4a,5,5a,6,11,12a-octahydro-3,6,10,12,12a-penta-hydroxy-6-methyl-1,11-dioxonaphthacene-2-carboxyamide
thiodan	I	see endosulfan
thiophanate-methyl	F	1,2-bis(3 methoxycarbonyl-2-thioureido) benzene
thiram	F	bis(dimethylthiocarbamyl)disulphide
thiouracil	D	2-mercapto-4-hydroxy pyrimidine
toxaphene	I	chlorinated camphene containing 67-69% chlorine
Trithion®	I	S-[[(p-chlorophenyl)thio] methyl] 0,0-diethyl phosphorodithioate
virginiamycin	A	mixture of antimicrobial substances produced by growth of a Streptomyces, related to S. virginiae
warfarin	P	3-(α-acetonylbenzyl)-4-hydroxycoumarin
zeranol	D	3,4,5,6,7,8,9,10,11,12-decahydro-7,14,16,tri-hydroxy-3-methyl-1H-2 benzoxyacyclo-tetradecin-1-one
zineb	F	zinc ethylene-1, 2-bis(dithiocarbamate)
ziram	F	zinc dimethyldithiocarbamate

REFERENCES

1. N. N. Melnikov, Residue Rev. 36 (1971).
2. F. A. Gunther, Pure Appl. Chem. 21:355 (1970).
3. T. Jukes, JAMA 232:292 (1974).
4. R. L. Metcalf, J. Agr. Food Chem. 21:511 (1973).
5. R. E. Duggan and P. Corneliussen, Pest. Monit. J. 5:331 (1972).
6. Verordnung zur Änderung der Höchstmengen-VO-Pflanzenschutz. Vom 14 December 1972 (BGBL.S.2459).
7. Verordnung über Höchstmengen von DDT und anderen Pestiziden in oder auf Lebensmitte In tierischer Herkunft. Vom 15 November 1973. (BGBL.1. S 1710).
8. D. M. Coulson, Gas Chromatogr. 3:134 (1965).
9. R. C. Hall, J. Chromatogr. Sci. 12 (1974).
10. A. O. Donoho, W. S. Johnson, R. F. Sieck, and W. L. Sullivan, J. Assoc. Offic. Anal. Chemists 56:785 (1973).
11. H. Maier-Bode, Naturwissensch. 55:470 (1968).
12. P. J. Polen, Pesticide Terminal Residues, Butterworth Scientific Publications, London, 1971, p. 138.
13. E. P. Lichtenstein, F. R. Myrdal, and K. R. Schulz, J. Agr. Food Chem. 13:126 (1965).
14. W. Ebeling, Residue Rev. 3 (1963).
15. D. C. Abbott, S. Crisp, K. R. Tarrant, and J. O'G. Tatton, Pestic. Sci. 1:10 (1970).
16. K. Kaemmerer and S. Buntenkotter, Residue Rev. 46 (1973).
17. N. Ganon, J. Agr. Food Chem. 7:829 (1959).
18. S. Williams, P. A. Mills, and R. E. McDowell, J. Assoc. Offic. Agr. Chemists 47:1124 (1964).
19. J. G. Cummings, M. Eidelman, V. Turner, D. Reed, K. T. Zee, and R. E. Cook, J. Assoc. Offic. Anal. Chemists 50:418 (1967).
20. A. G. Johnels and T. Westermark, in: Chemical Fallout (M. W. Miller and G. C. Berg, eds.), Charles C. Thomas, Springfield, Ill. 1969, p. 234.
21. V. Zitko, Bull. Environ. Contam. Toxicol. 6:464 (1971).
22. S. Jenson, A. G. Johnels, M. Olsson, and G. Otterlind, Nature 224:247 (1969).
23. R. L. Carr, C. E. Finsterwalder, and M. J. Schibi, Pestic. Monit. J. 6:23 (1972).
24. D. J. Sissons and A. Stanton, unpublished work.
25. P. Koivistoinen and A. Karinpää, J. Agr. Food Chemists 13:459 (1965).
26. R. E. Baldwin, K. Gonnerman Sides, and D. D. Hemphill, Food Technol. 22:126 (1968).
27. P. Koivistoinen, A. Karinpää, M. Könönen, and P. Roine, J. Agr. Food Chemists 13:468 (1965).

28. E. R. Elkins, R. P. Farrow, and E. S. Kim, J. Agr. Food Chemists 20:268 (1972).
29. D. D. Hemphill, R. E. Baldwin, A. Deguzman, and H. K. Deloach, J. Agr. Food Chemists 15:290 (1967).
30. E. R. Elkins, F. C. Lamb, R. P. Farrow, R. W. Cook, M. Kawai, and J. R. Kimball, J. Agr. Food Chemists 16:962 (1968).
31. J. E. Fahey, G. E. Gould, and P. E. Nelson, J. Agr. Food Chemists 17:1204 (1969).
32. R. P. Farrow, F. C. Lamb, E. R. Elkins, R. W. Cook, M. Kawai, and A. Cortes, J. Agr. Food Chemists 17:75 (1969).
33. L. Kilgore and F. Windham, J. Agr. Food Chemists 18:162 (1970).
34. J. E. Fahey, P. E. Nelson, and D. L. Balley, J. Agr. Food Chemists 18:866 (1970).
35. R. C. Blinn, R. W. Dorner, J. H. Barkley, L. R. Jeppson, F. A. Gunther, and C. C. Cassil, J. Econ. Entomol. 52:723 (1959).
36. C. A. Anderson, D. MacDougal, J. W. Kesterson, R. Hendrickson, and R. F. Brooks, J. Agr. Food Chemists 11:422 (1963).
37. A. Benvenue, J. N. Ogata, F. H. Haramoto, and J. E. Brekke, J. Agr. Food Chemists 16:863 (1968).
38. R. E. Menzer, P. P. Burbutis, R. N. Hofmaster, and L. P. Ditman, J. Econ. Entomol. 53:662 (1960).
39. F. C. Lamb, R. P. Farrow, E. R. Elkins, R. W. Cook, and J. R. Kimball, J. Agr. Food Chemists 16:272 (1968).
40. J. W. Ralls, D. R. Gilmore, A. Cortes, G. M. Schutt, and W. A. Mercer, Food Technol. 21:92 (1967).
41. A. F. Carlin, E. T. Hibbs, and P. A. Dohm, Food Technol. 20:80 (1966).
42. F. C. Lamb, R. P. Farrow, E. R. Elkins, J. R. Kimball, and R. W. Cook, J. Agr. Food Chemists 16:967 (1968).
43. J. E. Fahey, P. E. Nelson, and G. E. Gould, J. Agr. Food Chemists 19:81 (1971).
44. A. J. B. Powell, T. Stevens, and K. A. McCully, J. Agr. Food Chemists 18:224 (1970).
45. R. P. Farrow, F. C. Lamb, R. W. Cook, J. R. Kimball, and E. R. Elkins, J. Agr. Food Chemists 16:65 (1968).
46. J. G. Fair, J. L. Collins, M. R. Johnston, and D. L. Coffey, J. Food Sci. 38:189 (1973).
47. C. D. Usher and G. E. Denton, unpublished work.
48. D. F. Lee, J. Sci. Food Agr. 19:701 (1968).
49. R. P. Farrow, E. R. Elkins, and R. W. Cook, J. Agr. Food Chemists 14:430 (1966).
50. F. Acree, M. Beroza, and M. C. Bowman, J. Agr. Food Chemists 11:278 (1963).
51. W. H. Johnson, J. J. Domanski, T. J. Sheets, and C. S. Chang, J. Agr. Food Chemists 23:118 (1975).

52. M. Eerola, P. Varo, and P. Koivistoinen, Acta Agr. Scand. 24:286 (1974).
53. A. Sinios and W. Wodsak, Deut. Med. Wochschr. 90:1856 (1965).
54. E. H. Marth and B. E. Ellickson, J. Milk Food Technol. 22:179 (1959).
55. L. Richou-Bac, Lait, 523-524, 117 (1973).
56. G. M. Telling and G. E. Denton, unpublished work.
57. R. A. Ledford and J. H. Chen, J. Food Sci. 34:386 (1969).
58. C. F. Li, R. L. Bradley, Jr., and L. H. Schultz, J. Assoc. Offic. Anal. Chemists 53:127 (1970).
59. B. E. Langlois, B. J. Liska, and D. L. Hill, J. Milk Food Technol. 27:264 (1964).
60. H. R. H. Wendt, private communication.
61. A. M. Parsons, Progr. Chem. Fats Lipids 11:245 (1970).
62. K. J. Smith, P. Polen, D. M. de Vries, and F. B. Coon, J. Am. Oil Chemists' Soc. 45:866 (1968).
63. J. G. Saha, M. A. Nielsen, and A. K. Sumner, J. Agr. Food Chemists 18:43 (1970).
64. H. R. H. Wendt, private communication.
65. R. F. Addison and R. G. Ackman, J. Am. Oil Chemists Soc. 51:192 (1974).
66. Reports of the Laboratory of the Government Chemist, H.M.S.O., London 1964-1969.
67. S. J. Ritchey, R. W. Young, and E. O. Essary, J. Food Sci. 32:238 (1967).
68. S. J. Ritchey, R. W. Young, and E. O. Essary, J. Agr. Food Chem. 20:293 (1972).
69. B. J. Liska, A. R. Stemp, and W. J. Stadelman, Food Technol. 21:435 (1967).
70. R. Ferrando, M. R. Laurent, B. L. Terlain, and M. C. Caude, J. Agr. Food Chem. 19:52 (1971).
71. R. H. Carter, Ind. Eng. Chem. 40:716 (1948).
72. R. H. Carter, P. E. Hubanks, H. D. Mann, L. M. Alexander, and G. E. Schopmeyer, Science, 107:347 (1948).
73. M. K. Yadrick, K. Funk, and M. E. Zabik, J. Agr. Food Chemists 19:491 (1971).
74. E. G. Hill and R. H. Thompson, Pestic. Sci. 4:41 (1973).
75. FMBRA Report No. 54, May, 1972.
76. R. H. Thompson, E. G. Hill, and R. B. Fishwick, Pestic. Sci. 1:93 (1970).
77. G. M. Telling, unpublished work.
78. F. A. Gunther, Residue Rev. 28 (1969).
79. T. E. Archer and D. G. Crosby, Residue Rev. 29 (1969).
80. A. C. Waldron and E. C. Nabor, Poultry Sci. 53:1359 (1974).
81. Siebente VO Durchführung des Gesetzes zur Änderung Futtermittelrechtlicher Verschriften vom 28 Marz 1974.
82. D. D. Manske and P. E. Corneliussen, Pest. Monit. J. 8:110 (1974).

CHAPTER 11

PRESERVATIVES ADDED TO FOODS

Anthony J. Sinskey

Department of Nutrition and Food Science
Massachusetts Institute of Technology
Cambridge, Massachusetts

I. INTRODUCTION

Food preservatives can be defined as agents that extend the storage life of foods by retarding or inhibiting changes in flavor, nutritive value, odor, color, texture, and other organoleptic properties. Many of these undesirable changes are the result of microbial spoilage and/or enzymatic destruction. Enzymatic deterioration may be endogenous, deriving from the food itself, or exogenous, initiated by an extraneous source such as a microorganism. Traditionally, a variety of food processing techniques based on physical principles are employed to prevent such deterioration from taking place. The principles of food preservation by thermal processing, freezing, drying, concentration, and irradiation are well documented in such texts as Karel, Fennema, and Lund [1] and Nickerson and Sinskey [2].

Chemical methods of food preservation may also be employed. Although chemicals may be used to inhibit or control a variety of deteriorative reactions in foods, such as nonenzymatic browning and lipid oxidation, this chapter is devoted to a discussion of chemicals added to foods primarily to control microbial spoilage.

Many food preservatives, such as salt, vinegar, and sugar, have a long tradition, having been used for centuries to inhibit microbial growth and spoilage. The use of such preservatives was developed on an empirical basis. Although a more scientific approach has been used in the development and application of more modern food preservatives, such as sorbic acid and its derivatives, considerable basic information in still lacking on the antimicrobial mechanisms of food preservatives.

In many cases, there is also a lack of appreciation of how the physical and chemical properties of a given food system may influence the efficacy of a food preservative. This point is supported by the findings of Roberts [3], which show that the efficacy of nitrite in inhibiting the outgrowth of Clostridium botulinum spores is influenced by the pH, salt concentration, heat treatment, storage temperature, and initial spore concentration. Thus, the effectiveness of a given chemical preservative is a function of the intrinsic and extrinsic parameters governing the behavior of a food system. Mossel [4] has described the role of both intrinsic and extrinsic parameters in preventing food infections and intoxications, as well as in controlling microbial spoilage.

The addition of chemicals to foods in the United States is regulated by the Food and Drug Administration, Department of Health, Education, and Welfare. The Federal Food, Drug, and Cosmetic Act designates as a chemical preservative any chemical which tends to prevent or retard deterioration when added to foods. The regulations exclude natural preservatives such as table salt, sugar, and vinegars. A detailed discussion of the regulatory aspects of chemical preservatives is given by Furia [5].

Chemical preservatives have to satisfy a number of requirements before they may be used in foods [6], such as:

1. Exhibiting no carcinogenic properties
2. Being nontoxic at appropriate and realistic concentrations
3. Being soluble and not imparting off-odors or flavors
4. Being economical and practical
5. Exhibiting antimicrobial properties over a given range of pH

A variety of chemical agents are available for the specific purpose of inhibiting or retarding the growth of various types of microorganisms in foods. Most are acids and salts.

II. SALTS

A. Sodium Chloride

Sodium chloride is one of the oldest known preservatives and is used in a variety of ways. For example, it can be used in low concentrations (2-4%), in combination with refrigerated storage or with acid, or inhibit the growth of spoilage organisms. Higher concentrations may also be employed, such as in the method of brining.

Table 1 summarizes the types of microorganisms, i.e., halophilic, spoilage, and pathogenic, that are associated with various salted foods and seafoods. Note that the lightly salted products, such as fish containing 1-4% salt, may be readily spoiled by a variety of proteolytic and pathogenic microorganisms such as *C. botulinum*, *Clostridium perfringens*, *Staphylococcus aureus*, and *Vibrio parahaemolyticus*.

The effectiveness of salt as a preservative is complicated by the intrinsic and extrinsic parameters of the food system. Whether or not salt is to be used as the main method of preservation, such treatment of foods must be carried out under controlled temperature conditions. If the temperature of the product immediately after salting is not below 15.6°C, or preferably 4.4°C, spoilage organisms including putrefactive bacteria may invade the tissues previous to penetration by the salt. This applies equally well to the growth of pathogenic bacteria such as *C. botulinum*.

Another factor involved in the inhibition of microbial growth by salting is available water, as microorganisms grow only in aqueous solutions. The term "water activity," A_w, has been coined to express the degree of availability of water in foods [7]. This term can be applied to all foods, with fresh foods having an A_w of about 0.99 - 0.96 at ambient temperatures. Low water activities, which limit the growth of microorganisms in foods, may be achieved by the addition of salt or sugar, by the physical removal of water by drying, or by freezing.

According to Raoult's law,

$$\frac{P}{P_o} = \frac{N_2}{N_1 + N_2}$$

where P = the vapor pressure of the solvent (water in foods)
P_o = the vapor pressure of pure water
N_1 = the number of moles of solute present
N_2 = the number of moles of solvent present

Furthermore,

$$\frac{P}{P_o} = A_w = \frac{\text{equilibrium relative humidity}}{100}$$

TABLE 1 Halophilic, Spoilage, and Pathogenic Microorganisms Associated with Various Salted Foods

Food type	Salt associated with food (%)	Halophilic types	Spoilage microorganisms	Pathogens
Brined meats[a]	brine, 1-7[b]	Halotolerant molds, yeasts, and gram-positive bacteria	Molds, Lactobacillus, Micrococcus, and Vibrio in bacon; Clostridium in packaged salted meats	Clostridium botulinum, Clostridium perfringens, Staphylococcus aureus, and pathogenic Enterobacteriaceae in meats containing low salt
Salted vegetables	1-15	Moderate and halotolerant molds, yeasts, and gram-positive bacteria	Lactic acid bacteria, yeasts, and molds; Bacillus, Enterobacteriaceae in foods with low salt content; Clostridium in packaged foods.	Dependent on level of salt in food; pathogenic members of Enterobacteriaceae in low salt foods; Staphylococcus aureus in highly salted food
Salted fish				
Lightly	1-10	Slight and halotolerant types	Pseudomonas in lightly salted fish; Clostridium in packaged fish; Micrococcus in fish containing 5-10% salt	Dependent on salt concentration, same as for salted vegetables; Vibrio parahaemolyticus may occur when salt is 1-7%

Heavily	brine, 75-80; salt in interior of fish, 10-15	Moderately and extremely halophilic types	*Halobacterium*, *Halococcus* (cause condition called "pink" in fish); also some members of the *Micrococcaceae*	*Staphylococcus aureus*

Source: Adapted from Ref. 104.

[a]Ham, bacon, beef, prepared meat, and sausage.

[b]Brine concentration $= \frac{\text{g salt}}{\text{g salt} + \text{g water}} \times 100$.

For example, a 1 M solution of a perfect solute in water would reduce the vapor pressure of the solution to 98.23% of that of pure water. The A_w would be 0.9823.

Microbial water relationships have been reviewed by Brown [8]. In general, molds grow at lower water activities than yeasts, and yeasts grow at lower water activities than bacteria, but there are some exceptions. Bacteria usually require water activities between 0.99 and 0.96, but some nonhalophiles will grow at water activities as low as 0.94. *Micrococcus halodenitrificans*, a xerophilic red halophilic organism, will grow at an A_w as low as 0.76, the A_w of a saturated sodium chloride solution [9]. *Halobacterium salinasium* requires approximately 3 M sodium chloride for growth, when this is the only solute. The physiology of these bacteria is dominated by an absolute requirement for sodium chloride.

The water relations of food-borne bacterial pathogens have been reviewed by Troller [10]. Data indicate that in general food-borne bacterial pathogens can grow at water activity levels of 0.99 to 0.83.

The mechanisms by which microorganisms adjust to changes or alterations in water activity have been reviewed by Brown [8]. The intracellular physiology of extreme halophils is dominated by massive accumulations of K^+ and Cl^- and by the effective exclusion of Na^+. With xerotolerant yeasts, polyhydric alcohols function as compatible solutes.

Some bacteria accumulate specific metabolites in response to water stress. Measures [11] has demonstrated an accumulation in bacteria of glutamate, γ-aminobutyrate, or proline, in response to increased salt concentration. A qualitative trend was observed in the type of amino acid accumulation, that is, glutamate predominated in the least salt tolerant bacteria studied, whereas proline predominated in the most salt tolerant. Thus, amino acids might well act as amino regulators under mild conditions and function as low-grade compatible solutes.

Thus, one result of adding salt or sugar to foods is to limit the growth of microorganisms by lowering water activity. But there are other factors which play a role in limiting the growth of microorganisms when salt or sugar is added. For instance, it has been shown [12] that when sodium chloride is used to adjust the A_w, *C. botulinum* types A and B will grow at water activities as low as 0.96, but that when glycerol is used instead of salt, these organisms will grow at water activities as low as 0.93. Thus, the inhibiting effect of sodium chloride is not due entirely to the lowering of water activity.

B. Nitrite

Nitrite is included in meat curing mixtures primarily to develop and fix color. Nitrites decompose to nitric oxide, which then reacts with heme pigments to form nitrosomyoglobin [13]. Ingram [14] reports that up to 20 mg/kg nitrite is necessary to provide commercially adequate color stability.

In addition to fixation of color, nitrite is essential for producing the characteristic flavor of cured meat. To produce the flavor of ham, which distinguishes it from salt pork, some 50 mg of nitrite per kg are thought to be necessary.

Nitrite alone or in combination with sodium chloride has important antimicrobial properties, as demonstrated in a number of publications [15-18].

The use of nitrite as a curing agent provides protection against botulism [19]. It may also be important in inhibiting the growth of other food-poisoning and food-infection microbes, such as C. perfringens, Bacillus cereus, S. aureus, and salmonellae. Nitrite is also effective against saprohytic bacterial spoilage.

The antibacterial effect demands greater additions of nitrite than is needed for the development of color and flavor. According to Ingram [14], more than 100 mg per kg nitrite is necessary to secure protection against botulism.

1. Mode of Action of Nitrite

The antimicrobial effectiveness of nitrite, a weak acid salt, is dependent upon the pH of the food system, and to a large extent on the presence of the undissociated form HNO_2. The concentration of HNO_2 is related to pH, with a pK_a value of 3.4.

Nordin [20] has also reported that nitrite disappearance in meat or culture media is dependent upon acidity and temperature. The exponential decay relationship, as reported by Nordin [20], is:

$$\log_{10}(\text{half-life of nitrite in hr}) = 0.65 - 0.025 \times \text{temp } (^\circ\text{C}) + 35 \times \text{pH}$$

From this, it is seen that acidity and high temperature both shorten the half-life of nitrite in complex products. The use of filter-sterilized nitrite in the cold is warranted, in view of the effect of temperature on the disappearance of nitrite.

Ingram [14] has indicated that the pH resulting in a balance between formation and disappearance of HNO_2, which is optimal for antimicrobial action, may be around 5.5. Experiments have tended to confirm this result. For example, Tarr [21,22] showed that the preservative action of nitrite in fish is increased with acidification, and markedly so at levels below pH 6.0. As the pH is further lowered, antimicrobial effects begin to diminish around pH 5.5 [23]. Shank et al. [24] also report an increase in antimicrobial effect with decreasing pH, with a maximum around pH 5.0.

Several investigators have attempted to determine the biochemical mechanism by which nitrite inhibits bacteria. Castellani and Niven [17] considered several theories, including the well-known Van Slyke reaction of HNO_2 on amino acids, which could interfere with dehydrogenase structures.

However, they favored the hypothesis that nitrites react with cellular sulfhydryl constituents and sulfhydryl-aldehyde condensations. The reaction products were not metabolizable under anaerobic conditions and were thought to create a nutritional defect.

Although the antimicrobial mechanism remains to be determined, it has long been known that antimicrobial properties of nitrites are complicated by a variety of factors. For instance, complex interactions of pH, NaCl, $NaNO_2$, $NaNO_3$, heat treatment (F_o value), and incubation or storage temperature are known to be significant for the bacterial stability of cured meat products. Recent investigations have attempted to determine the extent and significance of particular interactions [25-28]. A detailed study of the triple interaction of pH, NaCl, and $NaNO_2$, and the influence of incubation temperature, has been quantitatively studied in a laboratory medium using _C. botulinum_ types A, B, E, and F [29].

Although significant conclusions have been drawn from the above studies, including the fact that salt is the major inhibitory factor in cured meats and that heating is critical for development of a stable cured meat product, little knowledge of the inhibitory mechanisms of nitrite has been gained. Johnston et al. [26] listed four possible mechanisms for the inhibitory effect of nitrite on bacterial spores. They are:

1. Enhanced destruction of spores by heat
2. Stimulation of spore germination during heating, followed by heat inactivation of germinated spores
3. Inhibition of spore germination after heating
4. Production of more inhibitory substances from nitrite

The first two mechanisms are not considered to be significant [14]. Inhibition of spore germination after heating has been described by Duncan and Foster [28]. Inhibition was dependent upon pH, and the undissociated HNO_2 again appears to be the active agent.

The fourth mechanism has attracted much interest recently. The historical developments are described by Ingram [14]. Briefly, it was observed that in some studies higher concentrations of nitrite than normally employed in industry were required for inhibition of spore outgrowth. Upon examination of the protocol, Perigo et al. [27] observed that these studies used filter-sterilized nitrite. It was therefore concluded that an inhibitory compound may be formed as a result of heating materials to which nitrite has been added.

This possibility was examined by two approaches, one concerning the development of inhibitory compounds during heating of bacterial culture media, and the other with meat systems.

A variety of studies have been conducted with heating of microbial culture media [27, 30-32]. A general finding has been that, from nitrite and one or more components in the growth medium, a microbial growth inhibitor is formed, which is commonly called the "Perigo inhibitor." In a systematic study with C. perfringens as the test organism, Moran et al. [32] observed that only amino acids and mineral salts were involved in the production of the inhibitor, which was proven to be a compound formed from cysteine, ferrous sulfate, and sodium nitrite. The inhibitor was compared to several known compounds, including S-nitroso cysteine, Roussin red salt, and Roussin black salt. S-nitrosocysteine, although inhibitory, was not formed in sufficient quantities in the test system, whereas Roussin red salt was found to be unstable. Roussin black salt may have been formed. However, no single compound could be isolated, and the authors conclude that low levels of each compound may result in inhibition of microbial growth.

According to Ingram [14], the Perigo inhibitor is not produced during the heat treatment of meat containing nitrite and cannot explain the role of nitrite in the stability of canned cured meats. Thus, further studies are required.

Raevuori [18] has confirmed earlier reports that the use of erythorbate increases the inhibitory effect of nitrite.

Nitrites have recently been shown to be involved in the formation of nitrosamines in cured meat products. The chemistry and health implications of nitrosamines are further discussed by Wogan in Chapter 9 of this volume. Nitrosamines can be formed from the reaction of secondary or tertiary amines with N_2O_3, the active nitrosating reagent in most food products. Nitrosation of amines may take place in foods during processing or storage, and thus nitrosamines may be ingested in foods as such. In addition, it has been shown that nitrosation can take place in the strongly acidic conditions of the stomach. Therefore, ingestion of precursors can also lead to local formation of nitrosamines.

Studies are underway to determine the prevalence of this type of contamination in foods, and significant improvements in analytical methodologies for nitrosamine detection have recently been developed.

The fact that nitrosamine may be formed in cured meats has stimulated considerable research and concern among public officials. Active research programs are under way to determine the conditions responsible for nitrosation in vivo and in foods. The solution to the problem is not simple. However, one consequence is a movement toward the use of lower initial levels of nitrite in cured meat products.

C. Sulfur Dioxide Generating Compounds

Sulfur dioxide, and salts of sulfurous acid, are used in foods to prevent enzymatic and nonenzymatic browning and the growth of undesirable microorganisms. Sulfur dioxide also has antioxidant properties and is sometimes employed as a bleaching agent [33].

Sulfur dioxide generating compounds include:

1. Sulfurous acid, H_2SO_3
2. The salts of sulfurous acid, including Na_2SO_3 (sodium sulfite), $NaHSO_3$ (sodium bisulfite), and K_2SO_3 (potassium sulfite)
3. Hydrosulfurous acid, $H_2S_2O_4$, the salt $Na_2S_2O_4$ (sodium hydrosulfite) being a strong reducing agent
4. Pyrosulfurous acid, $H_2S_2O_5$, and salt $Na_2S_2O_5$ (sodium pyrosulfite or metabisulfite)

The physical and chemical properties of sulfur dioxide generating compounds are described by Chichester and Tanner [34] and are briefly summarized below.

1. Sodium sulfite, Na_2SO_3. A white to tan, or slightly pink, odorless, or nearly odorless powder. One gram dissolves in 4 ml water. Sparingly soluble in alcohol.
2. Potassium sulfite, K_2SO_3. White, odorless, granular powder. One gram dissolves in 3.5 ml water.
3. Sodium bisulfite, $NaHSO_3$. White, crystalline powder; SO_2 odor. One gram dissolves in 3.5 ml cold water, 2 ml boiling water and about 70 ml alcohol.
4. Sodium metabisulfite, $Na_2S_2O_5$. White crystals or powder with an odor of SO_2. Freely soluble in water; slightly soluble in alcohol.
5. Potassium metabisulfite, $K_2S_2O_5$. White crystals or powder having odor of SO_2. Freely soluble in water; slightly soluble in alcohol.

Sulfur dioxide is highly soluble in water (11.3 g/100 ml at 20°C), forming sulfurous acid (H_2SO_3), which can dissociate into the bisulfite (HSO_3^-) or the sulfite (SO_3^{-2}), depending upon pH.

$$SO_2 + H_2O \longrightarrow H_2SO_3$$

$$H_2SO_3 \longrightarrow H^+ + HSO_3^- \qquad pK = 1.76$$

$$HSO_3^- \longrightarrow H^+ + SO_3^{-2} \qquad pK_2^1 = 7.21$$

The specific concentrations of SO_3^{-2}, HSO_3^-, and H_2SO_3 are dependent upon the pH of the solution, with HSO_3^- dominant between pH 2 and 7. Since it has been shown that sulfur dioxide is most effective as an antimicrobial agent in acid media, the antimicrobial agent is thought to be the undissociated sulfurous acid, which inhibits yeasts, molds and bacteria. It has been reported that the undissociated acid is 1000 times more active than HSO_3^- for Escherichia coli, 100-500 times for Saccharomyces cerevisiae, and 100 times for Aspergillus niger [35].

The possible mechanisms by which sulfurous acid inhibits microorganisms have been summarized by Lindsay [33] as follows: (a) reaction of bisulfite with acetaldehyde in the cell; (b) reduction of essential disulfide linkages in enzymes; and (c) formation of bisulfite addition compounds, which interfere with respiratory reactions involving nicotinamide dinucleotide. It should be pointed out, however, that most of the studies on the antimicrobial mechanisms of sulfite and sulfur dioxide were conducted a long time ago. Modern approaches of genetics and molecular biology have not been systematically applied.

Several studies have reported that sulfur dioxide and sulfites are metabolized to sulfate and then excreted in the urine without any obvious pathological results. For example, rats given oral doses of a 2% sodium metabisulfite solution eliminated 55% of the sulfur as sulfate within the first 4 hr [36]. In another study, groups of rats were fed sodium bisulfite in dosages from 0.01 to 2.0% of the diet for periods of 1 to 2 years. Rats fed 0.05% sodium bisulfite (307 ppm as SO_2) for 2 years showed no toxic symptoms. Sulfite in concentrations of 0.1% (615 ppm SO_2) or higher inhibited growth, probably through the destruction of thiamine in the diet [37].

Treatment of foods with sulfites reduces thiamine content [38], and it has been suggested that the ingestion of SO_2 in a beverage, for example, may cause a reduction in the level of thiamine in the rest of the diet [39]. Studies with rats have shown that the addition of SO_2 greatly reduces the urinary output of thiamine, especially when both are given together [40]. Thus, sulfur dioxide and sulfite salts should not be used in foods which are substantial sources of thiamine. Typical foods and recommended levels of SO_2 are summarized in Table 2. Levels of 2000 ppm are used in the initial stages. Levels above 500 ppm give a noticeably disagreeable flavor [34].

As will be discussed in Sec. VIII, several food preservatives have been shown to be chemical mutagens. Bisulfite ion has been shown to have mutagenic effects on viruses, bacteria, and plants [41].

TABLE 2 Levels of SO_2 in Dried and Dehydrated Fruits and Vegetables at Start of Storage

Product	SO_2 (ppm)	Product	SO_2 (ppm)
Apricots	2000	Raisins	800
Peaches	2000	Apples	800
Pears	1000	Cabbage	750-1500
		Carrots	200-250

Source: Adapted from Ref. 34.

Summers and Drake [42] reported that treatment of bacteriophage T4rII mutants with 0.9 M bisulfite for 4 hr resulted in mutations at G:C reversion sites specifically. Bisulfite was judged to be about as effective as nitrous acid applied at pH 5.3, but about 100 times weaker than nitrous acid applied at pH 3.7, in terms of mutations per mole per hr [41].

In vitro studies [43] have reported on the conversion of cytosine to uracil in nucleic acid (1 M bisulfite). Bisulfite also adds to the 5:6 double bond of cytidine and uridine by means of an ionic reaction, forming 5,6-dehydrocytidine-6-sulfonate and 5,6-dehydrouridine-6-sulfonate, respectively.

Deamination of the cytidine derivative to the uridine derivative occurs readily, and under alkaline conditions the derivative is converted to uridine. In this manner, chemical modifications of transfer ribonucleic acid have been performed with bisulfite [44,45].

Low bisulfite concentrations (1-2 x 10^{-2} M) have been shown to inactivate Bacillus subtilis, transforming DNA in vitro [46]. The reactive species in this case was thought to involve free radicals generated by the aerobic oxidation of sulfite.

Although the DNA in eukaryotic cells is protected by a variety of defense mechanisms, the potential exists for bisulfite damage in hereditary materials [41-46].

All sources of bisulfite should be considered in the estimation of potential health risks. As pointed out by Fishbein [41], these include not only bisulfite salts and SO_2 used as food preservatives, but bisulfite which is present in small amounts in polluted environments.

III. STERILIZING AGENTS

A. Diethyl Carbonate

Diethyl carbonate, also called diethyl pyrocarbonate (DEPC), has been used, especially in Europe, for stabilizing fruit juices, carbonated beverages, and wines. Although its use was accepted for a number of years, it was banned by the Food and Drug Administration after Lofroth and Geguall [47] reported that ethyl carbonate was formed in wine thus treated.

It has been proposed that diethyl pyrocarbonate first hydrolyzes in water to form ethyl carbonate, and that this product, which quickly decomposes to ethyl alcohol and carbon dioxide, is the active form of the compound, in terms of the destruction of microorganisms.

$$(C_2H_5-O-C(=O))_2O \longrightarrow (C_2H_5O)_2=C=O + CO_2$$

$$(C_2H_5O)_2=C=O + H_2O \longrightarrow 2C_2H_5OH + CO_2$$

When added to beverages, diethyl pyrocarbonate will destroy limited numbers of microorganisms, especially yeasts. The mechanisms by which diethyl pyrocarbonate inactivates microorganisms are thought to result from the reaction of histoidyl and sulfhydryl residues in enzymes of microorganisms.

Since diethyl pyrocarbonate is subject to nucleophilic attack, compounds which contain an active hydrogen atom will readily react with it, forming carbethoxy derivatives and ethyl esters.

Reaction with ammonia leads to the formation of the carcinogen urethane.

$$C_2H_5O-\overset{O}{\overset{\|}{C}}-O-\overset{O}{\overset{\|}{C}}-C_2H_5 + NH_3 \longrightarrow C_2H_5O\overset{O}{\overset{\|}{C}}-NH_2 + C_2H_5OH + CO_2$$

Ough [48,49] reports that ethyl carbonate occurs naturally in fermented foods, although the levels are low. Thus, diethyl pyrocarbonate is no longer permitted in foods in the United States. Urethane also occurs naturally in beverages. Hydrolysis of yeast carbamyl phosphate with ethanol, which yields phosphoric acid and urethane, is the source of the naturally occurring urethane.

B. Ethylene and Propylene Oxides

Although the epoxides, ethylene and propylene oxides, are used for gaseous sterilization of spices, they are normally classified as antimicrobial agents [34]. Phillips [50] has reviewed the use of ethylene and propylene oxides as sterilizing agents.

Both epoxides are reactive cyclic esters that inactivate most microbial forms including bacterial spores. With regard to ethylene oxide it has been proposed that alkylation of essential intermediary metabolites with an hydroxyethyl group ($-CH_2-CH_2-OH$) accounts for the inactivation [50].

Since ethylene oxide can react with inorganic chlorides, formation of toxic chlorohydrin is of concern [50]. Propylene glycol can also form chlorohydrin [51].

The desirability of the use of oxides for control of microbial levels lies not in the speed, simplicity, or economy of treatment, but rather in the fact that, with certain types of foods, microbial levels are reduced with least damage to the food. Spices and nutmeats are prime examples. Also, since for effective use of the gases, the relative humidity of the products has to be low (30% for both ethylene oxide and propylene oxide), their use is limited to low-moisture items. The chemical and physical factors affecting sterilization by ethylene oxide are described by Ernst and Doyle [52].

Stringent safety procedures are required for the use of these compounds. Ethylene oxide is volatile and flammable. Normally it is supplied as diluted mixtures. Propylene oxide is less reactive and has a narrower explosive range (2-22%).

Concern regarding ethylene oxide sterilization of foodstuffs has a long history. Hawk and Mickelsen [53] demonstrated that rats did not grow when fed special purified diets treated with ethylene oxide. Reaction of ethylene oxide with various essential vitamins and amino acids was determined as the cause.

Gordon et al. [54] found that ethylene oxide alkylated cellulose in dry prunes, and glycols were also formed, as summarized by Phillips [50]. Concern over possible ethylene glycol formation led the FDA to ban the use of ethylene oxide treatment of foods for human consumption, while allowing the use of propylene oxide, since it, unlike ethylene glycol, is nontoxic. However, according to Phillips [50], if salt is present in the product being treated, the more probable hydrolysis product is chlorohydrin rather than glycol, and both ethylene and propylene chlorohydrin are more toxic than ethylene glycol.

IV. ACIDS

A. Introduction

A variety of acids and their salts are used as food preservatives. Table 3 lists their pKs and concentrations commonly used in foods. As with nitrite and sulfite, a higher antimicrobial activity is associated with the undissociated acid. Therefore, the pK of the acid basically defines the pH range over which it may be expected to be an effective antimicrobial. The other factor that plays a dominant role in the case of a given acid is its solubility.

Many of the acids used in foods are added to reduce the pH of a given product. Generally speaking, all microorganisms have a range of pH over which growth may occur, and an optimal pH for high growth rate. Usually it is considered that bacteria will grow best at a pH of 5.5 to 7.0. Yeast and molds will grow well at low pH values.

It is normally impossible to lower the pH of a food product to the point where no microorganisms will grow, since such products would have objectionable tastes. Therefore, when acids are used, other sublethal treatments are also employed, such as heat pasteurization, refrigeration, or salting.

For example, acetic acid as vinegar is added to pickled fish products along with sodium chloride, and microbial growth is inhibited in such products as long as the temperature is low. Vinegar is also added to products such as pickles and ketchup, which are then heat pasteurized. Citric acid is sometimes used in tomato juice to control spoilage by spore-forming bacteria [2].

B. Acetic Acid

Acetic acid, CH_3COOH, is used in a variety of products as described above, and its salts, sodium acetate, hydrous calcium acetate, and especially

TABLE 3 Common Food Acids

Acid	pK	Maximal concentration present in some foods (mmol/kg)	Food
Acetate (H, Na, K, di-)	4.76	500	Pickled food
Benzoate (H, Na)	4.2	8	Beverages
Parabens[a]			
Methyl	8.47	5	Beverages
Heptyl	8.47	?	Beer
Propyl	8.47	0.6	Cream
Propionate (Na, Ca)	4.87	100	Cheese, bread
Sorbate (H, K, Na)	4.8	30	Cheese

Source: Adapted from Ref. 66.
[a]Esters of p-hydroxybenzoic acid.

sodium diacetate, are used to control ropiness in bread and bakery products. Further information on properties of acetic acid and its salts, and on their use in food, is to be found in Chichester and Tanner [34].

C. Propionic Acid

The aliphatic monocarboxylic acid, propionic acid (CH_3-CH_2-COOH), and its sodium and calcium salts are also used as preservatives. They are commonly employed in bakery products as mold and bacterial inhibitors. Propionates are also active against the ropy bread organism Bacillus mesentericus and useful for preventing the molding of cheese. Propionic acid is also a microbial fermentation product. It is produced by Propionicbacterium shermanii in Swiss cheese, to a final concentration as high as 1% by weight. In white bread and rolls, concentrations of sodium and calcium propionate are limited to 0.32% of flour. The antimicrobial effectiveness of the propionates is due to the undissociated form of the acid, which is effective up to a pH of 5.0.

Hoffmon et al. [55] evaluated the fungistatic properties of the normal saturated fatty acids containing 1 to 14 carbon atoms. They found that antifungal properties of the fatty acids varied according to chain length, concentration, and pH. It was at this time that the use of propionic acid was patented as a mold inhibitor in bread. Higher homologs, with higher antimicrobial activity, have objectionable tastes and odors.

D. Sorbic Acid

An unsaturated straight-chain fatty acid that is an effective preservative is sorbic acid and its salts. Sorbic acid (2,4-hexadienoic acid) was first recommended as a food preservative by Gooding [56] and since then has been used as a fungistatic agent in foods and wrapping materials. Sorbic acid is said to be effective as an antimicrobial agent [57]. However, because of its limited solubility, its salts are more effective. Furthermore, since the pK of the acid is 4.2, it is most useful in low-pH foods, although it is reported to be active up to pH 6.5, which is considerably above the range for propionic and benzoic acids [33].

The antimycotic activity of sorbic acid and its salts is attributed to the fact that molds are unable to metabolize the α-unsaturated diene system of the alphatic chain. Since this mechanism is not operative in higher animals, sorbic acid is thought to be metabolized in a manner similar to longer-chain fatty acids[58]. In addition, molds resistant to sorbic acid are said to be able to metabolize it via β-oxidation. With _Penicillium roqueforte_ sorbic acid is decarboxylated directly to CO_2 and 1,3-pentadiene. The latter can then lead to gasoline or hydrocarbon-like off-flavors in products such as cheese, where sorbate is used as a preservative.

The reader is referred to the article by Chichester and Tanner [34] for more details on specific applications of sorbic acid.

E. Formaldehyde

Formaldehyde is used in Italy as a bacteriostatic agent in cheese milk for the manufacture of Grana cheese. According to Kosikowski [59], 40 ppm formaldehyde are routinely added to milk for Grana cheese to improve its texture and to prevent clostridia bacteria from forming gas holes in the cheese. After three weeks, the cheese contains less than 3 ppm residual formaldehyde, which is regarded as harmless by Italian authorities. An alternative practice is to add hexamethylenetetramine to cheese curds. This compound eventually breaks down to formaldehyde and gives the cheese protection against late gas-blowing clostridia.

Although in the United States citrus fruits may be washed with hexamethylenetetramine, no residue is allowed on or in the food product. In European countries, hexamine is allowed in foods up to 750 ppm. It is used either alone or with sodium benzoate to stabilize pickled fish products which are to be used without refrigeration.

F. Benzoic Acid and the Parabenzoates

Benzoic acid and its sodium salt and parahydroxybenzoic ("parabenzoic" acid) and its methyl, propyl, and butyl esters are used in a number of food products. Benzoic acid is not very soluble in water and hence the sodium salt is more frequently used. Parabenzoic acid and its esters are more soluble than benzoic acid and the sodium salts of the parabenzoates are even more soluble, although the parabenzoates are more inhibitive to microorganisms at lower pH. These compounds are much more effective over a wide range of pH and inhibit a broader spectrum of microorganisms than do the benzoates.

Benzoates and the paraben compounds are used in fruit juices, carbonated beverages, pickles, and sauerkraut. Concentrations of both the benzoates and parabens of 0.1% by weight are allowed in foods.

The benzoic acid and the paraben compounds are eliminated from the body after conjugation with glycine. With the parabens, the ester is hydrolyzed and metabolic conjugation follows.

G. Antimicrobial Mechanisms of Food Acids

Despite the fact that chemical preservatives of various kinds have been used for several decades, their mode of action remains largely unexplained. Least is known about antimicrobial agents with relatively simple chemical structures, in contrast to more complex entities such as antibiotics, which have been extensively studied.

Postulated mechanisms of microbial inhibition can be grouped into interference with (a) the cellular membrane, (b) genetic mechanisms, and (c) enzyme activity. The abundant indirect evidence for these mechanisms has been reviewed by Wyss [60]. More recent evidence has shown interference with cellular membranes to be the primary mechanism of microbial inhibition.

Sheu and Freese [61] added fatty acids of different chain lengths to cultures of *Bacillus subtilis* growing in nutrient sporulation medium, and the effects of these fatty acids on growth, oxygen uptake, adenosine triphosphate (ATP) concentration, and membrane protein composition were examined. All fatty acids inhibited growth, the effect being reduced in the presence of glycolytic compounds and reversed by transfer to medium without fatty acids. The fatty acids also reduced the rate of oxygen consumption and the amount of ATP per cell. However, with cell envelope preparations, fatty acids did not inhibit the NADH-coupled oxidation, which is affected by typical electron-transport inhibitors, such as KCN or 2-n-heptyl-4-hydroxyquinoline-N-oxide. Romano and Kornberg [62] found that high concentrations of acetate inhibit the uptake of sugars in aspergillus. Borst-Pauwels and Jager [63] and Samson et al. [64] report that short-chain fatty acids inhibit the uptake of phosphate by yeast.

Sheu et al. [65] found that acetate and other short-chain n-fatty acids (C_1-C_6) strongly inhibit the uptake of L-serine and other L-amino acids, but only weakly inhibit that of α-methylglucoside or fructose, whether measured in whole cells of *B. subtilis* or in energized membrane vesicles. It was concluded that fatty acids "uncouple" the amino acid carrier proteins from the cytochrome-linked electron-transport system. Later Freese et al. [66] generalized that the undissociated lipophilic acids uncoupled both substrate transport and oxidative phosphorylation from the electron-transport system.

These results suggest that fatty acids react reversibly with the cell membrane or proteins in it. They may either alter the membrane structure (i.e., reduce its fluidity) or uncouple the connection between the electron-transport chain and two types of proteins, namely those required for ATP regeneration and the carrier proteins required for transport of various compounds into the cell. Since the affinity of the fatty acids to the membrane should increase with increasing lipophilic portions of the molecules, it is not surprising that the concentration of fatty acids required to inhibit growth, oxygen consumption, or ATP synthesis decreases somewhat with increasing chain length.

Sheu and Freese [67] found, in contrast to the above mentioned conclusion, that effectiveness of inhibition of microbial growth increased with increasing fatty acid chain length, and that the inhibitory effect on E. coli only increased up to a fatty acid chain length of six. Hexonate was equally effective with E. coli as with B. subtilis. However, decanoate inhibited growth of E. coli at a concentration fifty times higher than with B. subtilis. Long chain fatty acids were ineffective.

For this discrepancy two explanations were considered. The first was that E. coli could utilize the long-chain fatty acids as a carbon source, and that any long-chain fatty acids that were attached to the membrane were metabolized rapidly, before they could exert any significant inhibitory effect. If this is true, mutants unable to take up or metabolize long-chain fatty acids should be inhibited. The second explanation for differences in antimicrobial sensitivity was that the lipopolysaccharide (LPS) layer surrounding the cell wall and membrane prevents entry of the compounds [67]. If the LPS prevents the entry of long-chain fatty acids, its partial opening, in mutants devoid of a complete LPS or by chemical treatments, might make the cells more accessible to fatty acid inhibition. Treatment of E. coli with ethylenediaminetetracetate (EDTA), which removes much of the lipopolysaccharide since it is bound ionically by salt bridges, renders the cell permeable to a variety of drugs. Leive [68] has shown that EDTA treatment renders E. coli permeable to actinomycin D and lysozyme.

Sheu and Freese [67] found that the resistance of gram-negative organisms was not correlated with the ability to metabolize fatty acids. However, they found that mutants of both E. coli and Salmonella typhimurium, in which the lipopolysaccharide layer does not contain residues beyond the 2-keto-3-deoxyactonate core, are inhibited by medium- (C_{10}), but not by long-chain (C_{18}) fatty acids. Furthermore, removal of a portion of the LPS by EDTA treatment rendered the organisms sensitive to long-chain fatty acids.

Thus, they concluded that the intact LPS layer of gram-negative organisms seems to screen the cells against medium- and long-chain fatty acids and prevents their accumulation on the inner cell membrane (site of amino acid transport) to inhibitory concentrations. Such findings probably explain a commonly observed finding in food applications, namely that food preservatives are more effective against a variety of gram-negative bacteria, including pseudomonas species, in the presence of EDTA.

Sheu et al. [69] extended their studies to inhibitory effects of lipophilic acids and related compounds on bacteria and mammalian cells. Most compounds inhibit the growth of Hela cells about as efficiently as of B. subtilis. However, butyrate and propionate, as well as acetaminophen, antipyrine, phenacetin, and salicylamide, inhibit Hela cells at millimolar concentrations, whereas at least 10 times higher concentrations are needed to inhibit B. subtilis. There is a decrease in the concentrations needed to inhibit growth by 50% with increasing octanol-water partition coefficients of the compound.

As long as the pH is higher than the pK value of the compound, the effectiveness of a given lipophilic acid to inhibit bacterial and membrane transport increases with decreasing pH. Thus, it appears that the nonionized molecules are responsible for the inhibitory effect. However, as discussed by Sheu et al. [69], when one calculates the concentration of uncharged molecules at a given pH, and plots that against the octanol-water partition coefficient, discrepancies arise in the linear relationship between partition coefficient and inhibitory concentration. Several reasons for the discrepancies are put forth, including:

1. All compounds do not act by the same mechanisms.
2. Uptake of compounds may be determined by passive diffusion for some compounds and by group translocation or other active transport for others.
3. Lipophilic molecules tend to associate in aqueous solutions in pairs or multiple units. At increasing concentration, molecules with long aliphatic chains can form micelles, whereas aromatic compounds form stacks. This reduces the concentration of molecules able to attach to the membrane individually.
4. Molecules with high partition coefficients form dimers or polymers in cellular membranes, thereby potentially increasing the number of molecules that can be incorporated into the membrane.

Ingram [70] has reported on changes in fatty acid and phospholipid composition of E. coli resulting from growth in the presence of a variety of sublethal concentrations of organic solvents and food additives. Methyl paraben and sodium sorbate cause an increase in synthesis of lipids containing saturated fatty acids. Growth in the presence of propionate results in the production of fatty acids with odd chain lengths. Both 15- and 17-carbon saturated fatty acids, as well as 17-carbon unsaturated fatty acids, were found. Synthesis of these unusual fatty acids probably results from the utilization of propionyl- instead of acetyl-ACP (3-amino-6-chloropyrazinoylguanidine), as a primer for fatty acid synthesis.

Ingram [70] attributes changes in the fatty acid composition to attempts by the cell to maintain a homeoviscous membrane. Changes in phospholipid components could be expected to alter the charge distribution of the membrane. Whether or not such changes occur in complex eukaryotic systems remains to be determined.

V. ANTIBIOTICS

Antibiotics are used in a number of countries for the preservation of perishable commodities and canned foods. The use of antibiotics in the food industry has been reviewed by Wrenshall [71] and by Chichester and Tanner [34].

Nisin is a polypeptide antibiotic produced by Streptococcus lactis in milk and cheese [72]. The purification and nature of nisin are described by Berridge et al. [73]. The antibiotic contains 34 amino acid residues, eight of which are rarely found in nature, including lanthionine (two alanines bound to sulfur at the 3-carbon) and β-methyllanthionine [74]. The antibiotic is similar in structure to subtilin but contains no tryptophan.

Because of its heat stability, the use of nisin in combination with subtilin in canned foods has been investigated [75]. Bacillus stearothermophilus has been found to be particularly sensitive to nisin. Spores of this thermophilic flat-souring organism possess great heat resistance. Gibbs and Hurst [76] reported that nisin effectively controlled thermophiles in peas.

Nisin as discussed by Goldberg and Barnes [77] cannot be used indiscriminately in any type of food system. It has no effect against common spoilage organisms and some strains of C. botulinum are resistant to it. Regulations in England and Wales permit nisin to prevent flat-sour spoilage in certain canned foods that are given an adequate chemical treatment to control C. botulinum outgrowth.

Pimaricin is a polyene antifungal antibiotic produced by Streptomyces natalensis. It is used in Norway and Sweden for coating hard cheeses and for surface coating of cheese and sausage in Belgium. It is not used in the United States.

Nystatin is a polyene antifungal antibiotic produced by Streptomyces noursei, S. aureus and other streptomyces species [78]. This antibiotic may be used on the skin of bananas in England and Wales.

The broad spectrum antibiotics, including tetracycline, chlortetracycline, and oxytetracycline, have been investigated and used primarily for extending the shelf life of perishable foods such as chicken. For example, Kohler, et al. [79] reported that the shelf life of eviscerated and cut-up poultry was increased from 7 to 14 days at 4.4°C as a result of dipping for 30 min in solutions containing 3-30 ppm chlortetracycline. Applications with fish and other seafoods have also been studied. The tetracyclines are not permitted in the United States. Chlortetracycline and oxytetracycline for fish continues to be permitted in England and Wales. Chlortetracycline for raw fish, peeled shrimp, and shucked scallops is permitted in Canada.

Of the antibiotics, only chlortetracycline and oxytetracycline have been used as chemical food preservatives in the United States. The limits for residues of chlortetracycline and oxytetracycline were revoked by the Food and Drug Administration in 1967 [6].

VI. POLYHYDRIC ALCOHOLS

Polyhydric alcohols are valuable ingredients that are used in a wide range of food products. Griffin and Lynch [80] have reviewed the role played by the polyols (propylene glycol, glycerol, sorbitol, and mannitol) in controlling viscosity crystallization, taste, and hygroscopicity or humectancy, as well as their function as rehydration aids and bulking agents. Also included in the list of applications is microbiological preservation.

Prevention of microbial growth by polyols is the result of two phenomena: one is the physical mechanism of lowering the water activity of the food system; the other is independent of changes in water activity. Use of polyols as humectants and their effectiveness in lowering the A_w of intermediate products has been reviewed by Karel [81].

The bactericidal activity of a variety of glycols has been recognized [82]. Olitzky [83] reported a distinct difference in susceptibility to propylene glycol of gram-positive and gram-negative bacteria. Marshall et al. [84] compared the resistance of bacteria to glycerol and sodium chloride at similar water activities. Glycerol was more inhibitory than sodium chloride to relatively salt-tolerant bacteria and less inhibitory to salt-sensitive species.

Plitman et al. [85] investigated the bacteriostatic and bactericidal activity of several diols in intermediate-moisture meat products, as well as in a bacterial growth medium, employing S. aureus as a test organism. They found 1,2-propanediol and 1,3-butanediol to be bactericidal at an A_w of 0.92 to S. aureus, whereas with glycerol growth was inhibited at an A_w of approximately 0.88, which is near the minimum for growth.

Recently, Akedo et al. [86] investigated the antimicrobial activity of a variety of aliphatic diols and ester derivatives. Inhibition of growth of B. subtilis 60015 was found to depend on chain length and position of the hydroxyl groups. The 1,2-diols were more effective than 1,3-diols (see Fig. 1). Esterification also affected antimicrobial activity. Studies with membrane vesicles prepared from B. subtilis indicated that inhibition of amino acid transport is a primary antimicrobial effect of diols and their esters. This effect is similar to that observed by Freese et al. [66] with the lipophilic food acids.

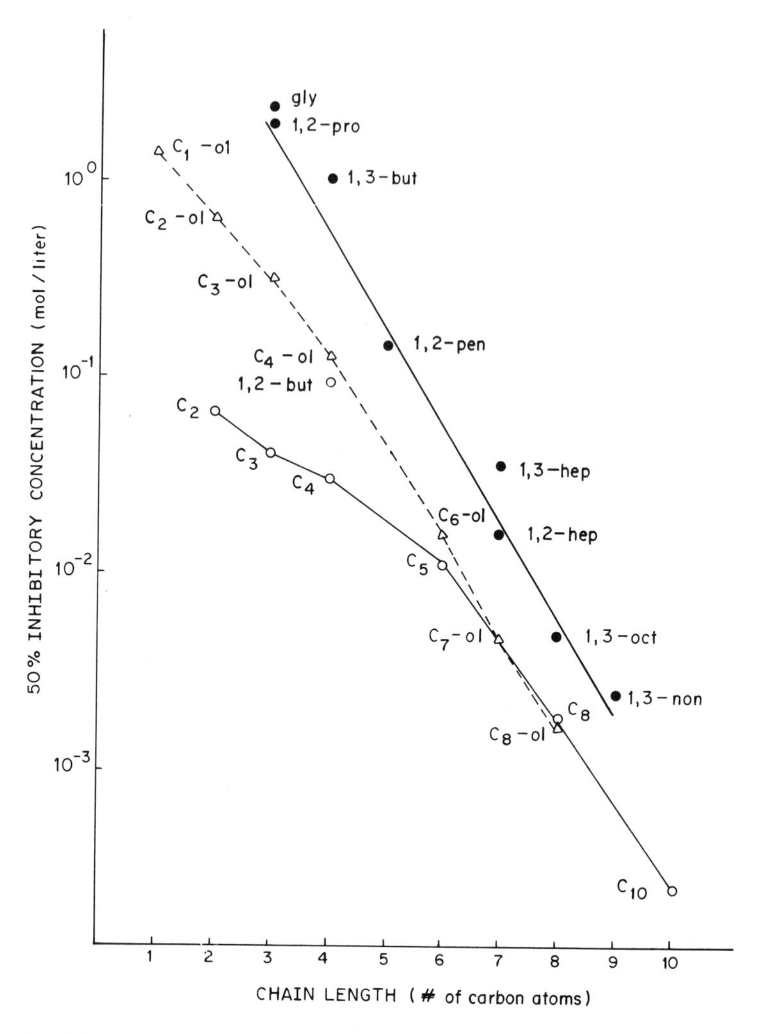

Fig. 1. Relationship between 50% growth inhibitory concentrations and chain length of compounds on Bacillus subtilis. Key: —○— acids; - -△- - alcohols; —●— diols. (Data on acids were taken from Sheu et al., Ref. 69.)

C_2:	acetic acid	C_1-ol:	methanol
C_3:	propionic acid	C_2-ol:	ethanol
C_4:	butyric acid	C_3-ol:	propanol
C_6:	hexanoic acid	C_4-ol:	butanol
C_8:	octanoic acid	C_6-ol:	hexanol
C_{10}:	decanoic acid	C_7-ol:	heptanol
		C_8-ol:	octanol

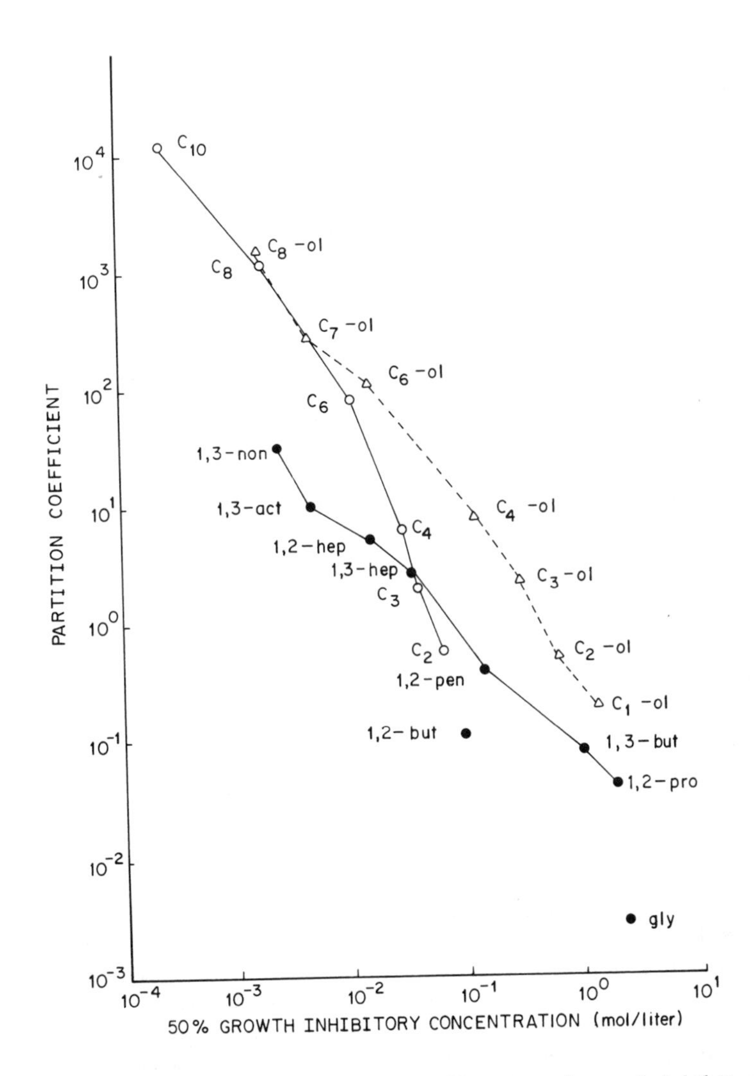

Fig. 2. Relationship between partition coefficient and growth inhibitory concentrations on Bacillus subtilis. Key: —○— acids; - - -△- - alcohols; —●— diols.

gly:	glycerol
1,2-pro:	1,2-propanediol
1,3-but:	1,3-butanediol
1,2-but:	1,2-butanediol
1,2-pen:	1,2-pentanediol
1,3-hep:	1,3-heptanediol
1,2-hep:	1,2-heptanediol
1,3-oct:	1,3-octanediol
1,3-non:	1,3-nonanediol

VII. ETHANOL

The straight-chain alcohol ethanol can also be considered a food preservative. Yeast metabolism is inhibited by alcohol as well as by sugar. It is reported that alcohol fermentation in wine musts is inhibited at about 18% vol/vol alcohol or at 80% wt/vol sugar. Therefore, alcohol is considered to be 4-5 times more inhibitory than sugar. This fact is expressed in Delle units (DU) in the following manner:

$$X + 4.5\ Y = DU$$

where X is the percentage of sugar (g of reducing sugar/100 ml), and Y is the volume percentage of alcohol (ml of ethyl alcohol/100 ml). Thus, a combination of sugar and alcohol concentrations which gives a DU of 80 should inhibit fermentation. A summary of the literature on the verification of the above equation and its practical significance has been made by Amerine and Kunkee [87]. Kunkee and Amerine [88] have applied the above relationships to stabilization of sweet wine.

A plot of the inhibitory concentrations of a variety of straight-chain alcohols, acids, and diols versus partition coefficients is presented in Fig. 2. The test organism was B. subtilis and, as can be seen, as chain length is increased, lower concentrations of all compounds are required for inhibition of growth by a standard amount. The straight-chain fatty acids appear to be more inhibitory than the alcohols or diols.

VIII. MUTAGENIC PROPERTIES OF ANTIMICROBIAL FOOD ADDITIVES

A number of powerful biological assay procedures have recently been developed for determining the mutagenic activities of a variety of chemicals. Ames [89] has discussed the validity of bacterial tests and their pertinence to human mutagenesis and carcinogenesis. Many of the procedures detect mutagenic properties by microbial assay systems [90-96]. Gene mutations that can be detected with microbial systems include base-pair substitutions and frame shifts. In addition, genetic changes such as large deletions, gene conversion, and mitotic recombinations can be observed. Listed below are microbial assay systems detecting genetic alterations and DNA-compound interactions.

1. Reverse mutations
2. Forward mutations
3. Transformations
4. Phage inductions
5. Repair tests

6. Mitotic recombinations
7. Gene conversions
8. Nondisjunctions

Many of these microbial systems can be used in conjunction with an activation system since some chemicals are actually metabolized to mutagens. Although in vitro chemical activation, whole animal, and tissue culture systems have been investigated, the most useful appears to be microsomal extracts prepared from induced rat liver preparations. For example, Gomez et al. [97] have shown that N-nitrosomorpholine and N-nitrosopyrillodine are activated to mutagenic chemicals by incubation with rat liver microsome preparation. The following food preservatives have been shown to be mutagenic [41]: nitrite (Na, K), sulfite (mono, bi, K, Na, SO_2), formaldehyde, propylene oxide, and ethylene oxide. A variety of such chemicals have been detected and most are alkylating agents.

Food preservatives may also react chemically within food systems to form mutagenic agents. As discussed earlier, an important example is the reaction of nitrite with secondary amines to form nitrosamines (Sec. II.B). Diethyl pyrocarbonate addition to foods results in urethane formation. Even ethanol has been shown to induce genetic damage in *Asperigillus nidulans* [98]. Hayatsu et al. [99] have recently shown that mutagenic compounds can be generated by a reaction between sorbic acid and nitrite.

The effect of various methods of food processing, i.e., drying, freezing, thermal processing, and irradiation, on the formation of mutagenic agents in foods will be an important subject of future research. The balance between the public health risks and benefits of such chemical mutagens in foods remains to be determined. The regulatory status of direct food additives is summarized by Furia [5].

One important aspect that should not be ignored is that environmental exposure to mutagenic compounds such as nitrite or sulfur dioxide comes not only from foods but also from other sources, such as air and water pollution. The total impact on human health of mutagenic chemicals needs to be evaluated.

IX. ANTIOXIDANTS, SEQUESTRANTS, AND INDIRECT ADDITIVES

Antioxidants are used to prevent rancidity in oils, fats, shortenings and other foods containing oxidizable lipids (see Chapter 4). They may be directly added to foods or incorporated in the packaging material, according to regulation. Furthermore, they may be used in combination with EDTA or citric acid as metal chelating agents, and with ascorbic acid as synergists.

The antioxidants are governed by specific provisions in the Code of Federal Regulations under the subpart for food additives. Those permitted for use in foods include BHA (butylated hydroxyanisole), BHT (butylated hydroxytoluene), TBHQ (tertiary butylhydroquinone), THBP (2,4,5-trihydroxybutyrophenone) and propyl gallate. Ethoxyquin is permitted in certain spices and as a residue in animal products.

The metabolism and toxicity of the phenolic antioxidants have been extensively investigated. The literature up until 1970 has been reviewed [100] and more recent publications have included detailed studies in man, rat, and dog [101]. There is little or no evidence for toxicity of these compounds at the levels used, and they are probably beneficial by virtue of their sporing action on vitamin E and reduction of the intake of oxidized lipids.

In addition, some antioxidants introduced into foods may have antimicrobial properties. For example, Trelase and Tompkin [102] demonstrate that butylated hydroxyanisole functions as an antimicrobial agent in food products, and was found to be bactericidal at a concentration of 0.05%, when tested for efficacy in controlling the normal spoilage flora of vacuum packaged franks. Butylated hydroxyanisole at 0.02% was borderline, whereas 0.01% had little or no effect in controlling the spoilage flora of franks. Butylated hydroxytoluene was found to be ineffective.

Some sequestrants have also been shown to inhibit the growth of microorganisms in foods. For example, EDTA has been shown to be an effective preservative for fresh fish fillets [103].

Finally, although this chapter has been primarily concerned with direct use of chemicals to control microbial proliferation in foods, there are also indirect uses of chemical preservatives, as discussed by Chichester and Tanner [34]. For example, the parabens, propionates, and sorbic acid are used as antimycotic agents in food packaging materials. In addition, a variety of chemicals are used as slimicides in paper and paper board. Many of the adhesives used in food packaging materials also contain preservatives [34].

X. CONCLUSIONS

Antimicrobial agents are employed in a variety of foods. Their effectiveness depends not only upon concentration and type of organism but on the intrinsic properties of the food system. Environmental and extrinsic parameters also play significant roles in maintaining the effectiveness of food additives.

Although many chemical food preservatives have long histories of use, the mechanisms by which they inhibit microbial proliferation are only recently being understood. Many of the lipophilic food additives interact with the cell membrane and interrupt transport properties. These recent findings are being extended to mammalian cells.

The fact that several of the commonly used food preservatives are mutagenic agents, or lead to formation of mutagenic agents, presents a complex problem. Only future research can provide the proper answers.

REFERENCES

1. M. Karel, O. R. Fennema, and D. B. Lund, in: Principles of Food Science (O. R. Fennema, ed.), Part II, Physical Principles of Food Preservation, M. Dekker, New York, 1975.
2. J. T. R. Nickerson and A. J. Sinskey, Microbiology of Foods and Food Processing, American Elsevier Publ. Co., New York, 1977.
3. T. A. Roberts, in: Proc. Int. Symp. Nitrite Meat Prod., Zeist, 1973, pp. 91-103.
4. D. A. A. Mossel, in: CRC Critical Reviews in Environmental Controls, CRC Press, Cleveland, 1975, pp. 1-1B9.
5. T. E. Furia (ed.), Handbook of Food Additives, 2nd ed., CRC Press, Cleveland, 1972, p. 783.
6. C. E. Kimble, in: Disinfection, Sterilization and Preservation (S. S. Block, ed.), 2nd ed., Lea & Febiger, Philadelphia, 1977, pp. 834-858.
7. W. J. Scott, Adv. Food Res. 7:83 (1957).
8. A. D. Brown, Bacteriol. Rev. 40:803 (1976).
9. J. H. B. Christian and J. A. Waltho, Biochim. Biophys. Acta 65:506 (1962).
10. J. A. Troller, J. Milk Food Technol 36:276 (1973).
11. J. C. Measures, Nature (London) 257:398 (1975).
12. A. C. Baird-Parker, and B. Freame, J. Appl. Bact. 30:420 (1967).
13. K. Möhler, in: Proc. Int. Symp. Nitrite Meat Prod., Zeist, 1973, pp. 13-18.
14. M. Ingram, in: Proc. Int. Symp. Nitrite Meat Prod., Zeist, 1973, pp. 63-74.
15. H. Riemann, Food Technol. 17:39 (1963).
16. J. H. Silliker, R. A. Greenberg, and W. R. Schack, Food Technol. 12:55 (1958).
17. A. G. Castellani and C. F. Niven, Jr., Appl. Microbiol. 3:154 (1955).
18. M. Raevuori, Zentralbl. Bakt. Hyg. I. Abt. Org. B. 161:280 (1975).
19. L. N. Christiansen, R. W. Johnston, D. A. Kautter, J. W. Howard, and W. J. Aunan, Appl. Microbiol. 25(3):357 (1973).
20. H. R. Nordin, Can. Inst. Food Technol J. 2(2):79 (1969).
21. H. L. A. Tarr, Nature (London) 147:417 (1941).
22. H. L. A. Tarr, J. Fish Res. Brd. Can. 5:265 (1941).
23. M. L. Henry, L. Joubert, and P. Goret, C-R Séanc-Soc. Biol. 148:819 (1954).

24. J. L. Shank, J. H. Silliker, and R. H. Harper, Appl. Microbiol. 10:185 (1962).
25. H. Pivnick, H. W. Barnett, H. R. Nordin, and J. L. Rubin, Can. Inst. Food Technol. J. 2:141 (1969).
26. M. A. Johnston, H. Pivnick, and J. M. Samson, Can. Inst. Food Technol, J. 2:52 (1969).
27. J. A. Perigo, E. Whiting, and T. E. Bashford, J. Food Technol. 2:377 (1967).
28. C. L. Duncan and E. M. Foster, Appl. Microbiol. 16:401 (1968).
29. T. A. Roberts, B. Jarvis, and A. C. Rhodes, J. Food Technol. 11:25 (1976).
30. T. A. Roberts and M. Ingram, J. Food Technol. 1:147 (1966).
31. W. E. Riha, and M. Solberg, J. Food Sci. 40:443 (1975).
32. D. M. Moran, S. R. Tannenbaum, and M. C. Archer, Appl. Microbiol. 30:838, (1975).
33. R. C. Lindsay, in: Principles of Food Science (O. R. Fennema, ed.), Part 1, Food Chemistry, M. Dekker, New York, 1976, p. 465.
34. D. F. Chichester and F. W. Tanner, Jr., in: Handbook of Food Additives (T. E. Furia, ed.), 2nd ed., CRC Press, Cleveland, 1972, p. 115.
35. H. J. Rehm and H. Wittmann, Z. Lebensm. Untersuch-Forsch. 118:413 (1962).
36. B. Bhaget and M. F. Locket, J. Pharmacol. 12:690, (1960).
37. D. G. Fitzhugh, L. Knudsen, and A. Nelson, J. Pharmacol. Exptl. Therap. 86:37 (1946).
38. M. C. Archer and S. R. Tannenbaum, "Vitamins" Chap. 3, this volume.
39. D. Hotzel, Verh. Deut. Ges. Inn Med. 67:868 (1962).
40. M. A. Joslyn and J. B. S. Braverman. Adv. Food Res. 5:97 (1954).
41. L. Fishbein, in: Chemical Mutagens: Principles and Methods for their Detection (A. Hollander, ed.), Vol. 4, Plenum, New York, 1976, p. 219.
42. G. A. Summers and J. W. Drake, Genetics 68:603 (1971).
43. R. Shapiro, R. E. Servis, and M. Welcher, J. Am. Chem. Soc. 92: 422 (1970).
44. U. Furuichi, Y. Wataya, H. Hayatsu, and T. Ukita, Biochem. Biophys. Res. Commun. 41:1185 (1970).
45. R. P. Singhal, J. Biol. Chem. 246:5848 (1971).
46. M. Inoue, H. Hayatsu, and H. Tandoka, Chem. Biol. Interactions 5:85 (1972).
47. G. Löfroth and T. Geguall, Science 174:1248 (1971).
48. C. S. Ough, J. Agr. Food Chem. 24:323, (1976).
49. C. S. Ough, J. Agr. Food Chem. 24:328, (1976).
50. C. R. Phillips, in: Disinfection, Sterilization and Preservation (S. S. Block, ed.), 2nd ed., Lea & Febiger, Philadelphia, 1977, pp. 592-610.

51. C. W. Bruch and M. K. Bruch, in: Disinfection (M. A. Bernarde, ed.), M. Dekker, New York, pp. 149-206.
52. R. R. Ernst and J. E. Doyle, Biotechnol. Bioengr. 10:1 (1968).
53. E. A. Hawk and D. Mickelsen, Science 121:442 (1955).
54. H. T. Gordon, W. W. Thornburg, and L. N. Werum, Agr. Food Chem. 7:196 (1959).
55. C. Hoffman, T. R. Schweitzer, and G. Dalby, Food Res. 4:539 (1939).
56. C. M. Gooding, U. S. Patent 2,379,294 (1945).
57. O. S. Strepanova, A. I. Chekurda, and N. Z. Prudnik, Mikol. Fitopatol. 4:543 (1970).
58. H. J. Deuel, Jr., C. E. Calbert, L. Anisfeld, H. McKeohan, and H. D. Blunden, Food Res. 19:13 (1954).
59. F. Kosikowski, in: Cheese and Fermented Milk Foods, 2nd ed., Edwards Brothers, Ann Arbor, Michigan, 1977, p. 213.
60. O. Wyss, Adv. Food Res. 1:373 (1948).
61. C. W. Sheu and E. Freese, J. Bacteriol. 111:516 (1972).
62. A. H. Romano and H. L. Kornberg, Proc. Roy. Soc. Ser. B 173:475 (1969).
63. G. W. F. H. Borst-Pauwels and S. Jager, Biochem. Biophys. Acta. 172:399 (1969).
64. F. E. Samson, A. Katz, and D. L. Harris, Arch. Biochem. Biophys. 54:406 (1955).
65. C. W. Sheu, W. N. Konings, and E. Freese, J. Bacteriol. 111:525 (1972).
66. E. Freese, C. W. Sheu, and E. Galliels, Nature 241:321 (1973).
67. C. W. Sheu and E. Freese, J. Bacteriol. 115:869 (1973).
68. L. Leive, J. Biol. Chem 243:2373 (1968).
69. C. W. Sheu, D. Salomon, J. L. Simmons, T. Steevalsan, and E. Freese, Antimicrob. Agents Chemother. 7:349 (1975).
70. L. O. Ingram, Appl. Environ. Microbiol. 33:1233 (1977).
71. C. L. Wrenshall, in: Antibiotics—Their Chemistry and Non-Medical Uses (H. S. Goldberg, ed), D. Van Nostrand Co., Princeton, 1959, p. 449-527.
72. A. T. R. Mattick and A. Hirsch, Nature, 154:551 (1944).
73. N. J. Berridge, G. G. F. Newton, and E. P. Abraham, Biochem. 52:529 (1952).
74. E. Gross and J. L. Morell, J. Am. Chem. Soc. 93:4634 (1971).
75. H. D. Michener, F. A. Thompson, and J. C. Lewis, Appl. Microbiol. 7:166 (1959).
76. B. M. Gibbs and A. Hurst, in: Microbial Inhibitors in Food, 4th Int. Symp. on Food Microbiol. (N. Main, ed.), Almquist & Wiksell, Stockholm, 1964.
77. H. S. Goldberg and E. M. Barnes, Antimicrob. Agents Ann. 576 (1960).
78. R. Brown and E. L. Hazen, Trans. NY Acad. Sci. Ser. II 19:447 1957.

79. A. R. Kohler, H. P. Broquist, and W. H. Miller, Food Technol. 8:19 (1954).
80. W. C. Griffin and M. J. Lynch, in: Handbook of Food Additives (T. E. Furia, ed.), 2nd ed., CRC Press, Cleveland, 1972, pp. 431-455
81. M. Karel, in: Intermediate Moisture Foods (R. Davies, G. G. Birch, and K. J. Parker, eds.), Applied Science Publ., London, 1976, pp. 4-28.
82. O. H. Robertson, E. M. Appel, T. T. Duck, H. M. Lemon, and M. H. Rittel, J. Infect. Dis. 83:124 (1948).
83. I. Olitzky, J. Pharm. Sci. 54:787 (1965).
84. B. J. Marshall, D. F. Ohye, and J. H. B. Christian, Appl. Microbiol. 21:363 (1971).
85. M. Plitman, Y. Park, R. Gomez, and A. J. Sinskey, J. Food Sci. 88:1004 (1973).
86. M. Akedo, A. J. Sinskey, and R. Gomez, J. Food Sci. 42:699 (1977).
87. M. A. Amerine and R. E. Kunkee, Vitis 5:187 (1965).
88. R. E. Kunkee and M. A. Amerine, Appl. Microbiol. 16:1067 (1968).
89. B. N. Ames, in: Chemical Mutagens: Principles and Methods for Their Detection (A. Hollander, ed.), Vol. 1, Plenum, New York, 1971, pp. 267-281.
90. B. N. Ames, F. D. Lee, and W. E. Durston, Proc. Natl. Acad. Sci. USA 70:782 (1973).
91. B. N. Ames, W. E. Durston, E. Yamasaki, and F. D. Lee, Proc. Natl. Acad. Sci. USA 70:2281 (1973).
92. B. N. Ames, J. McCann, and E. Yamasaki, Mut. Res. 31:347 (1975).
93. B. A. Bridges, Nature, 261:195 (1976).
94. R. Doll, Nature, 265:589 (1977).
95. J. W. Drake, Mut. Res. 33:65 (1975).
96. H. S. Rosenkranz, Ann. Rev. Microbiol. 27:383 (1973).
97. R. F. Gomez, M. Johnston, and A. J. Sinskey, Mut. Res. 24:5 (1974).
98. Z. Harsanyi, I. A. Granek, and D. W. R. MacKenzie, Mut. Res. 48:51 (1977).
99. H. Hayatsu, K. C. Chung, T. Kada, and T. Nakajima, Mut. Res. 30:417 (1975).
100. F. C. Johnson, Crit. Rev. Food Technol. 2:267 (1971).
101. J. W. Daniel, T. Green, and P. J. Phillips, Food Cosmet. Toxicol. 11:771, 781, 793 (1973).
102. R. D. Trelase, and R. B. Tompkin, U.S. Patent, 3, 955, 005 (1976).
103. R. E. Levin, J. Milk Food Technol. 30:277 (1967).
104. J. A. Baross, in: Compendium of Methods for the Microbiological Examination of Foods (M. L. Speck, ed.), Amer. Publ. Health Asso., Washington, 1976, pp. 196-197.

CHAPTER 12

IMMUNOLOGICAL ASPECTS OF FOODS AND FOOD SAFETY

Nicholas Catsimpoolas

Department of Nutrition and Food Science
Massachusetts Institute of Technology
Cambridge, Massachusetts

I. INTRODUCTION

A. Immunology and Food Safety

The availability of immunological methods for the identification and quantitation of biological compounds [1-12] has made possible the utilization of antigen-antibody interactions in evaluating several aspects of food safety. These involve primarily adulteration of food products with foreign protein, presence of toxic plant proteins, contamination with bacterial protein toxins, and development of allergic reactions to food components. In all of the above cases, immunology has been used as a specific and sensitive analytical tool in the service of food science. The successful interaction of these two distinct disciplines in various areas of research has been amply demonstrated [13].

B. Methodological Aspects

With the exception of food allergy where humans are the test subjects, the use of immunological methods requires the production of an antiserum in experimental animals. In general, rabbits are used in the laboratory for this purpose. The immunization procedure involves the injection of the antigenic material into the animal under a defined schedule of antigen dose, administration route (e.g., intraperitoneal, intravenous, subcutaneous, etc.), and time interval between challenges. The animals respond to the injected foreign material (usually proteins and complex carbohydrates) by producing specific antibodies to these macromolecules. Immunization with small molecules (haptens) is possible if these are first coupled covalently to a protein. The resulting antiserum is obtained by bleeding the animal, followed by removal of the blood cells and fibrin clot. The antiserum can be considered monospecific, i.e., recognizing one specific antigen only if a pure compound was injected. When a mixture of antigenic components is used for the immunization, the animals will produce antibodies to most of the antigens of different potency and specificity. The antisera, once produced, can be stored in the cold for many years. Under appropriate dilution for final use, hundreds of tests can be performed from the serum of even one animal.

The use of antisera in food research primarily involves the detection and quantitation of antigenic proteins. The most commonly used techniques to achieve this purpose involve antigen-antibody reactions in agarose gels where a specific immunoprecipitin line is formed for each immunologically distinct protein present in the sample. There are several variations of immunodiffusion and immunoelectrophoretic techniques which can be employed, according to the kind of information desired. The methods which combine electrophoresis and immunodiffusion, such as immunoelectrophoresis [2], quantitative bidimensional immunoelectrophoresis [8,11], and immunoelectrofocusing [14], offer the highest resolution possible and therefore are recommended when dealing with complex mixtures of proteins. The low-resolution method of double-gel immunodiffusion [4] provides a relatively simple means for detecting the presence of a few protein components and establishing the immunological cross-reactivity of structurally similar antigens. The techniques of single immunodiffusion [15], radial immunodiffusion [16], and rocket immunoelectrophoresis [17] are usually employed for the quantitation of a single protein component in a mixture, if a monospecific antiserum is available.

The availability of a monospecific antiserum and the corresponding homogenous protein also makes possible the use of radioimmunoassay procedures [18,19] which provide the ultimum in sensitivity and convenience of analysis.

The detection of allergenic components in food causing immediate hypersensitivity reaction has been traditionally performed by local skin testing (wheal-and-flare) of the suspected allergens on the sensitized person [20]. A new method which is now increasingly used for the detection of reaginic antibodies directed to foods is the radioallergosorbent test (RAST) [21]. A blood sample of the allergic subject is used for the in vitro quantitation of IgE antibody (reaginic antibody) directed toward specific allergens. The procedure involves the binding of the specific IgE antibodies present in the patient's serum with the allergen which is covalently bound on a solid support. The allergen-IgE antibody complexes are subsequently allowed to react with radioactively labeled human anti-IgE antibodies which bind to the IgE molecule. The quantity of the radioactivity bound is related to the concentration of reaginic antibodies present in the allergic serum. Another less frequently used method is the histamine release assay [22]. For measurement of allergen potency, various quantities of the extract are added to leukocytes obtained from the blood of the sensitized person. The amount of extract needed for release of 50% of total leukocyte histamine is determined from the dose-response curve, and therefore a comparison of the potency of various allergens can be made.

For detection of delayed hypersensitivity type of allergic reactions to food, which involves cellular immune responses, both skin testing (patch) [23] and in vitro techniques [24] can be employed. The patch testing is carried out by application of the suspected allergen source to the skin; the spot is covered and left undisturbed for 24-48 hr. The results are recorded as mm of erythema observed. The in vitro tests are performed by culturing peripheral lymphocytes of the sensitized person in the presence of the suspected allergen and tritiated thymidine. The uptake of the radioactive precursor by the cells is indicative of DNA synthesis in the S phase, which gives an estimate of lymphocyte proliferation. It takes from 4 to 7 days of incubation to detect optimal DNA synthesis in response to allergen stimulation. Lymphocytes from nonsensitized persons do not respond.

The above mentioned methodology represents the backbone of experimental approaches for the immunological assessment of food safety. Undoubtedly, other less known or still developing techniques could be found to be useful. Several variations of the basic methods are described in the literature [1-12] and can be applied for a specific problem at hand.

C. Scratching the Surface

Although considerable information exists on the use of immunological approaches to problems associated with the presence of undesirable components in food, this field of scientific research is still in its infancy. Therefore, the purpose of this chapter is to provide some awareness of the capabilities and restrictions of immunological methods in dealing with various aspects of food safety. An exhaustive review on the subject is not supplied here since a detailed treatise has been published elsewhere [13]. However, the many areas of promising applications are highlighted in the hope that further work will eventually provide objective means for determining the suitability of new food sources of plant, single cell, and animal origin.

II. FOOD ADULTERATION

A. A Food Safety Problem

Foreign protein adulteration of a food product can be considered a serious food safety problem because the consumer is unaware of the presence of the added material. For example, if infested pork is used as adulterant to hamburger which is not fully cooked, it may contribute to the possible contraction of trichinosis. In other cases, the consumer may be strongly allergic to the foreign protein, which he would normally avoid in his diet. Furthermore, the parent source of the adulterant may have been processed in such a manner as to result in the formation of toxic substances (e.g., carcinogens) which do not produce apparent clinical symptoms upon ingestion. Thus, food adulteration is not only a matter of concern in obtaining a nutritionally inferior product, but it can be considered also as a health problem of grave consequences. For these reasons, health inspection and regulatory agencies are anxious to use and develop rapid, simple, and reliable methods for the identification of foreign proteins. Differentiation of proteins from various species (e.g., addition of buffalo milk to cow milk, horse meat to beef) and heterologous sources (e.g., soy protein in meat) is highly desirable, especially if the analysis can be performed quantitatively.

B. Requirements for Immunological Detection and Quantitation

Immunological methods are now increasingly applied to food adulteration problems, especially those involving meat [25] and milk [26]. In addition, the availability of antisera for the detection of soybean [27], peanut [28], barley [29], wheat [30], bean [31,32], and egg [33] antigens makes possible the immunological identification of a large number of animal and plant proteins in foods. However, the utilization of these antisera is not free of experimental difficulties, which have to be overcome for each particular food product. The major problem seems to be associated with the possible denaturation of the adulterant proteins in the final product, especially if heat is involved. The denatured protein may not be recognized by the specific antiserum even if it still remains soluble. Antisera prepared to denatured proteins may interact very weakly and unpredictably with the denatured antigen. Often the heating process causes insolubility to the protein and interaction with other components in the system, which may block the antigenic sites. Other processing conditions such as acid, alkaline, and chemical treatment may also denature the antigens and render them immunologically unreactive. For all the above reasons, in the development of an immunochemical method to detect adulterant proteins, the probability of success is highest in a system where the antigen is likely to remain soluble, undenatured, and chemically unmodified. For such products, the immunological approach offers the best analytical solution.

III. PLANT PROTEINASE INHIBITORS AND AGGLUTININS

A. Proteinase Inhibitors

Proteinase inhibitors are proteins which occur naturally in many plants, especially in the seeds of legumes and grasses and in the tubers of potatoes and sweet potatoes [34,35]. Concern about their presence in foods in an undenatured state stems primarily from the fact that these proteins are unusually stable to proteolytic attack [36], and some of them (e.g., soybean inhibitors) have been shown to have growth-inhibitory and pathogenic properties in animals [37]. Very little is known about their possible effects on humans [37]. The particular structure of many of the inhibitor proteins allows them to resist inactivation by heat treatment, which is the most commonly relied upon procedure for their denaturation. Although cooking will probably destroy the inhibitory activity of many of these compounds, it should be mentioned that future engineered food products (e.g., imitation milk, protein-fortified beverages, etc.) may not be subjected to such a process. Therefore, in the absence of sufficient information, it would be safer to include the various proteinase inhibitors in the category of proteins which have potential for the induction of adverse physiological responses in man. As such they may represent a food safety problem.

B. Agglutinins

Another category of toxic plant proteins called hemagglutinins, agglutinins, or lectins are widespread in food materials of plant origin [38]. They have been shown to cause agglutination of red and white blood cells, blastic transformation and mitosis of lymphocytes, and interaction with specific receptor groups existing on the surface of mammalian cells [39,40]. Lectins represent a diverse group of proteins in regard to size, structure, and composition. Food scientists are particularly concerned about their toxic [41] and antigrowth [37,42] properties. Although presoaking of beans, followed by cooking, leads to the destruction of some agglutinins, dry heat may not be effective [37]. Since some protein products from oilseeds and beans may have been exposed only to semidry heat, the presence of active hemagglutinins in processed food cannot be excluded. Again, the available information is so meager as to exclude any general conclusions. For this reason, there is a great need for rapid methods of determining the quantity of active agglutinins in food ingredients and final products.

C. Potential Usefulness of Immunoassays and Restrictions

One of the most promising analytical approaches involves the use of immunological methods for the detection and quantitation of proteinase inhibitors and agglutinins. Both of these classes of proteins when injected into animals elicit the production of specific antibodies [36,43,44]. In the case of the Kunitz soybean proteinase inhibitor, it has been demonstrated [45] that as few as 0.5-10 μg of the protein can be accurately quantitated. The same level of sensitivity or better is expected for all the other proteinase inhibitors and agglutinins. The employment of radioimmunoassays may provide rapid and automated procedures for the routine evaluation of food ingredients for the presence of these potentially toxic factors. However, before these assays can be considered meaningful, a good correlation between biological and antigenic activity has to be demonstrated. In other words, the ability of the antibody to react with the particular proteinase inhibitor or agglutinin has to parallel closely the inhibitory and agglutinating or mitotic activity of the protein. This may be possible if both the antigenic and biologically active sites depend on a similar conformational stability of the structure and if any externally applied inactivating agent (e.g., heat) abolishes both of them to the same degree. It can be safely predicted that studies of this nature will be of great benefit in the overall effort of providing safe food products for long term and abundant consumption.

IV. BACTERIAL TOXINS

Detection and quantitation of protein toxins associated with severe or deadly food poisoning is undoubtedly one of the most important analytical tasks involved in assessing food safety. Among the various toxins, those of staphylococcal, *Clostridium botulinum*, pathogenic *E. coli*, and *Clostridium perfringens* origin are the most commonly encountered causative agents of food-borne poisoning. Although various bioassays exist for their detection, immunological methods are especially attractive because of their sensitivity and specificity. However, production of antibodies to the various enterotoxins involves particular protocols and precautions in immunization work and appropriate selection of immunoassays to be used for their detection in food. The subject has been covered extensively in recent reviews [46-48] where specific details can be found in regard to the preparation of antisera to specific toxins and applications to food examination. These studies point out the necessity for further work especially with unidentified toxins which require purification in order to be utilized in immunological analysis.

V. FOOD ALLERGENS

A. Allergy

The term allergy is generally used to describe an altered immunological reactivity to a foreign material, which is then called an allergen [49-51]. Most of the common allergens are proteins, although a variety of other compounds have been implicated. In immediate hypersensitivity type of reactions, the symptoms appear in a few minutes, usually no longer than one hour after exposure. In delayed hypersensitivity, symptoms can appear after 24 hr and up to 96 hr after ingestion of the allergen. Anaphylactic-type allergy, associated with violent often fatal symptoms, may occur in seconds to minutes. The affected organs include primarily the skin and mucus membranes, the respiratory tract, and the gastrointestinal tract. Less frequently the central nervous system, special sense organs, skeletal structures and the genitourinary system may be involved. Thus, the symptoms of the disease can be expressed as eczema, hives, angioedema, rhinitis, asthma, abdominal pain, vomitting, diarrhea, dizziness, headache, joint swelling, bladder inflammation, and others. The development of sensitivity to food allergens by a genetically predisposed person is presumed to occur from repeated exposure not only by ingestion, but also by inhalation and skin contact.

The symptoms of the disease are caused by compounds called mediators (such as histamine, kinins, heparin, and SRS-A and ECRF factors), which are released from special target cells, i.e., tissue mast cells and blood basophils. The reaction is triggered when an allergen to which a person has become sensitized combines with its IgE antibody (reagin), which is fixed on the surface of the target cells. The formation of the complex initiates a series of events in the target cell, which leads to the extracellular secretion of some of its granules containing the mediator substances. The action of the mediators can explain the clinical symptoms of immediate and anaphylactic types of allergies. The delayed allergy may involve the interaction of the allergen with T lymphocytes, which then undergo transformation to produce blast cells and soluble factors affecting the activity of other cells involved in the immunological response.

Methods for distinguishing immediate- from delayed-type hypersensitivity were described in Sec. I.B. The in vitro techniques, although not thoroughly explored in relation to food allergens, offer great promise for immunological studies of food safety. The assessment of the allergenicity of new unconventional proteins should be one of the primary objects of research, with regard to their anticipated incorporation in human diet.

B. Allergens in Foods

Food allergy is usually detected by subjective association of symptoms to ingested foods and is treated by dietary restriction. The employment of objective methods for the identification of the particular allergen involved is complicated by several factors: (a) the allergen may be present in minimal quantity; (b) it cannot be adequately extracted; (c) it is denatured during isolation; (d) it is not naturally present in the food tissue, i.e., it is a contaminant associated with chemical additives, bacteria, and fungi; (e) it is a natural constituent, but chemically modified by the food processing involved; (f) it is active as a complex between two different compounds; and (g) it is the product of an enzymatic reaction. Nevertheless, work on the isolation of food allergens has resulted in valuable information about the nature of these components. One important aspect that has been demonstrated is the inconsistency in cross-allergenicity among closely related foods on the basis of biological or botanical relationship. The shared allergenic components may be abundant in one food (e.g., cow's milk) and minimal in the other (e.g., goat's milk). Thus, it may not be necessary to remove an entire family of foods from the diet. Another troublesome aspect of allergenicity concerns the heat stability of certain allergenic determinants. Thus, cooking and baking of a food may not alter the allergenicity of certain constituents; in fact it may cause its enhancement.

Some of the major allergens in foods have been identified and shown to be proteins [20]. In milk, α-lactalbumin is the major allergen, whereas β-lactoglobulin and casein have been implicated but on a minor scale. Ovomucoid, an egg-white glycoprotein, causes strong allergenic reactions. Lesser allergenic activity has been obtained from ovalbumin, lysozyme, ovoinhibitor, and flavoprotein-apoprotein [33]. Other allergens of proteinaceous nature have been extracted from the "albumin" fraction of grain and grain cereals (e.g., barley, corn, oat, rice, rye, wheat) and nuts, seeds, and beans (e.g., walnut, brazil nut, filbert, sesame, peas, etc.). Other sources of allergens include fish and shellfish, meats, vegetables, fruits, and beverages. Food additives, bacteria and fungi (molds), and arthropods are responsible for the occurrence of a large variety of unexpected "food" allergies which have little to do with the nature of the food itself. These, of course, present a serious health hazard because of the inability of the allergic person to clearly identify the source of the offender. Furthermore, it is expected that the incorporation of unconventional protein products from new sources subjected to complex processing will contribute to the development of new allergic reactions in sensitive individuals. Identification and inactivation of highly active allergens may be necessary before these products can be considered safe for human consumption.

VI. CONCLUSIONS

The information presented here indicates several important areas of food safety where immunological analysis can play an important role. These are primarily the detection and quantitation, by immunological in vitro techniques, of food adulterants, toxic plant proteins, bacterial protein toxins, and food allergens. Some of the advantages and pitfalls of these methods have been hinted at when applied to complex processed food materials. The time seems ripe for the more extensive use of immunology in food research and regulation.

REFERENCES

1. E. A. Kabat and M. M. Mayer, Experimental Immunochemistry, 2nd ed., Charles C. Thomas, Springfield, Ill., 1961.
2. P. Graber and P. Burtin, Immunoelectrophoretic Analysis, Elsevier, New York, 1964.
3. D. M. Wier (ed.), Handbook of Experimental Immunology, 2nd ed., Blackwell Scientific Publications, Oxford, 1973.
4. Ö. Ouchterlony, Handbook of Immunodiffusion and Immunoelectrophoresis, Ann Arbor Science, Ann Arbor, Mich., 1968.
5. J. Clausen, Immunochemical Techniques for the Identification and Estimation of Macromolecules, Elsevier, New York, 1969.
6. M. Goldman, Fluorescent Antibody Methods, Academic Press, New York, 1968.
7. J. B. G. Kwapinski, Methodology of Immunochemical and Immunological Research, Wiley-Interscience, New York, 1972.
8. C.-B. Laurell (ed.), Electrophoretic and Electro-Immunochemical Analysis of Proteins, Universitets forlaget, Oslo, 1972.
9. A. J. Crowle, Immunodiffusion, 2nd ed., Academic Press, New York, 1973.
10. C. A. Williams and M. W. Chase (ed.), Methods in Immunology and Immunochemistry, Vols. 1-3, Academic Press, New York, 1967, 1968, 1971.
11. N. H. Axelsen, J. Krøll, and B. Weeke, Quantitative Immunoelectrophoresis, Universitets forlaget, Oslo, 1973.
12. N. R. Rose and H. Friedman (eds.), Manual of Clinical Immunology, American Society for Microbiology, Washington, 1976.
13. N. Catsimpoolas (ed.), Immunological Aspects of Foods, Avi Publishing Co., Westport, Conn., 1977.

14. N. Catsimpoolas, Ann. N.Y. Acad. Sci. 209:144 (1973).
15. J. Oudin, Methods Med. Res. 5:335 (1952).
16. G. Mancini, A. O. Carbonara, and J. F. Heremans, Immunochemistry 2:235 (1965).
17. C.-B. Laurell, Anal. Biochem. 10:358 (1965).
18. S. A. Berson and R. Yallow (eds.), Methods in Investigative and Diagnostic Endocrinology, North-Holland Publishing Co., Amsterdam, 1973.
19. B. M. Jaffe and H. R. Behrman (eds.) Methods of Hormone Radioimmunoassay, Academic Press, New York, 1974.
20. F. Perlman, in: Immunological Aspects of Foods (N. Catsimpoolas, ed.), Avi Publishing Co., Westport, Conn., 1977, p. 279.
21. L. Wide, in: Radioimmunoassay Methods (K. E. Kirkham and W. M. Hunter, eds.), E. and S. Livingstone, Ltd., Edinburgh, 1971, p. 405.
22. R. P. Siraganian, in: Manual of Clinical Immunology (N. R. Rose and H. Friedman, eds.), American Society for Microbiology, Washington, 1976, p. 603.
23. L. E. Spitler, in: Manual of Clinical Immunology (N. R. Rose and H. Friedman, eds.), American Society for Microbiology, Washington 1976, p. 53.
24. B. R. Bloom and P. Glade (eds.), In Vitro Methods of Cell-Mediated Immunity, Academic Press, New York, 1971.
25. R. G. Cassens, M. L. Greaser, and A. R. Handen, in: Immunological Aspects of Foods (N. Catsimpoolas, ed.), Avi Publishing Co., Westport, Conn., 1977, p. 1.
26. L. Å. Hanson and B. G. Johansson, in: Milk Proteins: Chemistry and Molecular Biology, Vol. 1 (H. A. McKenzie, ed.), Academic Press, New York, 1970, p. 45.
27. N. Catsimpoolas, E. Leuthner, and E. W. Meyer, Arch. Biochem. Biophys. 31:437 (1968).
28. J. Daussant, N. J. Neucere, and L. Y. Yatsu, Plant Physiol. 44:471 (1969).
29. R. J. Hill and R. Djurtoft, J. Inst. Brew. 70:416 (1964).
30. J. A. D. Ewart, J. Sci. Food Agr. 17:279 (1966).
31. J. Kloz, in: Chemotaxonomy of the Leguminosae (J. H. Harborne, D. Boulter, and B. L. Turner, eds.), Academic Press, New York, 1971, p. 309.
32. R. H. Falk and L. Bogorad, Plant Physiol. 44:1669 (1969).
33. H. J. Miller, in: Immunological Aspects of Foods (N. Catsimpoolas, ed.), Avi Publishing Co., Westport, Conn., 1977, p. 152.
34. R. Vogel, I. Trautshold, and E. Werle, Natural Proteinase Inhibitors, Academic Press, New York, 1966.
35. C. A. Ryan, Ann. Rev. Plant Physiol. 24:173 (1973).
36. C. A. Ryan, in: Immunological Aspects of Foods (N. Catsimpoolas, ed.), Avi Publishing Co., Westport, Conn., 1977.

37. I. E. Liener, in: Proteins in Human Nutrition (J. W. G. Porter and B. A. Rolls, eds.), Academic Press, New York, 1973, p. 481.
38. N. Sharon and H. Lis, Science 177:949 (1972).
39. N. Sharon, in: Mitogens in Immunobiology (J. J. Oppenheim and D. L. Rosenstreich, eds.), Academic Press, New York, 1976, p. 31.
40. H. Lis and N. Sharon, Ann. Rev. Biochem. 42:51 (1973).
41. W. G. Jaffé, in: Toxic Constituents of Plant Foodstuffs (I. E. Liener, ed.), Academic Press, New York, 1969, p. 93.
42. I. E. Liener, J. Agr. Food Chem. 22:17 (1974).
43. N. Catsimpoolas, in: Immunological Aspects of Foods (N. Catsimpoolas, ed.), Avi Publishing Co., Westport, Conn., 1977, p. 37.
44. W. G. Jaffé, in: Immunological Aspects of Foods (N. Catsimpoolas, ed.), Avi Publishing Co., Westport, Conn., 1977, p. 37.
45. N. Catsimpoolas and E. Leuthner, Anal. Biochem. 31:437 (1969).
46. M. S. Bergdoll, in: Immunological Aspects of Foods (N. Catsimpoolas, ed.), Avi Publishing Co., Westport, Conn., 1977, p. 199.
47. T. M. Jacks, in: Immunological Aspects of Foods (N. Catsimpoolas, ed.), Avi Publishing Co., Westport, Conn., 1977, p. 221.
48. C. E. Kimble, in: Immunological Aspects of Foods (N. Catsimpoolas, ed.), Avi Publishing Co., Westport, Conn., 1977, p. 233.
49. L. M. Lichtenstein, in: Clinical Immunobiology, Vol. 1 (F. H. Bach and R. A. Good, eds.), Academic Press, New York, 1976, p. 243
50. E. L. Becker and K. F. Austen (ed.), Second International Symposium on the Biochemistry of Acute Allergic Reactions, Blackwell, London, 1971.
51. J. M. Sheldon, R. G. Lovell, and K. P. Mathews, A Manual of Clinical Allergy, W. B. Saunders, Philadelphia, 1967.

AUTHOR INDEX

Numbers in brackets are reference numbers and indicate that an author's work is referred to although his name is not cited in the text. Underlined numbers give the page on which the complete reference is listed.

SUBJECT INDEX